S0-ADL-366

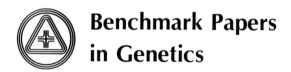

Benchmark Papers
in Genetics

Series Editor: David L. Jameson
University of Houston

PUBLISHED VOLUMES AND VOLUMES IN PREPARATION

GENETICS AND SOCIAL STRUCTURE / *Paul A. Ballonoff*
GENES AND PROTEINS / *R. P. Wagner*
DEMOGRAPHIC GENETICS / *Kenneth M. Weiss and Paul Ballonoff*
EUGENICS / *Carl Bajema*
MEDICAL GENETICS / *William Jack Schull*
POPULATION GENETICS / *James F. Crow and Carter Denniston*
ECOLOGICAL GENETICS / *W. W. Anderson*
QUANTITATIVE GENETICS / *R. E. Comstock*
GENETIC RECOMBINATION / *Rollin Hotchkiss*
REGULATION GENETICS / *Werner K. Maas*
ANIMAL BREEDING / *Robert C. Carter*
PLANT BREEDING / *D. F. Matzinger*
DEVELOPMENTAL GENETICS / *Antonie W. Blackler and Richard Hallberg*
CYTOGENETICS / *Ronald L. Phillips and Charles H. Burnham*

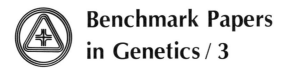

**Benchmark Papers
in Genetics / 3**

A BENCHMARK® Books Series

DEMOGRAPHIC GENETICS

Edited by
KENNETH M. WEISS
and PAUL A. BALLONOFF
Center for Demographic and Population Genetics
The University of Texas at Houston

Dowden, Hutchinson & Ross, Inc.

STROUDSBURG, PENNSYLVANIA

Distributed by

HALSTED
PRESS

A division of
John Wiley & Sons, Inc.

Copyright © 1975 by **Dowden, Hutchinson & Ross, Inc.**
Benchmark Papers in Genetics, Volume 3
Library of Congress Catalog Card Number: 75-31580
ISBN: 0-470-92698-8

77 76 75   1 2 3 4 5
Manufactured in the United States of America.

LIBRARY OF CONGRESS CATALOGING IN PUBLICATION DATA

Exclusive Distributor: **Halsted Press**
A Division of John Wiley & Sons, Inc.

ACKNOWLEDGMENTS
AND PERMISSIONS

ACKNOWLEDGMENTS

AMERICAN ASSOCIATION FOR THE ADVANCEMENT OF SCIENCE—*Science*
   Relation Between Birth Rates and Death Rates

DOVER PUBLICATIONS, INC.—*The Genetical Theory of Natural Selection*
   The Fundamental Theorem of Natural Selection
   Natural Selection and the Sex Ratio

GENETICAL SOCIETY OF GREAT BRITAIN—*Heredity*
   Contamination in Seed Crops: III. Relation with Isolation Distance

NATIONAL ACADEMY OF SCIENCES—*Proceedings of the National Academy of Sciences*
   Age-Specific Selection
   Demographic Approaches to the Measurement of Differential Selection in Human
      Populations

RANDOM HOUSE, INC.—*The Origin of Species and the Descent of Man*
   The Proportion of the Sexes in Relation to Natural Selection

TAYLOR & FRANCIS LTD.—*The Philosophical Magazine*
   A Problem in Age-Distribution

PERMISSIONS

The following papers have been reprinted with the permission of the authors and copyright holders.

AMERICAN ASSOCIATION FOR THE ADVANCEMENT OF SCIENCE—*Science*
   Social Subordination, Population Density, and Mammalian Evolution

THE BIOLOGICAL LABORATORY LONG ISLAND BIOLOGICAL ASSOCIATION, INC.—*Cold Spring Harbor Symposium on Quantitative Biology*
   The Problem of the Structure of Isolates and of Their Evolution Among Human Populations

CAMBRIDGE PHILOSOPHICAL SOCIETY—*Proceedings of the Cambridge Philosophical Society*
   A Mathematical Theory of Natural and Artificial Selection: Part IV

CAMBRIDGE UNIVERSITY PRESS—*Annals of Human Genetics, London*
   Natural Selection and the Sex Ratio

Acknowledgments and Permissions

CLARENDON PRESS FOR THE LONDON MATHEMATICAL SOCIETY—*Proceedings of the London Mathematical Society*
Natural Selection and Mendelian Variation

DUKE UNIVERSITY PRESS FOR THE ECOLOGICAL SOCIETY OF AMERICA—*Ecology*
Density-Dependent Natural Selection

GENETICS SOCIETY OF AMERICA—*Genetics*
Isolation by Distance

INDIA STATISTICAL INSTITUTE, CALCUTTA—*Sankhya*
On the Generation and Growth of a Population

REGENTS OF THE UNIVERSITY OF WISCONSIN—*Stochastic Models in Medicine and Biology*
Monte Carlo Simulation: Some Uses in the Genetic Study of Primitive Man

THE ROYAL SOCIETY, LONDON—*Proceedings of the Royal Society*
Concepts of Random Mating and the Frequency of Consanguineous Marriages

UNIVERSITY OF CHICAGO PRESS—*The American Naturalist*
Selection in Populations with Overlapping Generations: II. Relations Between Gene Frequency and Demographic Variables

THE UNIVERSITY OF TORONTO PRESS—*Proceedings of the 10th International Congress of Genetics 1958*
Some Data on the Genetic Structure of Human Populations

THE WAYNE STATE UNIVERSITY PRESS—*Human Biology*
Some Possibilities for Measuring Selection Intensities in Man

SERIES EDITOR'S PREFACE

The study of any discipline assumes the mastery of the literature of the subject. In many branches of science, even one as new as genetics, the expansion of knowledge has been so rapid that there is little hope of learning of the development of all phases of the subject. The student has difficulty mastering the textbook, the young scholar must tend to the literature near his own research, the young instructor barely finds time to expand his horizons to meet his class-preparation requirements, the monographer copes with a wider literature but usually from a specialized viewpoint, and the textbook author is forced to cover much the same material as previous and competing texts to respond to the user's needs and abilities.

Few publishers have the dedication to scholarship to serve primarily the limited market of advanced studies. The opportunity to assist professionals at all stages of their careers has been recognized by the publishers and by a distinguished group of editors knowledgeable in specific portions of the genetic literature. Some have contributed heavily to the development of that literature, some have studied with the early scholars, and some have and are in the process of developing entirely new fields of genetic knowledge. In many cases, the judgments of the editors become an historical document recording their opinion of the important steps in the development of the subject. These editors have selected papers and portions of papers that demonstrate both the development of knowledge and the atmosphere in which that knowledge was developed. There is no substitute for reading great papers. Here you can learn how questions are asked, how they are approached, and how difficult and essential it is to obtain definitive answers and clear writing.

My own pleasure in working with this distinguished panel is only exceeded by the considerable pleasure of reading their remarks and their selections. Their dedication and wisdom are impressive.

This volume presents a selection of literature that constitutes Benchmark papers in demographic genetics, a field that has been embryonic for some time but which has recently moved to the forefront of considerable research activity. This collection provides an example of one of the primary aims of the Benchmark series, that of stimulating research activity by providing ready and prompt access to the basic literature in an exciting and active area.

D. L. JAMESON

PREFACE

Choosing papers for this volume has been a difficult task. Not only is the literature of population genetics rich, but we have had to sort out subsets of that discipline to present, as a discrete area of investigation, what we call demographic genetics. Reality is not so neatly packaged, and we were forced to overlook and neglect many authors who could just as deservedly appear here. This book represents our own perspective, and we argue that demographic genetic thinking is crystallizing now in a more explicit way than it has in the past. Where relevant ideas have crossed the minds of certain investigators but have not been made central to the literature published by these investigators, the papers have been omitted from consideration. Also, of course, there is no doubt that we are ignorant of many works of which we should be aware.

We have tried to tie our view of the subject together with an editorial discourse which does not delve deeply into theoretical content or controversy, and which oversimplifies things, but which relates ideas to each other as we feel they should be. In addition, we have tried to add some historical or contextual commentary. As with our selection of papers, there will surely be disagreement with our presentation; we hope our actual errors are few. Our goal was to organize the material in a way that would provide an introduction to the subject and a stimulus to the reader to seek out more truth for himself. It was Socrates who said: "The unexamined life is not worth living." In this area of biology, the uncertainties which persist are such that each person must examine this life for himself. The most that we can do is to say why such an examination should be undertaken, and we hope that we have succeeded.

In a work such as this one, many acknowledgments are due. First, we should like to thank our two translators, Marianne Langenbucher Rowe and Janet Schreiber, both professionals in their own right, for taking the time and trouble to wade through abstruse academic German and Italian. Next in line, we thank Jeryl Silverman for her tireless and even cheerful typing of many drafts of this manuscript, a verve that we cannot expect to encounter more than a few times in a lifetime. Finally, acknowledgments are due to William J. Schull and Masatoshi Nei for taking the time to read and criticize our remarks. They have provided immeasurable assistance and have added quality to the final product with their additions and expurgations of our work. As must be the case, we have often insisted on having our own way, and they can neither be blamed for the results nor tarred with the brush of concurrence by association.

We have received financial support from the U.S. National Institutes of Health (Research Grant GM-19513-02). We also must thank those authors and publishers who have allowed us to republish their works, the many people who have helped us locate these authors and publishers, and Bernice Wisniewski of Dowden, Hutchinson & Ross for putting up with our many problems. To others whom we have carelessly forgotten, we offer both thanks and apologies.

KENNETH M. WEISS
PAUL A. BALLONOFF

CONTENTS

Acknowledgments and Permissions   v
Series Editor's Preface   vii
Preface   ix
Contents by Author   xv

Introduction   1

PART I: EARLY DEVELOPMENTS

Editors' Comments on Papers 1 Through 5   6

1   GRAUNT, J.:   Natural and Political Observations Mentioned in a
                   Following Index, and Made upon the Bills of Mortality   12
     The Economic Writing of Sir William Petty, Vol. 2, C. H. Hull, ed.,
                   Cambridge University Press, 1899, pp. 346–378, 383–388

2   LOTKA, A. J.:   Relation Between Birth Rates and Death Rates   52
                   Science, 26(653), 21–22 (1907)

3   SHARPE, F. R., and A. J. LOTKA: A Problem in Age-Distribution   54
                   Phil. Mag., Ser. 6, 21, 435–438 (1911)

4   VOLTERRA, V.:   A Mathematical Theory on the Struggle for Survival   58
                   Translated from Scientia, 41, 85–102 (1927)

5   LEWIS, E. G.:   On the Generation and Growth of a Population   73
                   Sankhya, 6, Pt. 1, 93–96 (1942)

PART II: ADDING GENETICS TO DEMOGRAPHIC CONCEPTS

Editors' Comments on Papers 6, 7, and 8   78

6   DARWIN, C.:   The Proportion of the Sexes in Relation to Natural
                   Selection   81
     The Origin of Species and The Descent of Man, Random House (The
                   Modern Library), pp. 608–611

7   FISHER, R. A.:   Natural Selection and the Sex Ratio   85
     The Genetical Theory of Natural Selection, Dover Publications, 1958, pp.
                   158–160

Contents

8   BODMER, W. F., and A. W. F. EDWARDS: Natural Selection and the Sex Ratio   88
Ann. Human Genetics, London, **24**(3), 239–244 (1960)

Editors' Comments on Papers 9, 10, and 11   94

9   HALDANE, J. B. S.:   A Mathematical Theory of Natural and Artificial Selection: Part IV   98
Proc. Cambridge Phil. Soc., **23**, 607–615 (1927)

10   NORTON, H. T. J.:   Natural Selection and Mendelian Variation   107
Proc. London Math. Soc., **28**(1639), 1–4, 25–45 (1928)

11   FISHER, R. A.:   The Fundamental Theorem of Natural Selection   132
The Genetical Theory of Natural Selection, Dover Publications, 1958, pp. 22–30, 37–41

Editors' Comments on Papers 12 Through 15   146

12   ROUGHGARDEN, J.:   Density-Dependent Natural Selection   153
Ecology, **52**(3), 453–468 (1971)

13   CHRISTIAN, J. J.:   Social Subordination, Population Density, and Mammalian Evolution   169
Science, **168**, 84–90 (Apr. 1970)

14   ANDERSON, W. W., and C. E. KING: Age-Specific Selection   176
Proc. Nat. Acad. Sci., **66**(3), 780–786 (1970)

15   CHARLESWORTH, B., and J. T. GIESEL: Selection in Populations with Overlapping Generations: Part II. Relations Between Gene Frequency and Demographic Variables   183
Amer. Naturalist, **106**(949), 388–401 (1972)

Editors' Comments on Papers 16 and 17   197

16   CROW, J. F.:   Some Possibilities for Measuring Selection Intensities in Man   200
Human Biol., **30**(1), 1–5, 13 (1958)

17   BODMER, W. F.:   Demographic Approaches to the Measurement of Differential Selection in Human Populations   205
Proc. Nat. Acad. Sci., **59**(3), 690–699 (1968)

PART III. DISPERSAL AND POPULATION DISTRIBUTION

Editors' Comments on Papers 18, 19, and 20   216

18   WAHLUND, S.:   Composition of Populations and of Genotypic Correlations from the Viewpoint of Population Genetics   224
Translated from *Hereditas,* **11**, 65–106 (1928)

19   WRIGHT, S.:   Isolation by Distance   264
Genetics, **28**, 114–121, 136–138 (Mar. 1943)

20   BATEMAN, A. J.: Contamination in Seed Crops: III. Relation with
Isolation Distance   274
Heredity, **1,** 303–311, 315–321, 324–326, 333–336 (1947)

Editors' Comments on Papers 21 Through 24   295

21   SUTTER, J., and TRAN-NGOC-TOAN: The Problem of the Structure of
Isolates and of Their Evolution Among Human
Populations   302
Cold Spring Harbor Symp. Quant. Biol., **22,** 379–383 (1957)

22   CAVALLI-SFORZA, L. L.: Some Data on the Genetic Structure of
Human Populations   307
Proc. 10th Intern. Congr. Genetics 1958, Vol. 1, University of Toronto Press,
1959, pp. 389–407

23   HAJNAL, J.: Concepts of Random Mating and the Frequency of
Consanguineous Marriages   326
Proc. Roy. Soc., **B159,** 125–174 (1963)

24   SCHULL, W. J., and B. R. LEVIN: Monte Carlo Simulation: Some Uses
in the Genetic Study of Primitive Man   376
Stochastic Models in Medicine and Biology, J. Gurland, ed., The University
of Wisconsin Press, 1964, pp. 179–196

References   397
Author Citation Index   407
Subject Index   411
About the Editors   416

CONTENTS BY AUTHOR

Anderson, W. W., 176

Bateman, A. J., 274

Bodmer, W. F., 88, 205

Cavalli-Sforza, L. L., 307

Charlesworth, B., 183

Christian, J. J., 169

Crow, J. F., 200

Darwin, C., 81

Edwards, A. W. F., 88

Fisher, R. A., 85, 132

Giesel, J. T., 183

Graunt, J., 12

Hajnal, J., 326

Haldane, J. B. S., 98

King, C. E., 176

Levin, B. R., 376

Lewis, E. G., 73

Lotka, A. J., 52, 54

Norton, H. T. J., 107

Roughgarden, J., 153

Schull, W. J., 376

Sharpe, F. R., 54

Sutter, J., 302

Toan, T-N., 302

Volterra, V., 58

Wahlund, S., 224

Wright, S., 264

IN THE FIRST PLACE: WHAT IS DEMOGRAPHIC GENETICS?

No specific discipline of science is traditionally known as "demographic genetics." The field of population genetics in general deals with those aspects of genetic processes which involve consideration of a population instead merely of individuals or single families. In the sense that demography is the study of properties of populations, almost every aspect of traditional population genetics could conceivably be considered "demographic." Yet the science of demography is usually thought of as a more restricted study of populations.

Our purpose here is to identify several demographic aspects of populations whose nature is important to the full understanding of genetic processes, especially as they relate to human populations (for which most data exist). This is not an exercise in territoriality, in which we claim for demographic genetics all aspects of traditional population genetics that are of interest to us. Rather, it is an attempt to note the traditional areas of population genetics which relate to the demography of real populations, and to show how these areas are connected. In a sense this is a personal perspective on a growing area of research.

In our opinion, the age distribution, growth, age-specific birth and death rates, migration pattern, size, sex ratio, subdivision, and isolation of populations are within what most people think of as demography. Some aspects of these topics would also be included in an "ecological-genetics" subset of population genetics. We shall try to trace the interrelated development of certain ideas in these areas, and the papers we have chosen to

1

reprint are a combination of earliest works, important empirical studies, and recent developments likely to lead to productive future work.

Since we are dealing with only certain aspects of these topics, our commentary on the papers and their context cannot be a complete survey of population genetics. Similarly, we simply cannot reprint every important paper, nor can we cite every important paper that is omitted. In most aspects of population genetics, an historic survey crediting the first development of ideas need probably do little more than list the relevant papers of Haldane, Fisher, Wright, Kimura, and a few others. They seem to have thought of everything at one time or another. We have tried to cite their major works in which they dealt in a serious way with the aspects of demographic genetics that we are discussing, but surely citation of many relevant papers of theirs and of other authors has been omitted.

Some topics that are a part of demographic genetics are not covered at all; severe space limitations inevitably forced us to focus on the aspects most interesting to us. These topics include the detailed studies of effective population size, the distribution and heritability of fertility and family size, aspects of subdivision and isolation of local populations, the details of demographic ramifications of social behavior, and topics generally included in ecological genetics. Also, stochastic processes are not covered; they represent such a growing area of present population genetics that to begin dealing with them would double the length of our book. To avoid prejudicing the reader's knowledge of the subject due to this kind of omission, we have repeatedly referred to the best and most comprehensive textbooks on the subject, where full development of the ideas is undertaken. Space limitations have also forced us to reprint only sections of some of the papers included here; we have tried always to present enough so that the continuity and relevant concepts are clear.

Our commentary is intended to be an introduction to the reader who does not already know the field. Those who do will not need such comments, and it is the reprints themselves that are of value. But for the reader who is curious about the demographic aspects of population genetics, we hope to provide a sufficiently cogent introduction to the subject for the papers to make sense. To do this, we have separated papers, ideas, and even authors into a somewhat excessively rigid series of categories; by reading the papers, the degree of overlap should become clear. Of course, we do not imply any rigidity or lack of insight on the part of the authors.

A survey of the development of demographic genetics must treat two fundamental ideas. First, we must explain the evolution of concepts of *population* itself, in which it was realized that living things may be productively considered in natural aggregates as well as singly. Second, we must trace the generally later development of the idea that populations can be viewed as *genetically structured aggregates;* that is, various individuals within the aggregate can be separated in a way that relates to their genetic makeup. In this book, we are interested in demographic ways in which individuals are differentiated. Both the idea of populations as meaningful aggregates and their internal differentiation have long histories as rudimentary concepts, but their crystallization into demographic genetics has come only recently.

Explicitly evolutionary, if not genetic, population concepts can of course be traced back no further than to Darwin, although evolutionary concepts were in the air long before he put them together. Indeed, the central problem in pre-Darwinian biology was the categorization of the variety of natural things and its explication. The *development* of such variation was not an important problem, since "origin" explanations were rather firmly entrenched and unquestioned in the Western ethos until Darwin, and the natural forms were considered permanent . In addition, Linnaean classification was fundamentally rooted in the venerable Platonic concept of types, through the use of type specimens. Little attention was paid to the variation within species except to see how that variation helped to focus on the essence of the type. The classical concept of the *Scala Natura,* or the Great Chain of Being, as modified by Biblical creation, was the organizing matrix for biology, and such attention as was directed to natural variation was directed to completing the description of the Great Chain. If all the required forms of living things had been created by a perfect Creator in a single perfect Creation, there was no reason to think that these forms were not permanent or that they changed one into the other.

Only after the nineteenth-century demonstration of geological succession, largely due to Charles Lyell, and of the discovery of fossils of extinct species grading from the archaic to the modern in form did the concept of the Great Chain of Being come under serious attack. The *species* was still the aggregate of interest; the variation of interest was in the fossil intermediates that were increasingly being found in the nineteenth century. The biological concept of an aggregate, or population, within a species did not reach maturity until well after Darwin, nor in fact did demography. The history of the discovery of evolution is a fascinating, well-documented chapter in the annals of science (see Lovejoy, 1936; Greene, 1959; Eiseley, 1961; Provine, 1971).

REFERENCES

Eisely, L. (1961) *Darwin's Century*. Garden City, N.Y.: Doubleday/Anchor Books.

Greene, J. C. (1959) *The Death of Adam*. Ames, Iowa: Iowa State University Press.

Lovejoy, A. O. (1936) *The Great Chain of Being*. Cambridge, Mass.: Harvard University Press.

Provine, W. B. (1971) *The Origins of Theoretical Population Genetics*. Chicago: University of Chicago Press.

Part I
EARLY DEVELOPMENTS

Editors' Comments
on Papers 1 Through 5

1 GRAUNT
Natural and Political Observations Mentioned in a Following Index, and Made upon the Bills of Mortality

2 LOTKA
Relation Between Birth Rates and Death Rates

3 SHARPE and LOTKA
A Problem in Age-Distribution

4 VOLTERRA
A Mathematical Theory on the Struggle for Survival

5 LEWIS
On the Generation and Growth of a Population

THE CONCEPT OF POPULATION

The concept of a population as a meaningful unit has existed as long as man himself. Primitive bands clearly distinguish themselves from their neighbors; tribal unity fundamentally differentiates mankind and usually includes concepts of difference in quality as well as merely descriptive differences. Thus, although people have probably been interested in other biological species (mainly food species) since earliest times, the evolution of our ideas about populations within species clearly has centered on ourselves. The history of the evolution of civilizations can be seen as a history of gradual social and economic stratification of society. Such a history naturally produced an interest in the various strata; in particular, the upper stratum desired to keep records on the availability of manpower, taxes, and so on, which derived from lower strata. Add to this the ubiquitous primeval interest in population structure as it relates to the availability of mates and the network of kinship and social obligation, and one has a reasonable concept of the early interest in human populations.

The founding of modern demography, however, is generally attributed to John Graunt in 1662. Graunt was a London cloth merchant with an amateur scientist's interest in the nature of the English population. For the better exercise of government, he felt that the use of population statistics (then called "political arithmetic") was necessary. The raising of taxes and of armies and better political administration of the population were important reasons for this, but the occurrence of the great plagues also fueled his interest. Records of deaths in London were kept at that time, which included age and cause of death. These Bills of Mortality were studied by Graunt, who extracted from them demographic statistics that clearly foreshadowed modern census demography. Therefore, we begin our collection of papers with extracts from Graunt's *Natural and Political Observations Mentioned in a Following Index, and Made upon the Bills of Mortality.*

The sections we have included from Graunt as Paper 1 begin with an enumeration of the causes of death and a discussion of their relative impact and constancy over time. He considers the causes of death at differing ages, and in relation to other stratifications of mortality, such as wealth, and country versus city living. In dealing with the effects of the plagues, he details his ideas on the speed of demographic recovery in the "population" (London) and its source (migration from outlying districts). He examines changes in birth and death rates and the rate of growth of the city. He looks at cultural causes of mortality, such as air pollution, considers aspects of the sex ratio, and concludes with the first attempt ever made to construct a life table.

Even at this early stage Graunt has really anticipated many aspects of modern demographic genetics. Differential mortality and fertility, population size and growth rate, sex ratios, migration, and even mating patterns are all discussed in seventeenth-century terms. Finally, he deals with age structure and survivorship, clearly treating London as an aggregate, and clearly understanding that *statistical constructs can be abstracted from a population* (such as the death "rates" by age and cause). Neither a biologist nor a geneticist, Graunt nonetheless anticipates the concepts of differing viability among different individuals in a defined aggregate of individuals.

After Graunt, the most influential early worker on demographic concepts was Thomas Malthus, who sowed the seeds for Darwin himself, as Darwin acknowledges. The *Essay on the Principle of Population* of Malthus is an explicit treatment of population growth, carrying capacity, and density-dependent self-regulation. Differential mortality is seen as a vital force in the processes of population control.

Malthus observed that human populations have a growth potential which would make them rapidly exceed their source of sustenance; therefore, means exist to regulate the population by eliminating the excess individuals. These can be either "positive" checks (such as war or famine) or "preventive" checks (birth prevention). This concept was used directly by Darwin in the theory of evolution by natural selection: species overproduce, and only the "fit" are allowed to survive. In fact, Darwin used only the differential mortality Malthusian theory (positive checks) and basically ignored the idea of regulated differential fertility. He observed that only human societies produced "artificial" increases in food and he (wrongly) felt only humans practice "restraint from marriage" (*The Descent of Man,* p. 53), that is, social damping of reproduction. Malthus had recognized the concept of a changing carrying capacity; in clearly acknowledging his debt to Malthus, Darwin still knew that in nature most populations do not grow. It was the *potential for growth in the absence of actual growth that is the key to natural selection.*

The concept of growth has a central role in the ethos, philosophy, and social thought of Western industrial civilizations. Demography as a science has developed only because nations were aware of their growth and wished to understand and keep track of it. Repeatedly counting a static population would have proved a dull and most uninspiring exercise, but keeping count of an increasing tax and military conscription base, of changing political and church districts, and so on, posed a problem of immense social and practical interest to ruler and scientist alike.

In the following papers (when it applies), we wish to use the concept of population growth as an organizing and interpretive point, and to trace its impact on demographic genetics through a path of scientific evolution. Our organizing thesis will be that *the concept of population growth has had a vital, and inordinate, role to play in the history of demographic genetics, which is due to its role in our society as a whole.* Also, we shall try to show that, although it generally holds itself to be a Darwinian science, much that has been done with demographic genetics has been conceptually different from Darwin. In many cases this may merely be the artifact of mathematical convenience and may not result in substantially different conclusions; in other cases, however, it has hindered progress, at least conceptually.

EARLY CONCEPTS OF POPULATION GROWTH

Historically, two basic lines of approach have been taken in regard to population growth. We can examine them here from their original works. As we progress it will be clear how these works have influenced the papers that followed them. Although we consider interpopulation relations, population size, and density to be "demographic" aspects of a population, we can generally ascribe one line of approach, which dealt with these concepts, as an *ecological* line. This is because the people who examined these questions usually did so with an ecologist's viewpoint relative to the other line of thought. That other line was a more purely *demographic* one, so called because it was developed and used mainly by demographers whose main consideration was age structure and vital rates within a single population, topics generally of interest to national census demographers.

The demographic line of inquiry into population growth concepts was due to Alfred J. Lotka, and it would be inexcusable to assemble a collection of benchmark demographic genetics papers without including his work. Lotka is the acknowledged father of theoretical demography; although he is also involved with the origins of population ecology (see below), it is his detailed investigations on age structure that form the foundation of most present-day theoretical demography. It was Lotka who, in 1907, first proposed the *stable population*. Assuming a population closed to migration, with constant age-specific rates of death and birth, he showed that the age distribution of a population would approach a fixed distribution, with the total population size growing ultimately at a fixed rate. Lotka called this rate the "intrinsic rate of natural increase." This terminology has caused a great deal of subsequent confusion, as many have taken this rate to be the maximal growth rate of a population in an unlimited environment. Although Lotka's model is certainly density-free in its formal formulation, the growth rate is *not* the maximum possible for the species or population.

Lotka's theory contains the nongrowing population as a special case, yet it is clear that this was only of minor interest to users of the stable population model. Lotka was a demographer for most of his career, in that he concerned himself with the empirical effects of vital rate schedules in assessing human population behavior. Empirically, populations with recently fixed vital rates grow at rate $r(t)$; Lotka and his followers have spent much time examining the determination of the "true" rate of increase—that which would ultimately result, as intrinsic to the vital rates. The approach of $r(t)$ to r has only recently been well examined and summarized (Coale, 1972; Keyfitz, 1968). It cannot be denied that *nonzero* growth rates were Lotka's principal interest; this is due to the growing nature of recent European populations and the interest demographers attached to assessing that growth and the manner in which vital rates determined it. Until Keyfitz's book in 1968, the only adequate summary of stable demographic theory was the 1939 two-volume set by Lotka in French, although his most famous work *Elements of Mathematical Biology* (in 1924), is a statement of general population biologic theory that remains one of the most influential books on this topic in the twentieth century. Papers 2 and 3 are Lotka's first statements of the theory (Lotka, 1907; Sharpe and Lotka, 1911), which are less readily available than his book. We should also mention another good recent treatment of stable population theory by Bourgeois-Pichat (1968), and the fact that Euler (1760) anticipated many of these concepts in rudimentary fashion in a recently reprinted paper.

The "ecological" line of investigation has pre-Darwinian origins in the demonstration by Verhulst (1838) that populations in a limited environment follow a "logistic" pattern of growth. But the first important biological treatment of this topic is the work of Volterra from 1926 onward (with a parallel development of some of it by Lotka, 1924). Basically a mathematician, Volterra became interested in population biology when, at the time of World War I, he was asked by his colleague Umberto D'Ancona to explain the relative abundances of predator and prey fishes in catches from Adriatic waters in that period. Volterra approached this problem by first constructing general equations for the relations between animal populations and then restricting these to specific cases for empirical or qualitative analysis. His life and work have been summarized by Whittaker (1959) and Scudo (1971), and his theories are presented in full in (among other sources) Chapman (1931) and D'Ancona (1954). Rather than reprint the more extensive treatments, we have chosen Paper 4 to give a short, early expression of Volterra's fundamental ideas on the subject of "a mathematical theory on the struggle for survival" (1927).

Volterra deals with the mechanics of change in total population numbers. Some of his results were independently derived by Lotka (1924); however, Volterra expands on the kinds of interactions that can occur. In so doing, he treats the general problem of the effect of various kinds of population interaction on population growth. This approach still finds widespread use in population biology (e.g., May, 1973), and the logistic-growth model is the first that is used when density effects are included in a model. These topics have been reviewed in Emlen (1973), Pielou (1969), Keyfitz (1968), Slobodkin (1961), MacArthur (1972), Watt (1968), Levins (1968), and May (1973), and in other extensive treatments in the ecological literature.

The ecological and demographic approaches to the problem of analyzing population growth have formed the threads of much thinking on demographic genetics. Before proceeding to document that genetic work, however, we must mention an immensely important method of studying the stable population —the discrete age–class matrix projection method generally attributed to Leslie (1945, 1948, 1959). Leslie derived a simple transition-matrix analog of Lotka's continuous age distribution model, which we have already reviewed. The method allows one to analyze real biological populations and has become the foundation of much of demography. Leslie's first two papers appeared in *Biometrika,* and hence are readily available. However, in 1942, three years before Leslie's work, E. G. Lewis published a short but full anticipation of Leslie's matrix (Paper 5). We therefore include Lewis's work in this volume (see also Bernardelli, 1941). Further development of Leslie's treatment may be found in Keyfitz (1968) or other formal demography works.

1

Reprinted from *The Economic Writings of Sir William Petty*, Vol. 2, C. H. Hull, ed., Cambridge University Press, London, 1899, pp. 346–378, 383–388

NATURAL AND POLITICAL OBSERVATIONS MENTIONED IN A FOLLOWING INDEX, AND MADE UPON THE BILLS OF MORTALITY

John Graunt

[*Editors' Note:* In the original, material precedes this excerpt.]

[C H A P. II.][1]

General Observations upon the Casualties.

IN my Discourses upon these *Bills*, I shall first speak of the *Casualties*, then give my Observations with reference to the *Places* and *Parishes* comprehended in the *Bills*; and next of the *Years* and *Seasons*.

---

[1] This line, omitted from the fifth edition, occurs in the first four.

1. There seems to be good reason, why the *Magistrate* should himself take notice of the ‖ numbers of *Burials* and (18) *Christnings, viz.* to see whether the City increase or decrease in People; whether it increase proportionably with the rest of the Nation; whether it be grown big enough, or too big, *&c.* But why the same should be made known to the People, otherwise than to please them, as with a curiosity, I see not.

2. Nor could I ever yet learn (from the many I have asked, and those not of the least *Sagacity*) to what purpose the distinction between *Males* and *Females* is inserted, or at all taken notice of? or why that of *Marriages* was not equally given in? Nor is it obvious to every body, why the Accompt of *Casualties* (whereof we are now speaking) is made? The reason, which seems most obvious for this later, is, That the state of health in the City may at all times appear.

3. Now it may be Objected, That the same depends most upon the Accompts of *Epidemical Diseases*, and upon the chief of them all, the *Plague*; wherefore the mention of the rest seems only matter of curiosity.

4. But to this we Answer, That the knowledge even of the numbers which dye of the *Plague*, is not sufficiently deduced from the meer Report of the *Searchers*, which only the Bills afford; but from other Ratiocinations, ‖ and com- (19) parings of the *Plague* with some other *Casualties*.

5. For we shall make it probable[1], that in the Years of *Plague*, a quarter part more dies of that *Disease* than are set down; the same we shall also prove by other *Casualties*. Wherefore, if it be necessary to impart to the world a good Accompt of some few *Casualties*, which since it cannot well be done without giving an Accompt of them all, then is our common practice of so doing very apt and rational.

6. Now, to make these Corrections upon the, perhaps, ignorant and careless *Searchers* Reports, I considered first of what Authority they were of themselves, that is, whether any credit at all were to be given to their Distinguishments: and finding that many of the *Casualties* were but matter of sense, as whether a Child were *Abortive* or *Stilborn*; whether men

[1] See p. 365.

were *Aged*, that is to say, above sixty years old, or thereabouts when they died, without any curious determination ; whether such *Aged* persons died purely of *Age*, as for that the *Innate heat* was quite extinct, or the *Radical moisture* quite dried up (for I have heard some Candid *Physicians* complain of the darkness which themselves were in hereupon[1]) I say, that (20) these Distin- guishments being but matter of sense, I concluded the Searche s Report might be sufficient in the Case.

7. As for *Consumptions*, if the *Searchers* do but truly Report (as they may) whether the dead Corps were very lean and worn away, it matters not to many of our purposes, whether the Disease were exactly the same, as *Physicians* define it in their Books. Moreover, In case a man of seventy five years old died of a *Cough* (of which had he been free, he might have possibly lived to ninety) I esteem it little errour (as to many of our purposes) if this Person be in the Table of *Casualties*, reckoned among the *Aged*, and not placed under the Title of *Coughs*.

8. In the matters of *Infants* I would desire but to know clearly, what the *Searchers* mean by *Infants*, as whether Children that cannot speak, as the word *Infant* seems to signifie, or Children under two or three years old, although I should not be satisfied, whether the *Infant* died of *Wind*, or of *Teeth*, or of the *Convulsion*, &c. or were choaked with *Phlegm*, or else of *Teeth*, *Convulsion*, and *Scowring*, apart, or together, which, they say, do often cause one another ; for, I say, it is somewhat to know how many die usually before they can speak, or how many live past any assigned number of years ||.

(21)   9. I say, it is enough, if we know from the *Searchers* but the most predominant Symptoms ; as that one died of the *Headach*, who was sorely tormented with it, though the *Physicians* were of Opinion, that the Disease was in the *Stomach*. Again, if one died *suddenly*, the matter is not great, whether it be reported in the Bills, *Suddenly*, *Apoplexy*, or *Planet-strucken*, &c.

---

[1] " For both the common phrases of physicians concerning Radical Heat and Natural Moisture are deceptive." Bacon, X 11

10. To conclude, In many of these Cases the *Searchers* are able to report the Opinion of the *Physician*, who was with the Patient, as they receive the same from the Friends of the Defunct : and in very many Cases, such as *Drowning, Scalding, Bleeding, Vomiting, making away themselves, Lunaticks, Sores, Small-pox, &c.* their own senses are sufficient, and the generality of the World are able pretty well to distinguish the *Gout, Stone, Dropsie, Falling sickness, Palsie, Agues, Pleuresie, Rickets,* one from another.

11. But now as for those Casualties, which are aptest to be confounded and mistaken, I shall in the ensuing Discourse presume to touch upon them so far, as the Learning of these Bills hath enabled me.

12. Having premised these general Advertisements, our first Observation upon the *Casualties* shall be, That in Twenty Years[1] ‖ there dying of all Diseases and Casualties 229250, (22) that 71124[2] died of the *Thrush, Convulsion, Rickets, Teeth* and *Worms*; and as *Abortives, Chrysomes, Infants, Livergrown,* and *Overlaid* ; that is to say, that about $\frac{1}{3}$ of the whole died of those Diseases, which we guess did all light upon Children under four or five years old.

13. There died also of the *Small Pox, Swine Pox,* and *Measles,* and of *Worms* without *Convulsions,* 12210[3], of which number we suppose likewise, that about $\frac{1}{2}$ might be Children under six years old. Now, if we consider that sixteen[4] of the said 229250 died of that extraordinary and grand Casualty, the *Plague,* we shall find that about thirty six *per Centum* of all quick conceptions died before six years old.

14. The second Observation is, That of the said 229250 dying of all Diseases, there died of *acute* Diseases, (the *Plague* excepted) but about 50000, or $\frac{2}{9}$ parts. The which proportion

---

[1] The years are 1629—1636, and 1647—1658, see the Table of Casualties, p. 406.

[2] These figures do not correspond to Graunt's table (p. 406) which gives thrush 211, convulsion 9,073, rickets 3,681, teeth and worms 14,236, abortive and still-born 8,559, chrisoms and infants 32,106, liver-grown, spleen, and rickets 1,421, overlaid and starved at nurse 529, or in all but 69,816.

[3] According to the table (p. 406) there died of swine-pox 57, of flox and small-pox 10,576, of measles 757, of worms (without convulsions) 830, or in all 12,220.

[4] That is, sixteen thousand; according to the table (p. 406), 16,384.

doth give a measure of the State, and disposition of this *Climate* and *Air* as to health; these *acute* and *Epidemical* Diseases happenning suddenly and vehemently, upon the like corruptions and alterations in the *Air.* ||

(23)   15.   The third Observation is, That of the said 229250, about seventy[1] died of *Chronical* Diseases, which shews (as I conceive) the State and Disposition of the Country (including as well its *Food* as *Air*) in reference to health, or rather to *longevity* : for as the proportion of *acute* and *Epidemical* Diseases shews the aptness of the *Air* to sudden and vehement Impressions ; so the *Chronical* Diseases shew the ordinary temper of the place : so that upon the proportion of *Chronical* Diseases seems to hang the judgment of the fitness of the Country for *long life*.   For, I conceive, that in Countries subject to great *Epidemical* sweeps, men may live very long, but, where the proportion of the *Chronical* distempers is great, it is not likely to be so ; because men being long sick, and alwaies sickly, cannot live to any great Age, as we see in several sorts of *Metal-men*, who, although they are less subject to *acute* Diseases than others, yet seldom live to be old, that is, not to reach unto those years, which *David* says is the Age of Man.

    16.   The fourth Observation is, That of the said 229250, not 4000 died of outward Griefs, as of *Cancers, Fistula's, Sores, Ulcers, broken and bruised Limbs, Imposthumes, Itch, King's*

(24) *Evil, Leprosie, Scald-head,* || *Swine Pox, Wens,* &c. *viz.* not one in sixty.

    17.   In the next place, whereas many persons live in great fear and apprehension of some of the more formidable and notorious Diseases following ; I shall only set down how many died of each : that the respective numbers, being compared with the total 229250, those persons may the better understand the hazard they are in.

---

[1] That is, seventy thousand.   The German translator of the *Observations* writes "70 vom hundert."

16

Table of notorious Diseases.

Apoplex	1306
Cut of the Stone	38
Falling Sickness	74
Dead in the Streets	243
Gout	134
Head-ach	51
Jaundice	998
Lethargy	67
Leprosie	6
Lunatick	158
Overlaid and Starved	529
Palsie	423
Rupture	201
Stone and Strangury	863
Sciatica	5
Suddenly	454 ‖

Table of Casualties. (25)

Bleeding	69
Burnt and Scalded	125
Drowned	829
Excessive drinking	2
Frighted	22
Grief	279
Hanged themselves	222
Kill'd by several accidents	1021
Murdered	86
Poysoned	14
Smothered	26
Shot	7
Starved	51
Vomiting	136

18. In the foregoing Observations we ventured to make a Standard of the healthfulness of the *Air* from the proportion of *acute* and *Epidemical* Diseases, and of the wholsomness of

17

the food, from that of the *Chronical.* Yet, for as much as neither of them alone do shew the *longevity* of the Inhabitants, we shall in the next place come to the more absolute Standard and Correction of both, which is the proportion of the Aged, (26) *viz.* 15757 to the Total 229250. That ‖ is, of about 1 to 15, or 7 *per Cent.* Only the question is, What number of years the *Searchers* call *Aged,* which I conceive must be the same that *David* calls so, *viz.* 70. For no man can be said to die properly of *Age,* who is much less. It follows from hence, That if in any other Country more than seven of the 100 live beyond 70, such Country is to be esteemed more healthful than this of our City.

19. Before we speak of particular *Casualties,* we shall observe, That among the several *Casualties* some bear a constant proportion unto the whole number of *Burials* ; such are *Chronical* Diseases, and the Diseases whereunto the City is most subject ; as for Example, *Consumptions, Dropsies, Jaundice, Gout, Stone, Palsie, Scurvy, Rising of the Lights* or *Mother, Rickets, Aged, Agues, Fevers, Bloody Flux* and *Scowring* : nay, some Accidents, as *Grief, Drowning, Men's making away themselves,* and being *Kill'd by several Accidents, &c.* do the like ; whereas *Epidemical* and *Malignant* Diseases, as the *Plague, Purples, Spotted Fever, Small Pox* and *Measles* do not keep that equality : so as in some Years, or Months, there died ten times as many as in others. ‖

---

(27) ## CHAP. III.

Of Particular Casualties.

1. MY first Observation is, that few are *starved.* This appears, for that of the 229250, which have died, we find not above fifty one to have been *starved,* excepting helpless *Infants* at Nurse, which being caused rather by carelessness, ignorance, and infirmity of the Milch-women, is not properly an effect or sign of want of food in the Country, or of means to get it.

2. The Observation which I shall add hereunto, is, That the vast number of *Beggars*, swarming up and down this City, do all live, and seem to be most of them healthy and strong; whereupon I make this question, Whether, since they do all live by begging, that is, without any kind of labour; it were not better for the State to keep them, even although they earned nothing? that so they might live regularly, and not in that Debauchery, as many Beggars do; and that they might be cured of their bodily Impotencies, || or taught to (28) work, &c. each according to his condition and capacity; or by being imployed in some work (not better undone) might be accustomed and fitted for labour?

3. To this some may Object, That *Beggars* are now maintained by voluntary Contributions, whereas in the other way the same must be done by general Tax; and consequently, the Objects of Charity would be removed and taken away.

4. To which we Answer, That in *Holland*, although no where fewer Beggars appear to charm up commiseration in the credulous, yet no where is there greater or more frequent Charity: only indeed the Magistrate is both the *Beggar*, and the *Disposer* of what is got by *begging*; so as all Givers have a Moral certainty that their Charity shall be well applyed.

5. Moreover, I question, Whether what we give to a Wretch that shews us lamentable sores and mutilations, be alwaies out of the purest Charity? that is, purely for God's sake; for as much as when we see such Objects, we then feel in our selves a kind of pain and passion by consent, of which we ease our selves, when we think we ease them, with whom we sympathised; or else we bespeak aforehand the like commiseration in || others towards our selves, when we shall (29) (as we fear we may) fall into the like distress.

6. We have said, *'Twere better the Publick should keep the Beggars, though they earned nothing*, &c. But most men will laugh to hear us suppose, That any able to work (as indeed most *Beggars* are, in one kind of measure or another) should be kept without earning any thing. But we Answer, That if there be but a certain proportion of work to be done, and

that the same be already done by the *non-Beggars*, then to imploy the *Beggars* about it, will but transfer the want from one hand to another; nor can a Learner work so cheap as a skilful practised Artist can. As for example, a practised *Spinner* shall spin a pound of Wool, worth two shillings, for six pence; but a Learner, undertaking it for three pence, shall make the wool indeed into yarn, but not worth twelve pence.

7. This little hint is the model of the greatest work in the World, which is the making of *England* as considerable for Trade as *Holland*; for there is but a certain proportion of Trade in the World, and *Holland* is prepossessed of the greatest part of it, and is thought to have more skill and experience to manage it; wherefore, to bring *England* into (30) *Holland*'s condition, as to this particular, ‖ is the same, as to send all the *Beggars* about *London* into the *West Country* to Spin, where they shall only spoil the Clothiers Wool, and beggar the present Spinners at best; but, at worst, put the whole Trade of the Country to a stand, until the *Hollander*, being more ready for it, have snapt that with the rest.

8. My next Observation is, That but few are *Murthered*, *viz.* not above 86 of the 229250, which have died of other Diseases and Casualties; whereas in *Paris* few nights scape without their *Tragedy*.

9. The Reasons of this we conceive to be *Two*: One is the *Government* and *Guard* of the City by *Citizens* themselves, and that alternately. No man setling into a Trade for that employment. And the other is, The natural and customary abhorrence of that inhuman *Crime*, and all *Bloodshed*, by most *English men*: for of all that are *Executed*, few are for *Murther*. Besides the great and frequent Revolutions and Changes in Government since the Year 1650, have been with little *bloodshed*; the *Usurpers* themselves having *Executed* few in comparison, upon the Accompt of disturbing their Innovations.

10. In brief, when any dead Body is found in *England*, (31) no *Algebraist*, or *Uncypherer* of ‖ Letters, can use more subtile suppositions and variety of conjectures to find out the Demonstration or Cipher, than every common unconcerned person

doth to find out the Murtherers, and that for ever, until it be done.

11. The *Lunaticks* are also but few, *viz.* 158 in 229250, though I fear many more than are set down in our *Bills*, few being entred for such, but those who die at *Bedlam*; and there all seem to dye of their *Lunacy*, who died *Lunaticks*; for there is much difference in computing the number of *Lunaticks*, that die (though of *Fevers* and all other Diseases, unto which *Lunacy* is no *Supersedeas*) and those that dye by reason of their *Madness*.

12. So that, this *Casualty* being so uncertain, I shall not force my self to make any inference from the numbers and proportions we find in our Bills concerning it : only I dare ensure any man at this present, well in his Wits, for one in a thousand, that he shall not dye a *Lunatick* in *Bedlam* within these seven years, because I find not above one in about one thousand five hundred have done so.

13. The like use may be made of the Accompts of men that made away themselves, ‖ who are another sort of Mad (32) men, that think to ease themselves of pain by leaping into *Hell*; or else are yet more Mad, so as to think there is no such place; or that men may go to rest by death, though they dye in *Self-murther*, the greatest Sin.

14. We shall say nothing of the numbers of those that have been *Drowned, Killed by falls from Scaffolds*, or by *Carts running over them, &c.* because the same depends upon the casual Trade and Employment of men, and upon matters which are but circumstantial to the Seasons and Regions we live in, and affords little of that Science and Certainty we aim at.

15. We find one *Casualty* in our Bills, of which, though there be daily talk, there is little effect, much like our abhorrence of *Toads* and *Snakes* as most poisonous Creatures, whereas few men dare say upon their own knowledge they ever found harm by either ; and this *Casualty* is the *French Pox*, gotten, for the most part, not so much by the intemperate use of *Venery* (which rather causeth the *Gout*) as of many common Women.

16. I say, the *Bills* of *Mortality* would take off these Bars, which keep some men within bounds, as to these (33) extravagancies : for in ‖ the aforementioned 229250, we find not above 392 to have died of the *Pox*.   Now, forasmuch as it is not good to let the World be lulled into a security and belief of Impunity by our *Bills*, which we intend shall not be only as *Deaths heads* to put men in mind of their *Mortality*, but also as *Mercurial Statues* to point out the most dangerous waies that lead us into it and misery ; We shall therefore shew, that the *Pox* is not as the *Toads* and *Snakes* afore-mentioned, but of a quite contrary nature, together with the reason why it appears otherwise.

17. Forasmuch as by the ordinary discourse of the World it seems a great part of men have, at one time or other, had some *species* of this Disease, I wondering why so few died of it, especially because I could not take that to be so harmless, whereof so many complained very fiercely ; upon enquiry, I found that those who died of it out of the Hospitals (especially that of *Kingsland*, and the *Lock* in *Southwark*) were returned of *Ulcers* and *Sores*.   And in brief, I found that all mentioned to dye of the *French Pox* were returned by the *Clerks* of Saint *Giles*'s and Saint *Martin*'s *in the Fields* only, in which place I understood that most of the vilest and (34) most miserable Houses of Un-‖cleanness were : from whence I concluded, that only *hated* persons, and such, whose very *Noses* were eaten off, were reported by the *Searchers* to have died of this too frequent *Malady*.

18. In the next place, it shall be examined, under what Name or *Casualty* such as die of these Diseases are brought in : I say, under the *Consumption* ; forasmuch as all dying thereof dye so emaciated and lean (their *Ulcers* disappearing upon Death) that the Old-women *Searchers*, after the mist of a Cup of *Ale*, and the bribe of a Two-groat fee, in stead of one given them[1], cannot tell whether this emaciation or leanness

---

[1] Cromwell's act of 24 August, 1653, provided for the election by each parish of a parish registrar, who might take " for every Birth of Childe, Four pence and no more ; and for every Death, Four pence and no more : And for Publications, Marriages, Births or Burials of poor people who live upon Alms, nothing shal

were from a *Phthisis*, or from an *Hectick Fever*, *Atrophy*, &c. or from an Infection of the *Spermatick* parts, which in length of time, and in various disguises hath at last vitiated the habit of the Body, and by disabling the parts to digest their nourishment, brought them to the condition of leanness above-mentioned.

19. My next Observation is, That of the *Rickets* we find no mention among the *Casualties*, until the Year 1634, and then but of 14 for that whole Year.

20. Now the Question is, Whether that Disease did first appear about that time ; or whether a Disease, which had been long be-‖fore, did then first receive its Name ? (35)

21. To clear this Difficulty out of the Bills (for I dare venture on no deeper Arguments) I enquired what other Casualtie before the Year 1634, named in the Bills, was most like the *Rickets*; and found, not only by Pretenders to know it, but also from other Bills, that *Livergrown* was the nearest. For in some years I find *Livergrown*, *Spleen*, and *Rickets*, put all together, by reason (as I conceive) of their likeness to each other. Hereupon I added the *Livergrowns* of the Year 1634, *viz.* 77, to the *Rickets* of the same Year, *viz.* 14, making in all 91 ; which Total, as also the Number 77 it self, I compared with the *Livergrown* of the precedent Year 1633, *viz.* 82 : All which shewed me, that the *Rickets* was a new Disease over and above.

22. Now, this being but a faint Argument, I looked both forwards and backwards, and found, that in the Year 1629, when no *Rickets* appeared, there were but 94 *Livergrowns*; and in the Year 1636 there were 99 *Livergrown*, although there were also 50 of the *Rickets* : only this is not to be denied, that when the *Rickets* grew very numerous (as in the Year 1660, *viz.* 521) then there appeared not above 15 of *Livergrown*. ‖

23. In the Year 1659 were 441 *Rickets*, and 8 *Livergrown*. (36) In the Year 1658 were 476 *Rickets*, and 51 *Livergrown*. Now,

be taken," Scobell, II. 236. In most cases the old parish clerk was elected registrar (Christie, 140), and in London the parish clerks may have collected their fees through the searchers.

though it be granted that these Diseases were confounded in the Judgment of the *Nurses*, yet it is most certain, that the *Livergrown* did never but once, *viz. Anno* 1630 exceed 100; whereas *Anno* 1660, *Livergrown* and *Rickets* were 536.

24. It is also to be observed, That the *Rickets* were never more numerous than now, and that they are still increasing; for *Anno* 1649, there were but 190, next year 260, next after that 329, and so forwards, with some little starting backwards in some years, until the Year 1660, which produced the greatest of all.

25. Now, such back-startings seem to be universal in all things; for we do not only see in the progressive motion of the wheels of *Watches,* and in the rowing of *Boats,* that there is a little starting or jerking backwards between every step forwards, but also (if I am not much deceived) there appeared the like in the motion of the *Moon,* which in the long *Telescopes* at *Gresham Colledge* one may sensibly discern[1]. ||

(37)   26. There seems also to be another new Disease, called by our Bills *The stopping of the Stomach,* first mentioned in the Year 1636, the which *Malady,* from that Year to 1647, increased but from 6 to 29; *Anno* 1655 it came to 145. In 57, to 277. In 60 to 314. Now these proportions far exceeding the difference of proportion generally arising from the increase of Inhabitants, and from the resort of *Advenæ* to the City, shews there is some new Disease, which appeareth to the Vulgar, as *A stopping of the Stomach.*

27. Hereupon I apprehended that this *Stopping* might be the *Green sickness,* forasmuch as I find few or none to have been returned upon that Account, although many be visibly stained with it. Now, whether the same be forborn out of shame, I know not: For since the World believes that Marriage cures it, it may seem indeed a shame, that any Maid should dye uncured, when there are more *Males* than *Females,* that is, an overplus of Husbands to all that can be Wives.

---

[1] "The author, going *ultra crepidam,* has attributed to the motion of the moon in her orbit all the tremors which she gets from a shaky telescope." De Morgan, *Budget of Paradoxes,* 68.

28. In the next place, I conjectured that this *stopping of the Stomach* might be the *Mother*, forasmuch as I have heard of many troubled with *Mother fits* (as they call them) ‖ although few returned to have died of them; which con- (38) jecture, if it be true, we may then safely say, That the *Mother-fits* have also increased.

29. I was somewhat taken off from thinking this *stopping of the Stomach* to be the *Mother*, because I ghessed rather the *Rising of the Lights* might be it. For I remembred that some Women, troubled with the *Mother-fits*, did complain of *a choaking in their Throats*. Now, as I understand, it is more conceivable, that the *Lights* or *Lungs* (which I have heard called *The Bellows of the Body*) not blowing, that is, neither venting out, nor taking in breath, might rather cause such a *Choking*, than that the *Mother* should rise up thither, and do it. For methinks, when a Woman is with Child, there is a greater rising, and yet no such Fits at all.

30. But what I have said of the *Rickets* and *stopping of the Stomach*, I do in some measure say of the *Rising of the Lights* also, *viz.* that these *Risings* (be they what they will) have increased much above the general proportion; for in 1629 there were but 44, and in 1660, 249, *viz.* almost six times as many. ‖

31. Now forasmuch as *Rickets* appear much in the *Over-* (39) *growing* of *Childrens Livers* and *Spleens* (as by the Bills may appear) which surely may cause *stopping of the Stomach* by squeezing and crowding upon that part. And forasmuch as these *Chokings* or *Risings of the Lights* may proceed from the same stuffings, as make the *Liver* and *Spleen* to over- grow their due proportion. And lastly, forasmuch as the *Rickets, stopping of the Stomach, and rising of the Lights*, have all increased together, and in some kind of correspondent proportions; it seems to me that they depend one upon another, And that what is the *Rickets* in Children, may be the other in more grown Bodies; for surely Children, which recover of the *Rickets*, may retain somewhat to cause what I have imagined: but of this let the Learned *Physicians* consider, as I presume they have.

32. I had not medled thus far, but that I have heard, the first hints of the circulation of the Blood were taken from a common Person's wondering what became of all the blood which issued out of the heart, since the heart beats above three thousand times an hour, although but one drop should be pump'd out of it at every stroke. ‖

(40)　　33. The *Stone* seemed to decrease : for in 1632, 33, 34, 35, and 36, there died of the *Stone* and *Strangury* 254. And in the Years 1655, 56, 57, 58, 59, and 1660, but 250, which numbers, although indeed they be almost equal, yet considering the Burials of the first named five Years were but half those of the later, it seems to be decreased by about one half.

34. Now the *Stone* and *Strangury* are Diseases which most men know that feel them, unless it be in some few cases, where (as I have heard *Physicians* say) a *Stone* is held up by the *Films* of the *Bladder*, and so kept from grating or offending it.

35. The *Gout* stands much at a stay, that is, it answers the general proportion of Burials ; there dies not above one of 1000 of the *Gout*, although I believe that more dye *Gouty*. The reason is, because those that have the *Gout*, are said to be *long livers*; and therefore, when such dye, they are returned as *Aged*.

36. The *Scurvy* hath likewise increased, and that gradually from 12, *Anno* 1629, to 95, *Anno* 1660.

37. The *Tyssick* seems to be quite worn away, but that it is probable the same is entred as *Cough* or *Consumption*. ‖

(41)　　[38]. *Agues* and *Fevers* are entred promiscuously, yet in the few Bills wherein they have been distinguished, it appears that not above 1 in 40 of the whole are *Agues*.

39. The *Abortives* and *Stilborn* are about the twentieth part of those that are *Christned,* and the numbers seemed the same thirty Years ago as now, which shews there were more in proportion in those years than now : or else that in these later years due Accompts have not been kept of the *Abortives,* as having been buried without notice, and perhaps not in *Church-yards*.

40. For that there hath been a neglect in the Accompts of the *Christnings*, is most certain, because until the Year 1642, we find the *Burials* but equal with the *Christnings*, or near thereabouts, but in 1648, when the differences in *Religion* had changed the Government, the *Christnings* were but two thirds of the *Burials*. And in the Year 1659, not half, *viz.* the *Burials* were 14720 (of the *Plague* but 36) and the *Christnings* were but 5670; which great disproportion could be from no other Cause than that abovementioned, forasmuch as the same grew as the Confusions and Changes grew. ‖

41. Moreover, although the Bills give us in *Anno* 1659, (42) but 5670 *Christnings*, yet they give us 421 *Abortives*, and 226 dying in *Child-bed*; whereas in the Year 1631, when the *Abortives* were 410, that is, near the number of the Year 1659, the *Christnings* were 8288. Wherefore by the proportion of *Abortives*, *Anno* 1659, the *Christnings* should have been about 8500: but if we shall reckon by the Women dying in *Childbed*, of whom a better Accompt is kept than of *Stilborns* and *Abortives*, we shall find *Anno* 1659, there were 226 *Child-beds*; and *Anno* 1631, 112, *viz.* not ½: Wherefore I conceive that the true number of the *Christnings*, *Anno* 1659, is above double to the 5690 set down in our Bills; that is, about 11500, and then the *Christnings* will come near the same proportion to the *Burials*, as hath been observed in former times.

42. In regular Times, when Accompts were well kept, we find that not above three in 200 died in *Childbed*, and that the number of *Abortives* was about treble to that of the Women dying in *Childbed*: from whence we may probably collect, that not one Woman of an hundred (I may say of two hundred) dies in her Labour; forasmuch as there be other Causes of a Womans dying with-‖in the Month, than (43) the hardness of her Labour.

43. If this be true in these Countries, where Women hinder the facility of their *Child-bearing* by affected straitening of their Bodies; then certainly in *America*, where the same is not practised, Nature is little more to be taxed as to Woman, than in *Brutes*, among whom not one in some thousands do

dye of their Deliveries: what I have heard of the *Irish women* confirms me herein.

44.   Before we quite leave this matter, we shall insert the Causes, why the Accompt of *Christnings* hath been neglected more than that of *Burials*: one, and the chief whereof, was a Religious Opinion against *Baptizing of Infants*, either as unlawful, or unnecessary.   If this were the only reason, we might by our defects of this kind conclude the growth of this Opinion, and pronounce, that not half the People of *England*, between the years 1650 and 1660, were convinced of the need of *Baptizing*.

45.   A second Reason was, The scruples which many publick *Ministers* would make of the worthiness of Parents to have their Children Baptized, which forced such questioned Parents, who did also not believe the necessity of having their (44) Children *baptized* ‖ by such Scruplers, to carry their Children unto such other *Ministers*, as having performed the thing, had not the Authority or Command of the *Register* to enter the Names of the *baptized*.

46.   A third Reason was, That a little Fee was to be paid for the *Registry*[1].

47.   Upon the whole matter it is most certain, That the number of *Heterodox* Believers was very great between the said year 1650 and 1660; and so peevish were they, as not to have the Births of their Children *Registred*, although thereby the time of their coming of Age might be known, in respect of such Inheritances as might belong unto them; and withal, by such *Registring* it would have appeared unto what *Parish* each Child had belonged, in case any of them should happen to want its relief.

48.   Of *Convulsions* there appeared very few, *viz.* but 52 in the year 1629, which in 1636 grew to 709, keeping about that stay till 1659, though sometimes rising to about 1000.

49.   It is to be noted, That from 1629 to 1636, when the *Convulsions* were but few, the number of *Chrysoms* and *Infants* was greater: for in 1629, there were of *Chrysoms* and *Infants* (45) 2596, and of the *Convulsion* 52, ‖ *viz.* of both 2648.   And in

[1] See p. 356 note.

1636 there were of *Infants* 1895, and of the *Convulsions* 709; in both 2604, by which it appears, that this difference is likely to be only a confusion in the Accounts.

50. Moreover, we find that for these later years, since 1636, the total of *Convulsions* and *Chrysoms* added together are much less, *viz.* by about 400 or 500 *per Annum*, than the like Totals from 1629 to 36, which makes me think, that *Teeth* also were thrust in under the Title of *Chrysoms* and *Infants*, inasmuch as in the said years, from 1629 to 1636, the number of *Worms* and *Teeth* wants by above 400 *per Annum* of what we find in following years. ‖

CHAP. IV. (46)

Of the Plague.

1. B Efore we leave to discourse of the *Casualties*, we shall add something concerning that greatest *Disease* or *Casualty* of all, The *Plague*.

There have been in *London*, within this Age, four times of great *Mortality*, that is to say, the years 1592 and 1593, 1603, 1625 and 1636.

[1]There died *Anno* 1592. from *March* to *December*,	25886
Whereof of the *Plague*	11503
Anno 1593,	17844
Whereof of the *Plague*	10662
Christned in the said year	4021
Anno 1603, within the same space of time, were Buried	37294
Whereof of the Plague	30561
Anno 1625, within the same space	51758
Whereof of the *Plague*	35417
Anno 1636, from *April* to *Decemb.*	23359
Whereof of the *Plague*	10460 ‖

[1] On the trustworthiness of the following figures see the notes to the "Table shewing how many died weekly," p. 426.

(47)   2.   Now it is manifest of it self, in which of these years most died; but in which of them was the greatest *Mortality* of all Diseases in general, or of the *Plague* in particular, we discover thus.   In the Years 1592, and 1636, we find the proportion of those dying of the *Plague* in the whole to be near alike, that is, about 10 to 23, or 11 to 25, or as about 2 to 5.

3.   In the Year 1625, we find the *Plague* to bear unto the whole in proportion as 35 to 51, or 7 to 10, that is almost the triplicate of the former proportion; for the *Cube* of 7 being 343, and the *Cube* of 10 being 1000, the said 343 is not $\frac{1}{8}$[1] of 1000.

4.   In *Anno* 1603, the proportion of the *Plague* to the whole was as 30 to 37, *viz.* as 4 to 5, which is yet greater than the last of 7 to 20[2]: For if the year 1625 had been as great a *Plague* year as 1603, there must have died not only 7 to 10, but 8 to 10, which in those great numbers makes a vast difference.

5.   We must therefore conclude the year 1603 to have been the greatest *Plague* year of this Age.

6.   Now to know in which of these four was the greatest Mortality at large, we reason thus : ‖

(48) *Anno* (Buried   26490} or (6
1592 (Christned   4277} as (1

Anno (There died in the whole year of all   38244} or (8
1603 (Christned   4784} as (1

1 to 8, or) *Anno* (Died in the whole year   54265} or (8
1¼ to 10 ∫ 1625 (Christned   6983} as (1

Anno (There died, *ut supra,*   23359} or (5
1636 (Christned   9522} as (2

7.   From whence it appears, That *Anno* 1636, the Christnings were about $\frac{2}{5}$ parts of the Burials : *Anno* 1592 but $\frac{1}{6}$; but in the year 1603, and 1625, not above an eighth : so that the said two years were the years of greatest *Mortality*.   We said that the year 1603 was the greatest *Plague* year.   And

---

[1] 1st. ed., '$\frac{2}{8}$,' German transl., 'nicht $\frac{1}{3}$.'   [2] 20 is a misprint for 10.

now we say, that the same was not a greater year of *Mortality* than *Anno* 1625. Now to reconcile these two Positions, we must alledge, that *Anno* 1625, there was an errour in the Accompts or Distinctions of the *Casualties*; that is, more died of the *Plague* than were re-‖counted for under that (49) name. Which Allegation we also prove thus, *viz.*

8. In the said year 1625 there are said to have died of the *Plague* 35417, and of all other Diseases 18848; whereas in the years, both before and after the same, the ordinary number of Burials was between 7 and 8000; so that if we add about 11000 (which is the difference between 7 and 18) to our 35, the whole will be 46000, which bears to the whole 54000, as about 4 to 5, thereby rendring the said year 1625 to be as great a *Plague*-year as that of 1603, and no greater; which answers to what we proved before, *viz.* that the *Mortality* of the two years was equal[1].

9. From whence we may probably suspect, that about $\frac{1}{4}$ part more died of the *Plague* than are returned for such; which we further prove by noting, that *Anno* 1636 there died 10400 of the *Plague*, the $\frac{1}{4}$ whereof is 2600. Now there are said to have died of all other Diseases that Year 12959, out of which number deducting 2600, there remain 10359, more than which there died not in several years next before and after the said Year 1636.

10. The next Observation we shall offer is, That the *Plague* of 1603 lasted eight Years. ‖ In some whereof there (50) died above 4000, in others above 2000, and in but one fewer than 600: whereas in the Year 1624 next preceding, and in the Year 1626 next following the said great *Plague*-year 1625, there died in the former but 11, and in the later but 134 of the *Plague*. Moreover, in the said Year 1625, the *Plague*

---

[1] The report of a case of the plague in any family led to the "shutting up" of the house infected, and thus increased the danger of the other members of the household. This danger was probably avoided, in many cases, by bribing the searchers. Creighton, I. 312, 318, 663, 672, also in *Social England*, IV. 469. The probable concealment of the plague was noted at the time. *Salvetti's Correspondence*, 11 July, 1625, *Hist. MSS. Com.* XI. pt. I. p. 26—27; Rev. Joseph Mead to Sir Martin Stuteville, Birch, *Court and Times of Charles I.*, vol. I. p. 39.

decreased from its utmost number 4461 a week, to below 1000 within six weeks.

11. The *Plague* of 1636 lasted twelve Years, in eight whereof there died 2000 *per annum* one with another, and never under 300. The which shews, that the Contagion of the *Plague* depends more upon the Disposition of the *Air*, than upon the *Effluvia* from the Bodies of men.

12. Which also we prove by the suddain jumps which the *Plague* hath made, leaping in one Week from 118 to 927 ; and back again from 993 to 258 ; and from thence again the very next Week to 852. The which Effects must surely be rather attributed to change of the *Air*, than of the Constitution of Mens Bodies, otherwise than as this depends upon that.

13. It may be also noted, That many times other *Pestilential* Diseases, as *Purple Fevers, Small-Pox,* &c. do fore-run (51) the *Plague* a ‖ Year, two or three ; for in 1622 there died but 8000: in 1623, 11000: in 1624, about 12000: till in 1625 there died of all Diseases above 54000.

CHAP. V.

Other Observations upon the Plague, and Casualties.

1. THE *Decrease* and *Increase* of People is to be reckoned chiefly by *Christenings*, because few bear Children in *London* but *Inhabitants*, though others die there. The Accounts of *Christenings* were well kept, until differences in *Religion* occasioned some neglect therein, although even these neglects we must confess to have been regular and proportionable.

2. By the numbers and proportions of *Christenings* therefore we observe as followeth, *viz.*

First, That (when from *December* 1602, to *March* following, there was little or no *Plague*) then the *Christenings* at a (52) *Medium* were between 110 and 130 *per Week*, few ‖ *Weeks* being above the one, or below the other; but when from thence to *July* the *Plague* increased, that then the *Christenings* decreased to under 90.

Secondly, The Question is, Whether *Teeming-Women* died, or fled, or miscarried? The latter at this time seems most probable, because even in the said space, between *March* and *July*, there died not above 20 *per Week* of the *Plague*; which small number could neither cause the death or flight of so many Women, as to alter the proportion $\frac{1}{4}$ part lower.

3. Moreover, We observe from the 21 of *July* to the 12 of *October*, the *Plague* increasing reduced the *Christenings* to 70 at a *Medium*, diminishing the above proportion down to $\frac{2}{5}$. Now the cause of this must be flying, and death, as well as Miscarriages and Abortions; for there died within that time about 25000, whereof many were certainly *Women-with child*: besides, the fright of so many dying within so small a time, might drive away so many others, as to cause this Effect.

4. From *December* 1624, to the middle of *April* 1625, there died not above five a Week of the *Plague*, one with another. In this time, the *Christenings* were one with ano-|| ther 180. The which decreased gradually by the 22 of (53) *September* to 75, or from the proportion of 12 to 5, which evidently squares with our former Observation.

5. The next Observation we shall offer is, The time wherein the City hath been *Re-peopled* after a great *Plague*; which we affirm to be by the second year. For in 1627 the *Christenings* (which are our Standard in this Case) were 8408, which in 1624, next preceding the *Plague*-year 1625 (that had swept away above 54000) were but 8299; and the *Christenings* of 1626 (which were but 6701) mounted in one year to the said 8408.

6. Now the Cause hereof, forasmuch as it cannot be a supply by Procreations; *Ergo*, it must be by new Affluxes to *London* out of the Country.

7. We might fortifie this Assertion by shewing, that before the *Plague*-year 1603, the *Christenings* were about 6000, which were in that very year reduced to 4789, but crept up the next year 1604 to 5458, recovering their former

ordinary proportion in 1605 of 6504, about which proportion it stood till the year 1610.

8. I say, it followeth, that, let the *Mortality* be what it will, the City repairs its loss of Inhabitants within two years; (54) which Ob-||servation lessens the Objection made against the value of Houses in *London*, as if they were liable to great prejudice through the loss of Inhabitants by the *Plague.*

CHAP. VI.

Of the Sickliness, Healthfulness, and Fruitfulness of Seasons.

1. HAving spoken of *Casualties*, we come next to compare the Sickliness, Healthfulness, and Fruitfulness of the several Years and Seasons one with another. And first, having in the Chapters afore going mentioned the several years of *Plague*, we shall next present the several other sickly years ; we meaning by a *sickly Year* such wherein the *Burials* exceed those, both of the precedent and subsequent years, and not above two hundred dying of the *Plague*, for such we call *Plague-Years* ; and this we do, that the World may see, by what spaces and intervals we may hereafter expect such times again. Now, we may not call that a more sickly year, wherein more die, because such excess of (55) *Burials* || may proceed from increase and access of People to the City only.

2. Such sickly years were 1618, 20, 23, 24, 1632, 33, 34, 1649, 52, 54, 56, 58, 61, as may be seen by the Tables[1].

3. In reference to this Observation we shall present another, namely, That the more sickly the years are, the less fecund or fruitful of Children also they be. Which will appear, if the number of Children born in the said sickly years be less than that of the years both next preceding and next following : all which, upon view of the Tables, will be

---

[1] According to the table on p. 408 the years 1623, 1624, 1633 and 1634 fail to satisfy Graunt's definition of sickly years.

found true, except in a very few Cases, where sometimes the precedent, and sometimes the subsequent years vary a little, but never both together. Moreover, for the confirmation of this Truth, we present you the year 1660, where the *Burials* were fewer than in either of the two next precedent years by 2000, and fewer than in the subsequent by above 4000 : And withal, the number of *Christenings* in the said year 1660 was far greater than in any of the three years next afore-going.

4. As to this year 1660, although we would not be thought *Superstitious*, yet it is not to be neglected, that in the said year was the *King's Restauration* to His Empire over these three Nations, as if God Almighty had ‖ caused (56) the healthfulness and fruitfulness thereof to repair the *Blood-shed* and *Calamities* suffered in His absence. I say, this conceit doth abundantly counterpoise the Opinion of those who think great *Plagues* come in with *King's* Reigns[1], because it hapned so twice, *viz. Anno* 1603, and 1625 ; whereas as well the year 1648, wherein the present *King* commenced His Right to reign, as also the year 1660, wherein He commenced the exercise of the same, were both eminently healthful : which clears both *Monarchy*, and our present *King's Family*, from what seditious men have surmised against them.

5. The Diseases, which beside the *Plague* make years unhealthful in this City, are *Spotted-Fevers*, *Small-Pox*, *Dysentery*, called by some *The Plague in the Guts*, and the unhealthful Season is the *Autumn*. ‖

CHAP. VII. (57)

Of the difference between Burials and Christenings.

1. THE next Observation is, That in the said Bills there are far more *Burials* than *Christenings*. This is plain, depending only upon *Arithmetical* computation ; for,

---

[1] The outbreak of the Plague at times of coronation was perhaps in part due to the concourse of people to London.

in 40 years, from the year 1603, to the year 1644, *exclusive* of both years, there have been set down (as hapning within the same ground, space, or Parishes[1]) although differently numbred and divided, 363935 *Burials*, and but 330747 *Christenings* within the 97, 16, and 10 Out Parishes; those of *Westminster, Lambeth, Newington, Redriff, Stepney, Hackney,* and *Islington,* not being included.

2. From this single Observation it will follow, That *London* should have decreased in its People; the contrary whereof we see by its daily increase of Buildings upon new Foundations, and by the turning of great Palacious Houses (58) into small Tenements. It is there-fore certain, that *London* is supplied with People from out of the Country, whereby not only to supply the overplus differences of *Burials* above-mentioned, but likewise to increase its *Inhabitants* according to the said increase of housing.

3. This supplying of *London* seems to be the reason, why *Winchester, Lincoln,* and several other Cities have decreased in their Buildings, and consequently in their *Inhabitants.* The same may be suspected of many Towns in *Cornwal,* and other places, which probably, when they were first allowed to send *Burgesses* to the *Parliament,* were more populous than now, and bore another proportion to *London* than now; for several of those *Burroughs* send two *Burgesses,* whereas *London* it self sends but four, although it bears the fifteenth part of the charge of the whole Nation in all *Publick* Taxes and Levies[2].

4. But, if we consider what I have upon exact enquiry found true, *viz.* That in the Country[3], within ninety years, there have been 6339 *Christenings,* and but 5280 *Burials,* the increase of *London* will be salved without inferring the decrease of the People in the Country; and withal, in case all *England* (59) have but fourteen times more People than || *London,* it will appear, how the said increase of the Country may increase the People, both of *London* and it self; for if there be in the 97, 16, 10, and 7 Parishes, usually comprehended within our

[1] See Introduction.   [2] See *Verbum Sap.,* p. 107, note 3.
[3] See table, p. 415.

Bills, but 460000 Souls, as hereafter we shall shew[1], then there are in all *England* and *Wales* 6440000 Persons, out of which subtract 460000, for those in and about *London*, there remain 5980000 in the Country, the which increasing about ¼ part in 40 years, as we shall hereafter prove[2] doth happen in the Country, the whole increase of the Country will be about 854000 in the said time ; out of which number, if but about 250000 be sent up to *London* in the said 40 years, *viz.* about 6000 *per Annum*, the said *Missions* will make good the alterations, which we find to have been in and about *London*, between the years 1603 and 1644 above-mentioned : But that 250000 will do the same, I prove thus ; *viz.* in the 8 years, from 1603 to 1612, the *Burials* in all the Parishes, and of all Diseases, the *Plague* included, were at a *Medium* 9750 *per Annum.* And between 1635 and 1644 were 18000, the difference whereof is 8250, which is the Total of the increase of the *Burials* in 40 years, that is, about 206 *per Annum.* Now, to make the *Burials* increase 206 *per Annum*, there must ‖ be added to the City 30 times as many (according to the (60) proportion of 3 dying out of 11 Families)[3] *viz.* 6180 *Advenæ*, the which number multiplied again by the 40 years, makes the Product 247200, which is less than the 250000 above-propounded ; so as there remain above 600000 of increase in the Country within the said 40 years, either to render it more populous, or send forth into other Colonies, or Wars. But that *England* hath fourteen times more People, is not improbable, for the Reasons following.

1. *London* is observed to bear about the fifteenth proportion of the whole Tax.

2. There are in *England* and *Wales* about 39000 square Miles of Land, and we have computed that in one of the greatest Parishes in *Hantshire*, being also a Market-Town, and containing twelve square Miles, there are 220 Souls in every square Mile, out of which I abate ¼ for the over-plus of People more in that Parish than in other wild Counties. So as the ¾ parts of the said 220, multiplied by the Total

[1] See p. 331, note.   [2] Cf. p. 389.   [3] See p. 385.

of square Miles, produces 6400000[1] Souls in all *London* included.

3. There are about 10000 Parishes in *England* and *Wales*, the which, although they should not contain the $\frac{1}{3}$ part of (61) the Land, nor the $\frac{1}{4}$ of the People of that Country-Pa-‖rish, which we have examined, yet may be supposed to contain about 600 People, one with another: according to which Account there will be six Millions of People in the Nation. I might add, that there are in *England* and *Wales* about five and twenty Millions of Acres at $16\frac{1}{2}$ Foot to the Perch; and if there be six Millions of People, then there is about four Acres for every head, which how well it agrees to the Rules of Plantation, I leave unto others, not only as a means to examine my Assertion, but as an hint to their enquiry concerning the fundamental Trade, which is Husbandry, and Plantation.

4. Upon the whole matter we may therefore conclude, That the People of the whole Nation do increase, and consequently the decrease of *Winchester*, *Lincoln*, and other like places, must be attributed to other Reasons, than that of re-furnishing *London* only.

5. We come to shew, why although in the Country the *Christenings* exceed the *Burials*, yet in *London* they do not. The general Reason of this must be, that in *London* the proportion of those subject to die, unto those capable of breeding, is greater than in the Country; That is, let there be an hundred Persons in *London*, and as many in the (62) Country; we say, that, if there be sixty of them ‖ Breeders in *London*, there are more than sixty in the Country, or else we must say, that *London* is more unhealthful, or that it inclines Men and Women more to Barrenness, than the Country: which by comparing the Burials and Christenings of *Hackney*, *Newington*, and the other Country-Parishes, with the most *Smoky* and *Stinking* parts of the City, is scarce discernible in any considerable degree.

6. Now that the Breeders in *London* are proportionably

---

[1] In fact 6,435,000.

fewer than those in the Country, arises from these Reasons, *viz.*

1. All, that have business to the Court of the King, or to the Courts of Justice, and all Country-men coming up to bring Provisions to the City, or to buy Forein Commodities, Manufactures, and Rarities, do for the most part leave their Wives in the Country.

2. Persons coming to live in *London* out of curiosity and pleasure, as also such as would retire and live privately, do the same if they have any.

3. Such as come up to be cured of Diseases do scarce use their Wives *pro tempore.*

4. That many Apprentices of *London*, who are bound seven or nine years from Marriage, do often stay longer voluntarily. ‖

5. That many Sea-men of *London* leave their Wives (63) behind them, who are more subject to die in the absence of their Husbands, than to breed either without men, or with the use of many promiscuously.

6. As for unhealthiness, it may well be supposed, that although seasoned Bodies may, and do live near as long in *London*, as elsewhere, yet new-comers and Children do not: for the *Smoaks, Stinks,* and close *Air,* are less healthful than that of the Country ; otherwise why do sickly Persons remove into the Country-*Air?* And why are there more old men in Countries than in *London, per rata?* And although the difference in *Hackney* and *Newington,* above-mentioned, be not very notorious, yet the reason may be their vicinity to -*London,* and that the Inhabitants are most such, whose Bodies have first been impaired with the *London-Air,* before they withdraw thither.

7. As to the causes of Barrenness in *London*, I say, that although there should be none extraordinary in the Native *Air* of the place; yet the intemperance in feeding, and especially the Adulteries and Fornications, supposed more frequent in *London* than elsewhere, do certainly hinder Breeding. For a Woman, admitting ten Men, is so far from ‖ having ten times as many Children, that she hath none at all. (64)

8. Add to this, that the minds of men in *London* are more thoughtful, and full of business, than in the Country where their work is *corporal* Labour and Exercises; All which promote Breeding, whereas *Anxieties* of the mind hinder it.

CHAP. VIII.

Of the difference between the numbers of Males and Females.

THE next Observation is, That there be more *Males* than *Females*[1].

1. There have been Buried from the year 1628, to the year 1662, *exclusive*, 209436 *Males,* and but 190474 *Females:* but it will be objected, That in *London* it may be indeed so, though otherwise elsewhere; because *London* is the great Stage and Shop of business, wherein the *Masculine Sex* bears the greatest part. But we Answer, That there have been also *Christened* within the same time 139782 *Males,* and (65) but 130866 *Females,* and that ‖ the Country-Accounts are consonant enough to those of *London* upon this matter[2].

2. What the Causes hereof are, we shall not trouble our selves to conjecture, as in other Cases: only we shall desire that Travellers would enquire, whether it be the same in other Countries.

3. We should have given an Account, how in every Age these proportions change here, but that we have Bills of distinction but for 32 years, so that we shall pass from hence to some Inferences from this Conclusion; as first,

I. That *Christian Religion,* prohibiting *Polygamy,* is more agreeable to the *Law of Nature,* that is, the *Law of God,* than *Mahumetism,* and others, that allow it: for one Man his having many Women, or Wives, by Law, signifies nothing, unless there were many Women to one Man in Nature also.

[1] The Table of Males and Females is at p. 411.   [2] See p. 389.

II. The obvious Objection hereunto is, That one *Horse*, *Bull*, or *Ram*, having each of them many *Females*, do promote increase. To which I Answer, That although perhaps there be naturally, even of these *species*, more *Males* than *Females*, yet *artificially*, that is, by making *Geldings*, *Oxen*, and *Weathers*, there are fewer. From whence it will follow, That when by experience it is found how ma-‖ny *Ews* (suppose twenty) (66) one *Ram* will serve, we may know what proportion of *male-Lambs* to castrate or geld, *viz.* nineteen, or thereabouts : for if you emasculate fewer, *viz.* but ten, you shall, by promiscuous copulation of each of those ten with two *Females*, hinder the increase, so far as the admittance of two *Males* will do it : but, if you castrate none at all, it is highly probable, that, every of the twenty *Males* copulating with every of the twenty *Females*, there will be little or no conception in any of them all.

III. And this I take to be the truest Reason, why *Foxes*, *Wolves*, and other *Vermin Animals*, that are not gelt, increase not faster than *Sheep*, when as so many thousands of these are daily Butchered, and very few of the other die otherwise than of themselves.

4. We have hitherto said, There are more *Males* than *Females*; we say next, That the one exceed the other by about a thirteenth part. So that although more Men die violent deaths than Women, that is, more are *slain* in *Wars*, *killed* by *Mischance*, *drowned* at *Sea*, and die by the *Hand of Justice*; moreover, more Men go to *Colonies*, and travel into Forein parts, than Women ; and lastly, more remain unmarried than of Women, as *Fellows* of *Colleges*, and *Apprentices* above eighteen, ‖ *&c.* yet the said thirteenth (67) part difference bringeth the business but to such a pass, that every Woman may have an Husband, without the allowance of *Polygamy*.

5. Moreover, although a Man be *Prolifick* fourty years, and a Woman but five and twenty, which makes the *Males* to be as 560 to 325 *Females*, yet the causes above-named, and the later marriage of the Men, reduce all to an equality.

6. It appearing, that there were fourteen Men to thirteen

Women, and that they die in the same proportion also; yet I have heard *Physicians* say, that they have two Women Patients to one Man, which Assertion seems very likely; for that Women have either the *Green-sickness,* or other like Distempers, are sick of *Breedings, Abortions, Child-bearing, Sore-breasts, Whites, Obstructions, Fits of the Mother,* and the like.

7. Now from this it should follow, that more Women should die than Men, if the number of *Burials* answered in proportion to that of Sicknesses: but this must be salved, either by the alleging, that the *Physicians* cure those Sicknesses, so as few more die than if none were sick; or else that Men, being more intemperate than Women, die as much (68) by reason of their Vices, as Women do by the Infir-‖mity of their *Sex;* and consequently, more *Males* being born than *Females,* more also die.

8. In the year 1642 many *Males* went out of *London* into the Wars then beginning, insomuch as I expected in the succeeding year 1643 to have found the *Burials* of *Females* to have exceeded those of *Males,* but no alteration appeared; forasmuch, as I suppose, Trading continuing the same in *London,* all those, who lost their *Apprentices,* had others out of the Country; and if any left their Trades and Shops, that others forthwith succeeded them: for, if employment for hands remain the same, no doubt but the number of them could not long continue in disproportion.

9. Another pregnant Argument to the same purpose (which hath already been touched on) is, That although in the very year of the *Plague* the *Christenings* decreased, by the dying and flying of *Teeming-Women,* yet the very next year after they increased somewhat, but the second after to as full a number as in the second year before the said *Plague:* for I say again, if there be encouragement for an hundred in *London,* that is, a Way how an hundred may live better than in the Country, and if there be void Housing there to receive‖ (69) them, the evacuating of a fourth or third part of that number must soon be supplied out of the Country; so as the great *Plague* doth not lessen the Inhabitants of the City, but of

the Country, who in a short time remove themselves from thence hither, so long, until the City, for want of receipt and encouragement, regurgitates and sends them back.

10. From the difference between *Males* and *Females,* we see the reason of making *Eunuchs* in those places where *Polygamy* is allowed, the later being useless as to multiplication, without the former, as was said before in case of *Sheep* and other *Animals* usually gelt in these Countries.

11. By consequence, this practice of *Castration* serves as well to promote increase, as to meliorate the Flesh of those Beasts that suffer it. For that Operation is equally practised upon *Horses*, which are not used for food, as upon those that are.

12. In *Popish* Countries, where *Polygamy* is forbidden, if a greater number of *Males* oblige themselves to *Cœlibate*, than the natural over-plus, or difference between them and *Females* amounts unto; then multiplication is hindred: for if there be eight Men to ten Women, all of which eight Men are married to eight of the ten Women, then the other two ‖ bear no Children, as either admitting no Man at all, or else (70) admitting Men as Whores (that is, more than one;) which commonly procreates no more than if none at all had been used: or else such unlawful Copulations beget Conceptions, but to frustrate them by procured Abortions, or secret Murthers; all which returns to the same reckoning. Now, if the same proportion of Women oblige themselves to a single life likewise, then such obligation makes no change in this matter of increase.

13. From what hath been said appears the reason, why the Law is and ought to be so strict against Fornications and Adulteries: for, if there were universal liberty, the Increase of Mankind would be but like that of *Foxes* at best.

14. Now forasmuch as Princes are not only Powerful, but Rich, according to the number of their People (Hands being the Father, as Lands are the Mother and Womb of Wealth)[1] it is no wonder why States, by encouraging Marriage,

---

[1] This idea, which occurs in slightly different phraseology in Petty's *Treatise of Taxes* (p. 68), has been pronounced a "leading thought in his writings."

and hindering Licentiousness, advance their own Interest, as well as preserve the Laws of God from contempt and violation.

(71) 15.   It is a Blessing to Mankind, that by this over-plus of *Males* there is this natural ‖ Bar to *Polygamy* : for in such a state Women could not live in that parity and equality of expense with their Husbands, as now, and here they do.

16.   The reason whereof is, not, that the Husband cannot maintain as splendidly three, as one ; for he might, having three Wives, live himself upon a quarter of his Income, that is, in a parity with all three, as well as, having but one, live in the same parity at half with her alone : but rather, because that to keep them all quiet with each other, and himself, he must keep them all in greater aw, and less splendour ; which power he having, he will probably use it to keep them all as low as he pleases, and at no more cost than makes for his own pleasure ; the poorest Subjects, (such as this plurality of Wives must be) being most easily governed. ‖

[*Editors' Note:* Material has been omitted at this point.]

Ingram, *Hist. of Political Economy*, 51 ; the suggestion is followed by Bevan, *Sir W. Petty, a Study*, 53.   The figure in which the idea is expressed apparently reflects the current notion, at least as old as Aristotle, that the female is passive in generation.   Legouvé, *Moral history of Woman*, tr. Palmer, 216.   Even the form of expressing the analogy is, probably, older than either Graunt or Petty, for both place the words in brackets—a seventeenth century equivalent for marks of quotation—and Schulz, in his translation of Graunt, writes, "weil, nach dem Sprichwort, die hander der welt vater, und das land derselbten mutter ist."

C H A P. XI.

Of the number of Inhabitants.

I Have been several times in company with men of great experience in this City, and have heard them talk seldom under Millions of *People* to be in *London*[1]: all which I was apt enough to believe, until, on a certain day, one of eminent Reputation was upon occasion asserting, That there was in the year 1661 two Millions of People more than *Anno* 1625 before the great *Plague.* I must confess, that, until this provocation, I had been frighted, with that mis-understood

---

[1] *The Scots Scouts Discoveries* declared that in 1639 London contained 100000 Frenchmen and Dutchmen. Morgan, *Phoenix Britannicus,* 463. Howell estimated that in 1657 the various parts of London "with divers more which are contiguous and one entire piece with London herself" had a population of a million and a half. *Londonopolis,* 403.

Example of *David*[1], from attempting any computation of the People of this populous place; but hereupon I both examined the lawfulness of making such Enquiries, and, being satisfied thereof, went about the work it self in this manner: *viz.*

2.   First, I imagined, That, if the Conjecture of the worthy Person afore-mentioned had any truth in it, there must needs (81) be about six or seven Millions of People in *London* ‖ now; but, repairing to my Bills, I found, that not above 15000 *per Annum* were buried; and consequently, that not above one in four hundred must die *per Annum*, if the Total were but six Millions.

3.   Next considering, That it is esteemed an even lay, whether any man lives ten years longer[2], I supposed it was the same, that one of any ten might die within one year.   But when I considered, that of the 15000 afore-mentioned about 5000 were *Abortive* and *Still-born*, or died of *Teeth, Convulsion, Rickets,* or as *Infants,* and *Chrysoms,* and *Aged*; I concluded, that of Men and Women, between ten and sixty, there scarce died 10000 *per Annum* in *London,* which number being multi-plied by 10^2, there must be but 10000^3 in all, that is not the $\frac{1}{60}$ part of what the *Alderman* imagined.   These were but sudden thoughts on both sides, and both far from truth, I thereupon endeavoured to get a little nearer, thus: *viz.*

4.   I considered, that the number of *Child-bearing Women* might be about double to the *Births:* forasmuch as such Women, one with another, have scarce more than one Child in two years.   The number of *Births* I found, by those years wherein the *Registries* were well kept, to have been somewhat (82) less than ‖ the *Burials.*   The *Burials* in these late years at a *Medium* are about 13000, and consequently the *Christenings* not above 12000.   I therefore esteemed the number of *Teem-ing-Women* to be 24000: then I imagined, that there might be twice as many Families, as of such Women; for that there

[1] 2 Samuel, xxiv. 1—9 ; 1 Chronicles, xxi. 1—8.

[2] If it be "an even lay, whether any man lives ten years longer," Graunt's multiplier, seven lines lower, should be 20, not 10.

[3] 10000 is a misprint for 100000.

might be twice as many Women *Aged* between 16 and 76, as between 16 and 40, or between 20 and 44; and that there were about eight Persons in a Family, one with another, *viz.* the Man and his Wife, three Children and three Servants or Lodgers: now 8 times 48000 makes 384000.

5. Secondly, I find, by telling the number of Families in some Parishes within the Walls, that 3 out of 11 Families *per annum* have died: wherefore, 13000 having died in the whole, it should follow, there were 48000[1] Families according to the last-mentioned Account.

6. Thirdly, the Account, which I made of the *Trained-Bands* and *Auxiliary*-Souldiers doth enough justifie this Account.

7. And lastly, I took the Map of *London* set out in the year 1658 by *Richard Newcourt*[2], drawn by a Scale of Yards. Now I ghessed that in 100 Yards square there might be about 54 Families, supposing every House ‖ to be 20 Foot in the (83) front: for on two sides of the said square there will be 100 Yards of Housing in each, and in the two other sides 80 each; in all 360 Yards: that is, 54 Families in each square, of which there are 220 within the Walls, making in all 11880 Families within the Walls. But forasmuch as there die within the Walls about 3200 *per Annum*, and in the whole 13000; it follows, that the Housing within the Walls is $\frac{1}{4}$ part of the whole, and consequently, that there are 47520 Families in and about *London*, which agrees well enough with all my former computations: the worst whereof doth sufficiently demonstrate, that there are two Millions[3] of People in *London*, which nevertheless most men do believe, as they do, that there be three Women for one Man, whereas there

---

[1] More accurately 47,667.

[2] "An exact Delineation of the Cities of London and Westminster and the Suburbs Thereof, Together wth ye Burrough of Southwark And All ye Through-fares Highwayes Streetes Lanes and Common Allies w^thin ye same Composed by a Scale and Ichnographically described by Richard Newcourt of Somerton in the Countie of Somersett Gentleman. Willm Faithorne sculpsit."——Facsimile, London: E. Stanford, 1878.

[3] The first edition has, "that there are no Millions," the fourth, "that there are not two Millions."

are fourteen Men for thirteen Women, as elsewhere hath been said[1].

8. We have (though perhaps too much at Random) determined the number of the Inhabitants of *London*[2] to be about 384000: the which being granted, we assert, that 199112 are *Males*, and 184186 *Females*.

9. Whereas we have found[3], that of 100 quick Conceptions about 36 of them die before they be six years old, and (84) that perhaps but one surviveth 76[4]; we having seven *De-‖cads* between six and 76, we sought six mean proportional numbers[5] between 64, the remainder, living at six years, and the one, which survives 76, and find, that the numbers following are practically near enough to the truth; for men do not die in exact proportions, nor in Fractions, from whence arises this Table following.

Viz. Of an hundred there die within the first six years[6]   36

---

[1] See p. 374.

[2] Excluding Westminster and the six parishes enumerated at p. 345.

[3] See p. 349.

[4] From the bills Graunt calculates (p. 352) that seven in 100 survive 70. The grounds of his assumption that but one survives 76 are not evident.

[5] This method of constructing a table of mortality suggests Petty's *Discourse of Duplicate Proportion.*

[6] With this calculation of London's mortality may be compared the figures for Geneva in the seventeenth century. The following table, compiled from Édouard Mallet's *Recherches hist. et stat. sur la population de Genève* (*Annales d'hygiène publique et de médecine légale,* XVII. p. 30, Janv., 1837), gives the returns for all the persons whose age at death was recorded in the years 1601—1700. The table reveals a juvenile mortality even higher than Graunt's calculation for London.

Age in years.	Number of deaths.	Percentage.
1—6	22,967	42·6
7—16	4,949	9·3
17—26	4,052	7·6
27—36	3,761	7·1
37—46	3,938	7·4
47—56	4,026	7·6
57—66	3,800	7·2
67—76	3,273	6·4
77—86	2,436	4·7
87—120	581	0·1
	53,783	100

The next ten years, or *Decad*	24
The second *Decad*	15
The third *Decad*	9
The fourth	6
The next	4
The next	3
The next	2
The next	1

10.   From whence it follows, that of the said 100 conceived, there remain alive at six years end 64.

At sixteen years end	40
At twenty six	25
At thirty six	16
At fourty six	10
At fifty six	6
At sixty [six]	3
At seventy six	1
At eighty [six]	0 ‖

11.   It follows also, That of all which have been conceived, (85) there are now alive 40 *per Cent*. above sixteen years old, 25 above twenty six years old, *& sic deinceps*, as in the above-Table.   There are therefore of Aged between 16 and 56 the number of 40, less by six, *viz.* 34 ; of between 26 and 66 the number of 25, less by three, *viz.* 22 : *& sic deinceps*.

Wherefore, supposing there be 199112 *Males*, and the number between 16 and 56 being 34 ; it follows, there are 34 *per Cent*. of all those Males fighting Men in *London*, that is 67694, *viz.* near 70000 ; the truth whereof I leave to examination, only the $\frac{1}{5}$ of 67694, *viz.* 13539, is to be added for *Westminster, Stepney, Lambeth*, and the other distant Parishes ; making in all 81233 fighting Men.

12.   The next enquiry will be, In how long time the City of *London* shall, by the ordinary proportion of Breeding and dying, double its breeding People?[1]   I answer, In about seven

---

[1] Apparently Graunt has not expressed himself with entire accuracy.   The question which he put is, in how many years will 24000 pairs become 48000 pairs?   The question which he probably meant to put is, in how many years will 24000 pairs beget 48000 children?   He answers, in seven years, or, plagues

years, and ·(*Plagues* considered) eight. Wherefore, since there be 24000 pair of Breeders, that is $\frac{1}{8}$ of the whole, it follows, that in eight times eight years the whole People of the City shall double, without the access of Forreiners: the which (86) contradicts not ‖ our Account of its growing from two to five in 56 years with such accesses.

13. According to this proportion, one couple, *viz. Adam* and *Eve*, doubling themselves every 64 years of the 5610 years[1], which is the *Age* of the World according to the *Scriptures*, shall produce far more People than are now in it. Wherefore the World is not above 100 thousand years older[2], as some vainly imagine, nor above what the *Scripture* makes it.

[*Editors' Note:* Material has been omitted at this point.]

considered, in eight. If, then, eight years are necessary for the birth of 48000 persons, the birth of 384000—a number sufficient, together with those already living, to double the population of the City—will require sixty-four years. It is unnecessary to dwell on the defects of this calculation. On one hand it ignores the increase in the number of pairs during sixty-four years. On the other hand, it tacitly assumes that the 384000 now living, and likewise all those new-born within the sixty-four years, will live to the end of that period.

[1] According to the chronology of Scaliger (*De emendatione temporum*, pp. 431—432) which places the Creation in the year 3948 B.C.

[2] Previous editions, 'old.'

REFERENCES

Bacon, Sir Francis. Works, edited by J. S. Spedding. Boston, 1861–64. 15 vols. 322, 348.

Bevan, W. L. Sir William Petty, a study in English economic literature. New York, 1894. xxxix, xlvi, xlviii, lxi, lxii, 378, 625.

Birch, Thomas. Court and times of Charles I. [really ed. by R. F. Williams, not by Birch]. London, 1848. 2 vols. 365.

Christie. History of the company of parish clerks. London. lxxxii, lxxxxviii, 357.

Creighton, Charles. A history of epidemics in Britain. Cambridge, 1891–94, 2 vols. lxxx, lxxxi, lxxxvii–lxxxix, 336, 365, 417, 418, 426–429, 432.

De Morgan, Augustus. A budget of paradoxes. London, 1872. xxxix, xlvii, 358.

Holy Bible. 384, 466.

Howell, James. Londonopolis, an historical discourse. London, 1657. 383.

Ingram, John Kells. A history of political economy. New York, 1893. 378.

Legouvé, E. Moral history of woman, translated by J. W. Palmer. New York, 1860. 378.

Mallet, Édouard. Recherches historiques et statistiques sur la population de Genève. (*In* Annales d'hygiène publique et de médecine légale XVII., 30, Paris, January, 1837.) 386.

Morgan, J. Phoenix britannicus. London, 1732. xv, 383.

Newcourt, Richard. An exact delineation of the cities of London and Westminster and the suburbs thereof . . . [map] composed by a scale. London, 1658.—Facsimile, E. Stanford, 1878. 385.

Petty, Sir W. Discourse of duplicate proportion. London, 1674. xxvii, xli, 9, 386.

Reports of the Historical Manuscripts Commission. London. Viz: 3rd, lvi, 630; 4th, lvi, 237; 7th, xxix, lvi, 125, 630; 8th, 237, 461; 10th, lxxxi; 11th, xxxi, 365; 14th, xxiv, xxix; 15th, lvi, lviii, 212.

Scaliger, J. J. Opus novum de emendatione temporum. Lutetiae, 1583. 388.

Scobell, Henry. A collection of acts and ordinances of general use. London, 1658. 2 pts. 40, 81, 129, 178, 179, 357.

Traill, H. D. Social England. New York, 1894–96. 6 vols. 365.

2

Reprinted from *Science*, **26**(653), 21–22 (1907)

RELATION BETWEEN BIRTH RATES AND DEATH RATES

A SHORT notice appeared on page 641 of SCIENCE, 1907, of a paper read by C. E. Woodruff before the American Association for the Advancement of Science, on the relation between birth rates and death rates, etc.

In this connection, it may be of interest to note that a mathematical expression can be obtained for the relation between the birth rate per head b and the death rate per head d, for the case where the general conditions in the community are constant, and the influence of emigration and immigration is negligible.

Comparison with some figures taken from actual observation shows that these at times approach very nearly the relation deduced on the assumptions indicated above.

I give here the development of the formula, and some figures obtained by calculation by its aid, together with the observed values, for comparison.

Let $c(a)$ be such a coefficient that out of the total number N_t of individuals in the community at time t, the number whose age lies between the values a and $(a + da)$ is given by $N_t c(a) da$.

Now the $N_t c(a) da$ individuals whose age at time t lies between the values a and $(a + da)$, are the survivors of the individuals born in time da at time $(t - a)$.

If we denote by $B_{(t-a)}$ the total birth rate at time $(t - a)$, and by $p(a)$ the probability at its birth, that any individual will reach age a, then the number of the above-mentioned survivors is evidently $B_{(t-a)} p(a) da$.

Hence:

$$N_t c(a) da = B_{(t-a)} p(a) da$$

$$c(a) = \frac{B_{(t-a)}}{N_t} p(a)$$

Now if general conditions in the community are constant, $c(a)$ will tend to assume a fixed form. A little reflection shows that then both N and B will increase in geometric progression with time,[1] at the same rate $r = (b - d)$. We may, therefore, write:

$$B_{(t-a)} = B_t e^{-ra}$$

$$c(a) = \frac{B_t}{N_t} e^{-ra} p(a)$$

$$= b e^{-ra} p(a) \qquad (1)$$

Now from the nature of the coefficient $c(a)$ it follows that

$$\int_0^\infty c(a)\,da = 1$$

Substituting this in (1) we have:

$$\frac{1}{b} = \int_0^\infty e^{-ra} p(a)\,da \qquad (2)$$

Equation (1) then gives the fixed age-distribution, while equation (2) (which may be expanded into a series if desired), gives the relation between b, the birth rate per head, and r, the rate of natural increase per head, and hence between b and d, since $r = b - d$.

Applying these formulæ to material furnished by the Reports of the Registrar-General of Births, etc., in England and Wales, the following results were obtained:

ENGLAND AND WALES 1871–80 (MEAN)

		Observed[2]	Calculated
Birth-rate per head	b	.03546	.0352
Death-rate per head	d	.02139	.0211
Excess	$(b-d) = r$	.01407	(.0141)

$p(a)$ from Supplement to 45th Ann. Rep. Reg. Gen. Births, etc., England and Wales, pp. vii and viii, assuming ratio:

$$\frac{\text{male births}}{\text{female births}} = 1.04.$$

[1] Compare M. Block, "Traité théorique et pratique de statistique," 1886, p. 209.

[2] Mean b and d from 46th Ann. Rep. Reg. Gen. Births, etc., England and Wales, p. xxxi.

Age Scale.—1,000 individuals, in age-groups of 5 and 10 years

$a_1 a_2$	$1000 \int_{a_1}^{a_2} c(a)\,da$	
0 – 5	136	138
5 – 10	120	116
10 – 15	107	106
15 – 20	97	97
20 – 25	89	87
25 – 35	147	148
35 – 45	113	116
45 – 55	86	87
55 – 65	59	59
65 – 75	33	33
75 – ∞	13	13

It will be seen that in the above example the values calculated for the age-scale and especially for b and d, show a good agreement with the observed values.[2]

The above development admits of further extension. But this, as well as further numerical tests, must be reserved for a future occasion. In view of the recent note of the work by Major Woodruff, it appeared desirable to the writer to publish this preliminary note.

ALFRED J. LOTKA

3

Reprinted from *Phil. Mag.*, Ser. 6, **21**, 435–438 (1911)

A PROBLEM IN AGE-DISTRIBUTION

F. R. Sharpe and A. J. Lotka

THE age-distribution in a population is more or less variable. Its possible fluctuations are not, however, unlimited. Certain age-distributions will practically never occur; and even if we were by arbitrary interference to impress some extremely unusual form upon the age-distribution of an isolated population, in time the "irregularities" would no doubt become smoothed over. It seems therefore that there must be a limiting "stable" type about which the actual distribution varies, and towards which it tends to return if through any agency disturbed therefrom. It was shown on a former occasion † how to calculate the "fixed" age-distribution, which, if once established, will (under constant conditions) maintain itself.

It remains to be determined whether this "fixed" form is also the "stable" distribution: that is to say, whether a given (isolated) population will spontaneously return to this "fixed" age-distribution after a small displacement therefrom.

To answer this question we will proceed first of all to establish the equations for a more general problem, which may be stated as follows:—

"Given the age-distribution in an isolated population at any instant of time, the 'life curve' (life table), the rate of procreation at every age in life, and the ratio of male to female births, to find the age-distribution at any subsequent instant."

1. Let the number of males whose ages at time t lie between the limits a and $a+da$ be $F(a, t)da$, where F is an unknown function of a and t.

Let $p(a)$ denote the probability ‡ at birth that a male shall reach the age a, so that $p(0)=1$.

Further, let the male birth-rate (*i. e.* the total number of males born per unit of time) at time t be $B(t)$.

Now the $F(a, t)da$ males whose age at time t lies between a and $a+da$ are the survivors of the $B(t-a)da$ males born a units of time previously, during an interval of time da. Hence

$$F(a, t)da = B(t-a)p(a)da$$

$$F(a, t) = p(a)B(t-a). \quad . \quad . \quad . \quad . \quad . \quad (1)$$

---

† A. J. Lotka, Am. Journ. Science, 1907, xxiv. pp. 199, 375; 'Science,' 1907, **xxvi.** p. 21.

‡ As read from the life table.

2. Let the number of male births per unit time at time t due to the $F(a, t)da$ males whose age lies between a and $a + da$ be $F(a, t)\beta(a)da$.

If γ is the age at which male reproduction ends, then evidently

$$B(t) = \int_0^\gamma F(a, t)\beta(a)da$$

$$= \int_0^\gamma B(t-a)p(a)\beta(a)da. \quad . \quad . \quad (2)$$

Now in the quite general case $\beta(a)$ will be a function of the age-distribution both of the males and females in the population, and also of the ratio of male births to female births.

We are, however, primarily concerned with comparatively small displacements from the " fixed " age-distribution, and for such small displacements we may regard $\beta(a)$ and the ratio of male births to female births as independent of the age-distribution.

The integral equation (2) is then of the type dealt with by Hertz (*Math. Ann.* vol. lxv. p. 86). To solve it we must know the value of $B(t)$ from $t=0$ to $t=\gamma$, or, what is the same thing, the number of males at every age between 0 and γ at time γ. We may leave out of consideration the males above age γ at time γ, as they will soon die out. We then have by Hertz, *loc. cit.*,

$$B(t) = \sum_{h=1}^{h=\infty} \frac{\alpha_h^t \int_0^\gamma \left\{ B(a) - \int_0^a \beta(a_1)p(a_1)B(a-a_1)da_1 \right\} \alpha_h^{-a}da}{\int_0^\gamma a\beta(a)p(a)\alpha_h^{-a}da}, \quad (3)$$

where $\alpha_1, \alpha_2, \ldots$ are the roots of the equation for α;

$$1 = \int_0^\gamma \beta(a)p(a)\alpha^{-a}da. \quad . \quad . \quad . \quad . \quad (4)$$

The formula (3) gives the value of $B(t)$ for $t > \gamma$, and the age-distribution then follows from

$$F(a, t) = p(a)B(t-a). \quad . \quad . \quad . \quad . \quad . \quad (1)$$

55

4. From the nature of the problem $p(a)$ and $\beta(a)$ are never negative. It follows that (4) has one and only one real root r, which is $\gtreqless 1$, according as

$$\int_0^\gamma \beta(a)p(a)\,da \gtreqless 1. \quad \ldots \ldots \quad (5)$$

Any other root must have its real part less than r. For if $r_1(\cos\theta + i\sin\theta)$ is a root of (4),

$$1 = \int_0^\gamma \frac{\beta(a)p(a)}{r_1^a}\cos a\theta\,da. \quad \ldots \ldots \quad (6)$$

It follows that for large values of t the term with the real root r outweighs all other terms in (3) and $B(t)$ approaches the value

$$B(t) = Ar^t. \quad \ldots \ldots \ldots \quad (7)$$

The ultimate age-distribution is therefore given by

$$F(a,\,t) = Ap(a)r^{t-a} \quad \ldots \ldots \quad (8)$$
$$= Ap(a)e^{r'(t-a)}. \quad \ldots \ldots \quad (9)$$

Formula (9) expresses the "absolute" frequency of the several ages. To find the "relative" frequency $c\,(a,\,t)$ we must divide by the total number of male individuals.

$$c(a,\,t) = \frac{F(a,\,t)}{\displaystyle\int_0^\infty F(a,\,t)\,da} = \frac{Ap(a)e^{r'(t-a)}}{Ae^{r't}\displaystyle\int_0^\infty e^{-r'a}p(a)\,da} = \frac{p(a)e^{-r'a}}{\displaystyle\int_0^\infty e^{-r'a}p(a)\,da}$$

$$= be^{-r'a}p(a), \quad \ldots \ldots \quad (10)*$$

where

$$\frac{1}{b} = \int_0^\infty e^{r'a}p(a)\,da. \quad \ldots \ldots \quad (11)*$$

The expression (10) no longer contains t, showing that the ultimate distribution is of "fixed" form. But it is also "stable;" for if we suppose any small displacement from this "fixed" distribution brought about in any way, say by temporary disturbance *of the otherwise constant conditions*, then we can regard the new distribution as an "initial" distribution to which the above development applies : that is to say, the population will ultimately return to the "fixed" age-distribution.

* Compare Am. Journ. Science, xxiv. 1907, p. 201.

It may be noted that of course similar considerations apply to the females in the population. The appended table shows the age-distribution calculated according to formula (10) for England and Wales 1871–1880. The requisite data (including the life table) were taken from the Supplement to the 45th Annual Report of the Registrar General of Births, &c. The mean value of r' (mixed sexes) for that period was ·01401, while the ratio of male births to female births was 1·0382.

It will be seen that at this period the observed age-distribution in England conformed quite closely to the calculated "stable" form.

TABLE.

Age (Years).	MALES.		FEMALES.		PERSONS.	
	Calc.	Obs.	Calc.	Obs.	Calc.	Obs.
0- 5...	139	139	136	132	138	136
5-10...	118	123	115	117	116	120
10-15...	107	110	104	104	106	107
15-20...	97	99	95	95	96	97
20-25...	88	87	87	91	87	89
25-35. .	150	144	148	149	149	147
35-45...	116	112	116	115	116	113
45-55...	86	84	88	87	87	86
55-65...	57	59	62	61	59	59
65-75...	30	31	35	35	33	33
75-∞	11	12	15	15	13	13

4

A MATHEMATICAL THEORY ON THE STRUGGLE FOR SURVIVAL

Vito Volterra

A translation of the article
UNA TEORIA MATEMATICA SULLA LOTTA PER L'ESISTENZA
Scientia 41:85-102, 1927

NOTE ON TRANSLATION
This article was translated by Dr. Janet Mogg Schreiber, Assistant Professor of Human Ecology, University of Texas School of Public Health, Houston, Texas.

1. There have been many applications of mathematics to biology. There has been research on physiology of the senses, circulation of blood, and movement of animals, all of which can be regarded as part of optics, acoustics, hydro-dynamics or the mechanics of solid bodies and, therefore, have not given rise to the development of new methods outside of the sphere of classical physical mathematics. On the other hand, biometry, with its own procedures, has recourse to the application of the calculus of probabilities and has created a body of new and original studies.[1] The recent research on the geometry of the form and growth of organized beings is also of an original character. In these studies, geometry has been adopted to describe the forms themselves and their development, as has been done for some time in astronomy to describe the orbits and movements of celestial bodies.[2] It is also hoped the methods used in the analysis of heredity can be employed in questions pertaining to biology.[3]

Leaving other applications of mathematics aside, I believe it worthwhile to study and examine the applications which I will discuss in this article, which can clarify various points of actual interest to biologists.[4]

2. Biological associations (biocenosis) occur when more than one species lives in the same environment. Ordinarily the various individuals of such associations compete for the same nutrients, or some species live at the expense of others on which they feed. Nothing, however, precludes these being also of

[*Editors' Note:* Figures appear on pages 71 and 72.]

mutual benefit. All of this enters into the phenomenon generally called "the struggle for survival".

The quantitative character of this phenomenon is manifested in the variations in the number of individuals which constitute the various species. Under certain conditions such variations fluctuate around a mean value, while in others there is a continuing decrease or increase in the species.

The study of these variations and tendencies is important theoretically, but many times also has a notable practical importance, as in the case of the species of fish that live in the same seas and whose variation are of interest to the fishing industry.[5] Agronomy is also interested in the fluctuations of plant parasites particularly when these can be controlled by the introduction of species-specific predators on these parasites. Infectious diseases (malaria, etc.) also show fluctuations that are probably of an analogous nature.

The question presents itself in a very complex way. There are certain periodic environmental conditions such as those depending on seasonal variation and producing forced oscillations of an external character in the number of individuals of the various species. These external periodic actions especially merit study from the statistical viewpoint, but are there also others of an internal character, actual periodic events which would exist even if the external periodic influences should cease, and that are superimposed on them?

Observations lead to an affirmative reply and mathematical calculations confirm it, as we will see in this article. At first glance, the question can seem by its extreme complexity not amenable to mathematical treatment and it may seem that mathematical methods, being too delicate, might stress some peculiarities but hide the essentials of the question. To defend against this danger, it is convenient to begin from hypotheses which are gross but simple, and schematicize the phenomenon.

We will begin, therefore, by studying what could be called "purely internal phenomena", those related only to the reproductive potential and the voracity of the species, as if they were alone. Then we will study their relation with external actions or periodic forces which are results of the environment.

3. Which mathematical methods will be employed? Perhaps those founded in the calculus of probability come first to mind. I will say quickly that it is not these that lead us to our goal.

Permit me to indicate how the question can be considered: We will seek to express in words roughly how the phenomenon proceeds; then we will translate these words into mathematical language. This leads us to formulate differential equations. If we then employ methods of analysis, we are carried much farther than we could be carried by the language of ordinary reason and we can formulate precise mathematical laws. These do not contradict the results of observation. Rather the most important of these seem in perfect accord with statistical evidence.[6] The road to be followed is clearly indicated with these brief words. We will see before long how to overcome the difficulties encountered.

4. Let us take a given animal species, assuming that it increases and decreases at a continual rate and that the number of individuals of the species (N) is not an integer but a positive number that varies continuously. In general, births come in definite seasons which are separated from each other. We will neglect these circumstances and assume that births come at a continuous rate, on a parity with all other conditions, and occur proportionally to the number of individuals in the species. The same can be said of deaths, and whether the births prevail over the deaths or *vice versa*, there will be an increase or decrease in the number of individuals. We will thus assume homogeneity of the individuals of each species, neglecting variations of age and size.

If the species is alone or others do not influence it, it will not change until there are births or deaths. The velocity of growth of the species, or the number of individuals that will grow in a unit of time, will be:

$$V = nN - mN = (n - m)N,$$

n being the coefficient of natality and m that of mortality, both of which are constant. Letting $n - m = \varepsilon$ we have

$$V = \varepsilon N$$

from which we get the well known law of exponential growth of the species: as time increases arithmetically, the number of individuals in the species will vary geometrically.

ε will be called the coefficient of growth of the species and if this is positive, the geometric progression will be increasing; if ε is negative, it will be decreasing. We can give this a geometric representation. We have a first species which finds in the environment sufficient nutrients so that its coefficient of growth is constant and positive. If N_1 denotes the number of individuals, the exponential curve (Figure 1) in which the time, t, is the abscissa while N_1 is the ordinate, represents the changes in the number of individuals of the species when it exists alone.

The curve of Figure 1 represents the equation

$$V_1 = \varepsilon_1 N_1,$$

where V_1 is the velocity of growth of the species.

We can easily determine the time necessary for the species to double because the number of its individuals grows from N_1 to $2N_1$. This construction is indicated in Figure 1. The time, t_1, is independent of the value beginning from N_1 and depends only on the coefficient ε_1. This is, in fact, given by:

$$\frac{\text{log. nep. } 2}{\varepsilon_1} .$$

Let us consider a second species, one which does not find nutriment in the environment so that it has only the coefficient of growth, $-\varepsilon_2$, which would be constant and negative. (ε_2 will be called the coefficient of decrease.) If N_2 denotes the number of individuals, the exponential curve (Figure 2) will represent the variation of the number of individuals of the species when it is alone in the environment.

The slope of the curve shows the indefinite decrease of the species. This is represented geometrically by the equation

$$V_2 = -\varepsilon_2 N_2$$

where V_2 is the velocity of negative growth of the species.

When this has been done in an analogous way we can obtain the time in which the species will be reduced to one half. Figure 2 gives the time t_2 in which the individuals of the species N_2 becomes $N_2/2$.

Also t_2 is independent of the initial value and is

$$\frac{\text{log. nep. } 2}{\varepsilon_2} .$$

5. Suppose now that the two species coexist and that the individuals of the second species nourish themselves on those of the first. What will happen in a biological association constituted this way?

We shall try to express the direction of this phenomenon with words. It is certain that the coefficient of growth ε_1 of the first species will be modified; it will no longer be constant. It will be decreased the greater the number of individuals in the second species which feeds on the first, and can be negative. Also the number $-\varepsilon_2$ will no longer be constant but will be augmented (it can change sign) the greater the number of individuals in the first species because with the increase in this, the food base of the second species increases. We can now translate this into mathematical language saying that the constant quantity ε_1 must be substituted for a quantity that decreases with the growth of N_2. As a first approximation we can substitute ε_1 with $\varepsilon_1 - \gamma_1 N_2$ and $-\varepsilon_2$ with $-\varepsilon_2 + \gamma_2 N_1$ where γ_1 and γ_2 are two positive coefficients. If then V_1 and V_2 denote the velocities of growth of two species, or the change in the number of individuals of the two species in a unit of time, we have for the two preceding equations:

$$(1) \quad V_1 = (\varepsilon_1 - \gamma_1 N_2)N_1 \qquad (2) \quad V_2 = (-\varepsilon_2 + \gamma_2 N_1)N_2.^7$$

These cannot be considered separately, but simultaneously. In such a way one finds that the calculation defines two simultaneous differential equations. From these, the solution of a fixed relation between N_1 and N_2 can easily be obtained.

We can give a very simple idea of this relation with a geometric representation in which an auxiliary variable X is used, which is bound to N_1 by a relation represented by the curve on the left of Figure 3 and to N_2 by a relation represented by the curve on the same figure. N_1 and N_2 are the abscissas of the two curves, while X is the value in common of their ordinates. To find the values of N_2 which correspond to a single value of N_1 use the construction shown in Figure 3, from which it is clear that for each value of N_1 there correspond two values of N_2, and alternatively for each value of N_2 there are two values of N_1.

The construction itself permits us to obtain the curve which has for its ordinate N_1 and for abscissa N_2. This closed artificial curve is represented in Figure 4. The time in which the entire cycle is run is denoted by T after which the numbers N_1 and N_2 take the same values they had in the beginning; that is, the final conditions are identical to the initial ones. The phenomenon is, therefore, periodic and the period T can be obtained from the coefficients ε_1, ε_2 and the original data.

Approximately, when we are dealing with small flucuations, this is given by the formula

$$T = \frac{2\pi}{\sqrt{\varepsilon_1 \varepsilon_2}} = 9.06\sqrt{t_1 t_2}$$

that is *the period of the fluctuations is proportional to the geometric average of the two times in which the first species doubles and the second is reduced to 1/2, respectively.*

How N_1 and N_2 vary with time can also be represented with a graph. This is given in Figure 5 in which time is taken for the abscissa and N_1 and N_2 are the ordinates of the two curves.

If we change the initial conditions, the cycle changes. We have, therefore, infinitely many possible cycles which can be represented by many closed concentric curves as indicated in Figure 6. The unique point Ω internal to these infinite curves has as coordinates the relations

$$K_1 = \frac{\varepsilon_2}{\gamma_2} \ , \ K_2 = \frac{\varepsilon_1}{\gamma_1} \ .$$

We will see before long the significance of the numbers K_1 and K_2.

6. First we look at the significance of ε_1, ε_2, γ_1, γ_2. Evidently ε_1, is the coefficient of growth of the first species (when alone in the environment); ε_2 is the coefficient of diminution of the second species (when alone).

It is easily seen that γ_1 and γ_2 grow with the voracity of the second species while they diminish with improved methods of protection for the first species. These can be called "the coefficients of voracity".

As regards K_1 and K_2, it is easily seen that if $N_1 = K_1$, $N_2 = K_2$, then V_1 and V_2 are zero; therefore, K_1 and K_2 are numbers of individuals corresponding to a stationary state, in which the species neither increase or decrease. On the other hand, since the curves in Figure 6 are not symmetrical with respect to Ω, K_1 and K_2 are the mean of the values of N_1 and N_2 during the course of a single period.

Therefore: 1) *The fluctuations of the species are periodic.* 2) *The mean of the number of individuals of the two species does not depend on the initial conditions so long as the coefficients of growth and competition remain constant.* For any initial state in which there are few or many individuals, the means after a given period will always be the same. The periods, however, will change with variation in initial states.

Suppose now that the two species are artificially destroyed, for example, if we were dealing with fish, they could be fished. Now ε_1, the coefficient of growth of the first species, will diminish while ε_2, the coefficient of decrease of the second species, will grow. The voracity of the first species and the means of protection of the second will not change: K_1 will increase and K_2 decrease, that is, the mean of the eaten species will increase and the mean of the eating species will decrease. From this comes the third law; 3) *If you try to destroy both species, the mean number of individuals of the prey species will grow and the mean number of individuals of the predator species will decrease.*

7. As has been previously noted, this third law is in perfect accord with the results of the study of fish in the North Adriatic. D'Ancona has examined the fish statistics of the markets in Venice, Trieste and Fiume, which catch most of the fish in the North Adriatic, for the periods preceding and following World War I and has noted that towards the end of the war there was a greater abundance of the most predatory species, principally the selaci, and a relatively lesser amount of prey species. D'Ancona interprets this by positing that the fishing fleet in 1914-1918 temporarily destroyed the equilibrium between the diverse species in the Adriatic, favoring the most predatory at the expense of the less predatory, but economically more important, species. Fishing with nets favored the species with the least defenses. With the cessation of fishing during the war period the populations returned to their primitive conditions, that is, an increment in the predatory species.

8. One can easily see that this can be true only up to a certain limit, at which both species will be depleted. This limit, below which the destructive force favors the eating species, can easily be calculated and if it is surpassed the two species are depleted. The eating species will diminish but the eaten species will tend towards a limit that is inferior to the mean previously attained. In other words, there is an upper limit that is not a maximum.

9. We have assumed that when the two species coexist ε_1 and ε_2 diminish and grow respectively with the growth of N_2 and N_1 in a way given by $\varepsilon_1 - \gamma_1 N_2$ and $-\varepsilon_2 + \gamma_2 N_1$ as a first approximation. These linear functions of N_2 and N_1 not only serve as a first approximation of the phenomenon, but also show that the growth of the two species occurs proportionally to the probable number of encounters between individuals of them, and therefore to the product $N_1 N_2$, which is proportional to the number of these encounters. The form of the two equations is thus rigorously proven.

These same equations can be used to incorporate all possible hypotheses,

V. Volterra

depending on the signs of the coefficients ε_1, ε_2, γ_1, γ_2, taking into consideration the diverse cases in which the encounters between individuals of the two species are favorable or unfavorable, with these coefficients.

When one thinks that many of the events of interest to medicine can be related to phenomena which depend on encounters and reciprocal actions between diverse species (human species and pathogens, parasitic species and host species), one understands how the fluctuations of epidemics have a pattern similar to the one theoretically presented here.

10. To consider the coexistence of only two species limits the question unnecessarily. It is mathematically possible to treat a case in which a greater number of species coexist with reciprocal interaction.

Let us set their number as n and assume that an encounter between two individuals of diverse species will always have a result favorable to the species of which one is a member and unfavorable to that species of which the other is a member, or a result which is null for both. We take two of these species, for example the first and the second, and formulate a relationship between the number of individuals by which the first is augmented and the number of individuals by which the other diminishes as a consequence of their encounters, during which the first species devours the second producing a decrease in that species and an increase in the first species, proportional to the nutriment obtained. Let us say that this relationship is always expressed by γ_1/γ_2, in which γ_1, γ_2, ..., γ_n denote n positive numbers corresponding respectively to the first, second,..., to the nth species. Thus, we suppose that analogous relationships may be formulated regarding encounters between individuals of any two of the species. In such hypotheses, the numbers γ_1, γ_2, ..., γ_n constitute equivalents in terms of individuals of the various species. It is possible that the individuals of the first species, in virtue of their rapacity, could destroy γ_2 individuals of the second species, thus increasing their number γ_1 which means that γ_1 individuals of the first species are equivalent to γ_2 individuals of the second.

This can be expressed with other terms calling them, respectively,

$$\beta_1 = \frac{1}{\gamma_1} \, , \ \beta_2 = \frac{1}{\gamma_2} \, , \ \ldots, \ \beta_n = \frac{1}{\gamma_n}$$

the *values* of the single individuals of the first, of the second, ..., of the nth species, and if N_1, N_2, ..., are the corresponding numbers of individuals of the various species, we can call

$$W = \beta_1 N_1 + \beta_2 N_2 + \ldots + \beta_n N_n$$

the value of the biological association.

The preceding hypotheses are equivalent to assuming that encounters between individuals of the various species do not alter the value of the biological association.

We will call an association of this nature conservative. The variation in the number of individuals of the various species are in this case regulated by a system of simultaneous quadratic differential equations characterized by a symmetrical determinant. In virtue of the properties of these determinants, the analytical development in the case of an even number of pairs of species is different from that of an odd number. In both the cases, one can find an integral of the system of differential equations.

11. When the number of species is even, three laws can be established that are an extension of those previously formulated. The first law states:

If there exists a stationary state for a conservative biological association of n pairs of species, the numbers of individuals in each species is bounded between positive numbers, between which undamped fluctuations always occur.

With this extension it becomes evident that the property of periodicity is lost while that of fluctuation persists.

When the fluctuations are small and *n* is the number of species, the pattern of fluctuation can be approximated by the superposition of *n*/2 undamped fluctuations, each with its own period, which are independent of the initial conditions.

The second law remains unaltered *when one takes as a mean number of individuals of a single species the limit of the mean number of individuals over an infinitely long time period (the asymptotic mean).*

The third law takes the following form: *In a conservative association of an even number of species for which a stationary state exists, and in which the predatory and prey species may be distinguished, if there is a descruction of all species uniformly in proportion to their numbers, the asymptotic means of the number of individuals of some of the prey species (if not of all) will increase, while the asymptotic numbers of some of the predator species (if not of all) will diminish.*[8]

If the number of species in a conservative system is odd the number of individuals of each species cannot be limited between two positive numbers, and such a system does not have a stable equilibrium.[9]

V. Volterra

12.   The case of conservative systems can be considered as a limiting approximation to the associations in nature, but *dissipative associations* would seem to be even closer to systems which actually exist; the value of such associations decreases with each encounter between individuals when one is eaten by the other. Therefore, the fluctuations around a stationary state diminish and the system converges to that state.

13.   In this way we can study internal causes of fluctuations which are sufficient to explain various observed phenomena and to predict new phenomena which are probably amenable to experimental control and observation.

We have previously mentioned external periodic forces. We can take account of these if we assume that the coefficients of growth have small periodicities about constant means, whence for small fluctuations the principle of superimposing the inherent fluctuations on the externally-caused fluctuations applies, that is, the small fluctuations are obtained by superimposing on the inherent variations those which are external and have the period of the growth coefficients, when that period does not coincide with any of the periods of the inherent fluctuations.

14.   One case in particular which can be studied mathematically in a very complete way in virtue of the preceding results, is that in which there are three living species in a limited environment such as an island and in which the first species eats the second and the third but not *vice versa*. As an example, we can take a species of carnivorous animals that eats an herbivorous species which in turn eats a plant species, admitting that, for the last, one can use the same values and methods as are used for animals. The same procedures can be applied even to parasitic insects and to plants and plant parasites.

Diverse cases and subcases can be presented. For these, one can characterize the values of the coefficients in the relevant equations; that is, the coefficients of growth and voracity and the values of the individuals of each species.

First case: We assume that the plant species can increase indefinitely. The nutriment which it provides to the carnivores through the herbivores is not sufficient to maintain the carnivorous species and it is depleted, while the herbivores and the plants tend toward an undamped periodic fluctuation.

Second case: If the coefficient of growth of the plant species were constant, the number of individuals in this would grow indefinitely, hence we can assume that this coefficient decreases proportionally to the number of individuals.

Case 2, subcase a: The nutriment furnished by the plants is not sufficient to maintain the herbivores; therefore, the herbivorous species and the carnivorous die out while the plant species holds a constant value.

Case 2, subcase b: The plants are sufficient to maintain the herbivores but not sufficient nutriment for the carnivores as mediated through the herbivores; therefore, the carnivorous species is depleted while the herbivores and the plants tend to damped fluctuation, and finally to a stationary state.

Case 2, subcase c: The food is sufficient for all the species to live, and through asymptotic and damped variations all of these will tend towards a stationary state.[10]

15. It should be noted that from an analytic point of view, the study of fluctuations and oscillations in the number of individuals of coexisting species is beyond the limits of the ordinary study of oscillations when the general equations are not linear, whereas the classical scheme of the theories of oscillations is normally dealt with by linear equations.

In fact, the fluctuations studied are generally not small fluctuations. It is only when we have made the hypothesis of small fluctuations that we have gone beyond the terms of the second order and we have been able to make use of linear equations.

16. Before closing this article, we wish to forewarn the readers of objections which could be raised which might put the preceding results in a false light making them seem inexact and senseless. We have to foresee these, and present our response.

For example, in the case of the two species, one of which preys on the other, we find that this stabilizes a periodic cycle in which the two species oscillate about a mean value in reaction to each other. One could object that it is easy to imagine that the predatory species is so numerous and voracious that it destroys one by one all the individuals of the other species in a very short time and, therefore, renders the predicted oscillations impossible.

We can observe that the law of the cycle is deduced from the assumption that a species which lacks nutrients will not be depleted in an infinite time and this may seem far from reality and even farther from the law itself. This is dependent on the fact that among the fundamental hypotheses there is the hypotheses that the number of individuals is a continuously variable positive number, while in reality, it must be an integer and cannot descend below unity.

Therefore, we can say that if the number of individuals of a species is reduced
to a sufficiently small number, it must be assumed to be null and to prolong
the value is nothing other than a theoretical concession without any real significance. Thus, we return to the case of section 5, if the coefficient of voracity
γ_1 were much greater and the initial value of N_2 unchanged, N_1 can rapidly
become smaller than 1 which in practical terms means its extinction and, therefore, the cycle which would theoretically continue cannot resume but will cease
at this point.

There is, therefore, nothing peculiar about these mathematical expositions
as applied to biology, but an analog is presented to all other cases in which
the continuous is substituted for the discontinuous. It is necessary in most
cases to make such a substitution otherwise it would not be possible to apply
the most powerful instrument which mathematics provides, which is infinitesimal
calculation, and for the rest of the classical cases, the consequences do not
always have a practical application.

It is not only when this substitution is made, but it could also be said that
for any application of mathematics to natural phenomenon the type of question
to which we are alluding must be presented. Therefore, to apply mathematics
to almost any subject, it is necessary to build hypotheses and models that are
appropriate to the subject or event under consideration even if it only approximates the actual phenomenon. Thus, in dealing with the mechanics of solid
bodies it is assumed that, subject to whatever forces, they will not be deformed
which, of course, never is true for any material.

How can we proceed theoretically to overcome these difficulties? This can
be facilitated by distinguishing two phases: In the first, the problem is
solved by abandoning it. That is to say we proceed with our analyses by assuming
that the hypotheses we have made have been absolutely verified. Having thus
obtained a solution, we can in the second phase discuss it and if it appears
that certain limits have been exceeded in the solution, then the hypotheses made
are too far from reality, and it is necessary to reject the solution or modify
it.

Thus, we can calculate the forces that support the parts of an ideal structure assuming that these are infinitely resistant and absolutely rigid. But
once we have obtained this solution, we must in the second phase see if some
of these forces exceed certain limits because then an equilibrium would not be

possible, but the structure would disintegrate. This is extremely important to see.

In the case of the fluctuations, the species which we have considered as ideal species are those formed from any positive number of individuals. But if, after having made the calculations, we find that the number of individuals of one species have a value of less than one, we can say (as we have previously mentioned) that the variations of the species will be interrupted because the species cannot exist. The solution found was not, therefore, in vain, but reveals to us a circumstance of notable importance and utility.

The first phase of which we have spoken could be called the rational phase and the other the applied. We have built a mechanical rationale and an applied rationale. The researches in mathematical biology which have been briefly presented here pertain to the rational phase.

FOOTNOTES

1. See Volterra, *Saggi scientifici*. -I. Sui tentativi di applicazione delle matematiche alle Scienze biologiche e sociali, Bologna, Zanichelli, 1920. For biometry see the periodical *Biometrika* established in 1901 by Karl Pearson.

2. D'Arcy Thompson Wentworth, *On Growth and Form,* Cambridge, 1917.

3. Volterra, Ibid., VII. L'evolutzione delle idee fondamentali del calcolo infinitesimali; VIII. L'applicazione del calcolo ai fenomeni d'eredita.

4. I have published the complete work in *Memoria della R. Accademia Nazionale dei Lincei* classe di Scienze fisico-matematiche e naturali, Serie VI, Vol. II, Fasc. III with the title: *Variazioni e fluttuazioni del numero d'individui in specie animali conviventi*. After this publication I have been informed that in the parasitological questions relating to malaria the equations of Ross were used. I have learned also that Dr. Lotka in the volume: *Elements of Physical Biology,* New York, 1925, had considered the case of two species which I had developed in the third section of part I arriving by other methods at the integral, to his diagram and to the period of small oscillations. But the general laws I developed in the same section, the various cases I developed in other sections of the first part and other parts of the publication in which I consider the conservative and the dissipative associations which are new and treated for the first time. I regret that I was not able to site in this publication the interesting work of Dr. Lotka which contains other diverse applications of mathematics to chemical and biological questions, which will form the subject of a special review.

5. Dr. Umberto D'Ancona has many times discussed with me the statistics which he was making of fisheries in the period during the war and in periods before and after, asking me if it were possible to give a mathematical explanation

V. Volterra

of the results which he was getting on the percentages of the various species
in these different periods. This request has spurred me to formulate the problem
and solve it, establishing the laws which are set forth in section 6. Both
D'Ancona and I working independently were equally satisfied in comparing results
which were revealed to us separately by calculus and by observation, as these
results were in agreement, showing for example that fishing, by disturbing
the natural proportions of the two species, one of which preys on the other,
causes decline in the number of the predator species and an increase in the
numbers of the prey.

6. As we see in section 7, D'Ancona examines the market statistics from Trieste,
Venice and Fiume which were caught during the war in the high Adriatic. He finds
a displacement of the proportion of the individuals of the various species of
fish to the advantage of the selacians which we must consider among the most
voracious. This result agrees with the law of the disturbances of the means
which we will present further on in this article.

7. These two equations are much better justified by utilizing the probability
of encounter between individuals of the two species as is treated in section 9.
This is also treated as a general case of an undetermined number of coexisting
species in section 10.

8. As we have stated in section 8, this law is valid up to a certain limit
from which additional distruction could deplete all the species.

9. This result should not be suprising given the absolute character of conservative systems. (See section 12.)

10. We have not given the algebraic equations which the coefficients character-
izing the diverse cases and subcases must satisfy. These can be found in the
Memoria published by the Accademia dei Lincei, page 62.

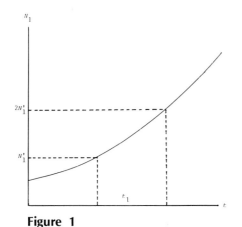

Figure 1

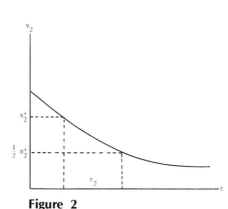

Figure 2

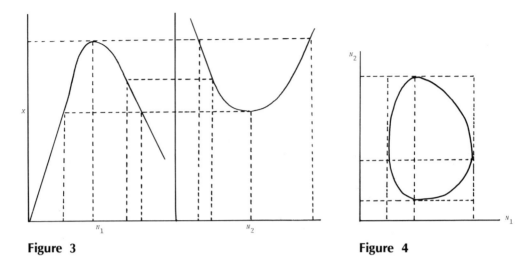

Figure 3

Figure 4

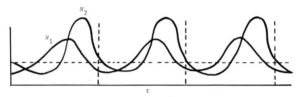

Figure 5

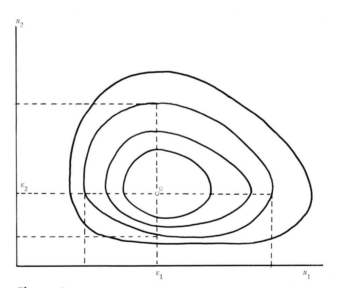

Figure 6

5

Reprinted from *Sankhya*, **6**, Pt. 1, 93–96 (1942)

ON THE GENERATION AND GROWTH OF A POPULATION

By E. G. LEWIS

University College, Rangoon

1. Some of the factors which complicate the theoretical and practical quantitative study of human populations may be negligible in the treatment of certain lower biological populations. The comparatively short lifetime and the regularity of breeding epochs is one simplification, which, incidentally suggests the interval between breeding epochs as a natural unit of age. It may also be that fertility and mortality rates at ages are more nearly constant in time - in any case, time fluctuations in these rates due to environment changes are more amenable to experimental control in colonies of lower animals or plants than in human populations.

2. Suppose a group y_1 of individuals to be 'born' at some epoch $(t = 0)$ and consider the population generated by this group. Assume that 'breeding' occurs at regular epochs $t = z, 2z, \ldots . N\,Z, \ldots$ and let us adopt z as the unit of age for individuals. Let f_r be the fertility factor per individual entering the rth age group $(r-1\,z$ to $rz)$ and let S_r be the survival factor per individual for survival from the rth to the $(r+1)^{th}$ age group. Both f_r and S_r are taken to be constant in time and we assume, further, that for reproductive purposes there is no interaction between different age groups.

Under these somewhat stringent conditions, the relaxation of which can be considered later, the numerical history of the population generated by the original y_1 individuals, will, just after the third breeding epoch (say), be as follows :

	After epoch $t=0$	After epoch $t=Z$	After epoch $t=2Z$	After epoch $t=3Z$
N° aged 0+	y_1	$f_1 y_1$	$(f_1^2 + f_2 s_1)y_1$	$(f^3_1 + 2f_1 f_2 s_1 + f_3 s_2 s_1 f_1')$
N° aged Z+	0	$s_1 y_1$	$f_1 s_1 y_1$	$(s_1 f_1^2 + f_2 s_2^2)y_1$
N° aged 2Z+	0	0	$s_2 s_1 y_1$	$f_1 s_2 s_1 y_1$
N° aged 3Z+	0	0	0	$s_3 s_2 s_1 y_1$

If we now suppose that no individual survives the age nZ (i. e. $S_n = 0$), the frequency distribution in the n age groups 0 to Z ; Z to $2Z \ldots, n-1\,Z$ to nz will, after the Nth breeding epoch be given by the Nth power of the linear substitution.

$$A \equiv (f_1 y_1 + f_2 y_2 + \ldots + f_n y_n, s_1 y_1 \ldots s_{n-1} y_{n-1}), \qquad \qquad .. \ (1)$$

the initial set of frequencies being $y_1 = y_1$; $y_2 = y_3 = \ldots y_n = 0$

3. Now consider the n age groups 0 to z, z to $2z$, etc. to be filled by an *arbitrary* set of $y_1, y_2, \ldots, y_n$ individuals and let the fertility and survival factors for age groups be $f_1 f_2 \ldots f_n$ and $s_1, s_2 \ldots s_{n-1}$, 0 as before. Then we may take the arbitrary column matrix

$$\{y\} = \{y_1, y_2 \ldots y_n\}$$

as an initial age distribution corresponding to the epoch $t = 0$. The age distribution after the first breeding epoch will now be given by

$$\{y'\} = \{y'_1, y'_2 \ldots y'_n\}$$

where

$$y_1' = f_1 y_1 + f_2 y_2 + \ldots + f_n y_n$$

$$y'_2 = s_1 y_1$$

$$y'_3 = s_2 y_2$$

$$y'_n = s_{n-1} y_{n-1}$$

which is, of course, the linear substitution A of equation (1). The matrix of this substitution is the square matrix

$$
M =
\begin{matrix}
f_1 & f_2 & \ldots & f_{n-1} & f_n \\
s_1 & 0 & \ldots & 0 & 0 \\
0 & s_2 & \ldots & 0 & 0 \\
0 & 0 & \ldots & s_{n-1} & 0
\end{matrix}
$$

all elements not in the first row or sub-diagonal being zero. The age frequency distribution after the Nth breeding epoch will be given by the Nth power of A or, what is the same thing, by the column matrix $M^N\{y\}$.

Now the characteristic equation of M is easily shown to be

$$\lambda^n - f_1 \lambda^{n-1} - \ldots - f_{n-1} s_1 s_2 \ldots s_{n-2} \lambda - f_n s_1 s_2 \ldots s_{n-1} = 0$$

where, of course $f_r \geqslant 0$ $(r = 1, 2, \ldots, n)$ and $0 \leqslant s_r \leqslant 1$ $(r = 1, 2, \ldots \overline{n-1})$, $s_n = 0$. This equation determines the form to which the age distribution of the population envisaged ultimately settles down under the action of the given system of fertility and survival factors.

4. Under certain conditions however the initial age distribution will be repeated exactly after the cycle of n breeding epochs. This follows at once since the matrix M satisfies its own characteristic equation

$$M^n - f_1 M^{n-1} - \ldots - f_{n-1} s_1 s_2 \ldots s_{n-2} M - f_n s_1 \ldots s_{n-1} I = 0$$

where I is the unit matrix of order n.

Now if $M^n - I = 0$ the initial distribution $\{y\}$ will be repeated exactly after n operations of M. We see that the condition for this is

$$f_1 = f_2 \ldots f_{n-1} = 0 \text{ and } f_n s_1 s_2 \ldots s_{n-1} = 1$$

If

$$f_1 = f_2 \ldots f_{n-1} = 0 \text{ and } f_n\, s_1\, s_2\, \ldots\, s_{n-1} = k$$

then the distribution after n breeding epochs will be

$$\{ky_1 \ldots, ky_n\}. \quad i.e., \quad k\{y\}$$

After $2n$ epochs it will be $k^2\{y\}$ and so on.

5. In general, however, the age distribution will display no pure periodicity but will settle down to a definite (stable) distribution depending on the dominant root (root of maximum modulus) of the characteristic equation.

In this equation, namely

$$\lambda^n - f_1\, \lambda^{n-1} - f_2\, \lambda^{n-2}\, s_1 \ldots - f_{n-1}\, s_1 . s_2 \ldots s_{n-2}\, \lambda - f_n\, s_1, \ldots s_{n-1} = 0$$

the f and s are all positive so that the equation has at most one positive root, and moreover, it always has one.   It can be shown that this positive root is also the dominant root* (a fact of itself not without interest).   The only exception which can arise yields the periodicity mentioned in para 4.

6. Now when the characteristic equation has a single (non repeated) real dominant root l(say) the ratio of an element in the $(m+1)^{th}$ power of M to the corresponding element in the mth power will, for large m, tend to l.   The column matrix giving the age distribution, being acted upon successively by the matrix M, will also tend to a form in which an element of $M^{m+1}\{y\}$ will bear to the corresponding element in $M^m\{y\}$ the ratio l.

Now let $Y_1, Y_2, \ldots Y_n$ be a set of frequencies such that

$$
\begin{aligned}
\lambda\, Y_1 &= f_1\, Y_1 + f_2\, Y_2 \ldots + f_n\, Y_n \\
\lambda\, Y_2 &= s_1\, Y_1 \\
\lambda\, Y_3 &= s_2\, Y_2 \\
&\vdots \\
\lambda\, Y_n &= s_{n-1}\, Y_{n-1}
\end{aligned}
\qquad \ldots (2)
$$

Eliminating $Y_1\, Y_2, \ldots Y_n$ we obtain the characteristic equation of M, the matrix of A.

The set $Y_1\, Y_2 \ldots Y_n$ constitutes a 'pole' of the substitution, A corresponding to the root λ of the characteristic equation.   We have seen that the frequencies in all age groups will ultimately increase in the ratio l, it follows that the ultimate ratios of the n frequencies one to another will be $Y_1, Y_2, \ldots Y_n$ where $Y_1 \ldots Y_n$ is a pole corresponding to the dominant root l.   Now from equation (2) with $\lambda = l$ we obtain

$$Y_1 : Y_2 : \ldots Y_n = 1 : \frac{s_1}{l} : \frac{s_1\, s_2}{l^2} : \ldots : \frac{s_1\, s_2 \ldots s_{n-1}}{l^{n-1}} \qquad \ldots (3)$$

---

* This theorem and its consequences, which lead to a simple upper bound to the modulus of the roots of any equation, have been investigated by Mr. D. B. Lahiri and the present author in a paper awaiting publication.

7. We see, therefore, that under the operation of the matrix M the age distribution column matrix $\{y\}$ will in general settle down to a form in which the relative frequencies in the different age groups are determined by the ratios (3) and that after each breeding epoch the numbers in each group (and so of course in the total population) will increase in geometric progression of ratio equal to the positive root of the characteristic equation of M. The population will increase, just maintain itself or decrease ultimately according as $l \gtrless 1$. The number l is related to Kuczynski's net reproduction rate, the analogue of which would here be

$$(f_1 + f_2\, s_1 + \ldots + f_n\, s_1 \ldots s_{n-1})$$

but l appears to be a better index of what can be called the true (ultimate) rate of increase. We have

$$1 = \frac{f_1}{l} + \frac{f_2\, s_1}{l^2} \ldots \ldots + \frac{f_n\, s_1 \ldots s_{n-1}}{l^n}$$

so that a net reproduction rate $\gtrless 1$ means $l \gtrless 1$ and *vice versa*, but the magnitudes of the two measures will be different.

8. The question of approximation to the case of pure periodicity discussed in para 4 may be of practical importance for some lower animals with few age groups. The rapidity with which an arbitrary initial age distribution settled down close to the 'ultimate' form depends on the difference between the dominant and sub-dominant root of the characteristic equation.

As for the relaxation of the conditions assumed in para 1, the question of interaction between age groups would in many cases be of no importance but to take account of time changes in f_r and s_r would mean assuming suitable laws of variation for the matrix elements. Such laws lead to much more complicated analysis but are being investigated.

Part II
ADDING GENETICS TO DEMOGRAPHIC CONCEPTS

Editors' Comments
On Papers 6, 7, and 8

6   DARWIN
     The Proportion of the Sexes in Relation to Natural Selection

7   FISHER
     Natural Selection and the Sex Ratio

8   BODMER and EDWARDS
     Natural Selection and the Sex Ratio

Our discussion so far has dealt with the development of the concept of a population as a meaningful aggregate of individuals. This aggregate is of no genetic interest unless its internal structure is related to differing genotypes within, or between, populations. When genotypes are related to demographic aspects of a population, we have demographic genetic phenomena. We shall now turn to an examination of various aspects of such demographic structure.

SEX RATIO

The most important and immediate genetically relevant structure of a biological population is its division according to sex. The sex ratio is presumably under genetic control, and it has ramifications for other aspects of population behavior and structure.

The first problem is to explain the nearly one to one ratio between males and females in a population of higher organisms. Relatively little is known about the biological details of how this has evolved. Darwin was the first to consider this as an evolutionary problem, and Paper 6 is a section from *The Descent of Man* (originally published in 1872) to illustrate his approach (in the full work, Darwin precedes this discussion with an extensive, one may say typically Darwinian, detailing of observations on animal sex ratios). Darwin was mainly concerned with the way the sex ratio relates to human populations and sexual selection. Clearly, a disparity in the sex ratio necessitates mate selection in the population, and this is the grist for the sexual-selection mill. This was the mechanism through which Darwin acknowledged the role of differential fertility in evolution, which otherwise he credits to differential mortality. In the section chosen here, Darwin concludes with his famous statement that the problem of the evolution of the sex ratio is too difficult and must be left "to the future."

The "future" in this case was R. A. Fisher, who tackled this problem in his *The Genetical Theory of Natural Selection* (1930, 1958). We have included the relevant section in this volume (Paper 7), for it, and not Darwin, forms the foundation of most later treatments. Fisher was not concerned with the evolution of parity as was Darwin; in fact, if Darwin had considered three generations and not simply two he would have found a priori argument for sex parity (see Crow and Kimura, 1970). Fisher was primarily interested in the way that differential mortality between the sexes can lead to selection for the production of a disparate sex ratio. He argues that this can occur if the amount of parental care needed in raising offspring of one sex exceeds that required by the other. The argument was vaguely phrased and caused some minor genetic controversy. In 1937, Crew argued that sex parity maximizes the probabilities of one sex encountering the other for reproduction, and hence is to be expected. In 1953, Shaw and Mohler pass over the parental expenditure problem and argue that, when all the members of a population mate equally and no care is involved, parity must result. Shaw (1958) goes on to argue that nonrandom mating will not affect parity. If parental care is involved, a form of reproductive compensation may act to produce a sex ratio at birth that differs from parity, but sex parity at the end of the parental-care period must result.

Paper 8 is an important recent treatment of the problem, which tries to incorporate the findings of other workers. Here the concept of parental care is rescued from vagueness. Equilibrium sex ratios are derived using Fisherian concepts; a problem with this treatment is that the sex ratio is treated as a quantitative trait, but it is not clear that significantly different conclusions would result from a more rigorous approach. Good discussions of this question are in Crow and Kimura (1970), Edwards (1962), Stern (1973), and Cavalli-Sforza and Bodmer (1971).

The sex ratio has effects on many aspects of the social biology of a species; this is especially true for human societies, a great deal of whose energy has traditionally been invested in marriage-related behavior, around which their basic cultural system is centered. Treating the sex ratio in this fully demographic way, Beiles (1974) has recently discussed the effect of different marriage ages of spouses as this relates and responds to the sex ratio at birth and other factors.

6

Reprinted from *The Origin of Species* and *The Descent of Man,* Random House, Inc. (The Modern Library), New York, pp. 608–611

THE PROPORTION OF THE SEXES IN RELATION TO NATURAL SELECTION

Charles Darwin

[*Editors' Note:* In the original, material precedes this excerpt.]

There is reason to suspect that in some cases man has by selection indirectly influenced his own sex-producing powers. Certain women tend to produce during their whole lives more children of one sex than of the other: and the same holds good of many animals, for instance, cows and horses; thus Mr. Wright of Yeldersley House informs me that one of his Arab mares, though put seven times to different horses; produced seven fillies. Though I have very little evidence on this head, analogy would lead to the belief, that the tendency to produce either sex would be inherited like almost every other peculiarity, for instance, that of producing twins; and concerning the above tendency a good authority, Mr. J. Downing, has communicated to me facts which seem to prove that this does occur in certain families of short-horn cattle. Col. Marshall[94] has recently found on careful examination that the Todas, a hill-tribe of India, consist of 112 males and 84 females of all ages—that is in a ratio of 133.3 males to 100 females. The Todas, who are polyandrous in their marriages, during former times invariably practised female infanticide; but this practice has now been discontinued for a considerable period. Of the children born within late years, the males are more numerous than the females, in the proportion of 124 to 100. Colonel Marshall accounts for this fact in the following ingenious manner. "Let us for the purpose of illustration take three families as representing an average of the entire tribe; say that one mother gives birth to six daughters and no sons; a second mother has six sons only, whilst the third mother has three sons and three daughters. The first mother, following the tribal custom, destroys four daughters and preserves two. The second retains her six sons. The third kills two daughters and keeps one, as also her three sons. We have then from the three families, nine sons and three daughters, with which to continue the breed. But whilst the males belong to families in which the tendency to produce sons is great, the females are of those of a converse inclination. Thus the bias strengthens with each generation, until, as we find, families grow to have habitually more sons than daughters."

That this result would follow from the above form of infanticide seems almost certain; that is if we assume that a sex-producing tendency is inherited. But as the above numbers are so extremely scanty, I have searched for additional evidence, but cannot decide whether what I have found is trustworthy; nevertheless the facts are, perhaps, worth giving. The Maories of New Zealand have long practised infanticide; and Mr. Fenton[95] states that he "has met with instances of women who have destroyed four, six, and even seven children, mostly females. However, the

---

[94] 'The Todas,' 1873, pp. 100, 111, 194, 196.
[95] 'Aboriginal Inhabitants of New Zealand; Government Report,' 1859, p. 36.

universal testimony of those best qualified to judge, is conclusive that this custom has for many years been almost extinct. Probably the year 1835 may be named as the period of its ceasing to exist." Now amongst the New Zealanders, as with the Todas, male births are considerably in excess. Mr. Fenton remarks (p. 30), "One fact is certain, although the exact period of the commencement of this singular condition of the disproportion of the sexes cannot be demonstratively fixed, it is quite clear that this course of decrease was in full operation during the years 1830 to 1844, when the non-adult population of 1844 was being produced, and has continued with great energy up to the present time." The following statements are taken from Mr. Fenton (p. 26), but as the numbers are not large, and as the census was not accurate, uniform results cannot be expected. It should be borne in mind in this and the following cases, that the normal state of every population is an excess of women, at least in all civilised countries, chiefly owing to the greater mortality of the male sex during youth, and partly to accidents of all kinds later in life. In 1858, the native population of New Zealand was estimated as consisting of 31,667 males and 24,303 females of all ages, that is in the ratio of 130.3 males to 100 females. But during this same year, and in certain limited districts, the numbers were ascertained with much care, and the males of all ages were here 753 and the females 616; that is in the ratio of 122.2 males to 100 females. It is more important for us that during this same year of 1858, the *non-adult* males within the same district were found to be 178, and the *non-adult* females 142, that is in the ratio of 125.3 to 100. It may be added that in 1844, at which period female infanticide had only lately ceased, the *non-adult* males in one district were 281, and the *non-adult* females only 194, that is in the ratio of 144.8 males to 100 females.

In the Sandwich Islands, the males exceed the females in number. Infanticide was formerly practised there to a frightful extent, but was by no means confined to female infants, as is shown by Mr. Ellis,[96] and as I have been informed by Bishop Staley and the Rev. Mr. Coan. Nevertheless, another apparently trustworthy writer, Mr. Jarves,[97] whose observations apply to the whole archipelago, remarks:—"Numbers of women are to be found, who confess to the murder of from three to six or eight children," and he adds, "females from being considered less useful than males were more often destroyed." From what is known to occur in other parts of the world, this statement is probable; but must be received with much caution. The practice of infanticide ceased about the year 1819. when idolatry was abolished and missionaries settled in the Islands. A careful census in 1839 of the adult and taxable men and women in the island of Kauai and in one district of Oahu (Jarves, p. 404), gives 4723 males and 3776 females; that is in the ratio of 125.08 to 100. At the same time the number of males under fourteen years in Kauai and under eight-

[96] 'Narrative of a Tour through Hawaii,' 1826, p. 298.
[97] 'History of the Sandwich Islands,' 1843, p. 93.

een in Oahu was 1797, and of females of the same ages 1429; and here we have the ratio of 125.75 males to 100 females.

In a census of all the islands in 1850,[98] the males of all ages amount to 36,272, and the females to 33,128, or as 109.49 to 100. The males under seventeen years amounted to 10,773, and the females under the same age to 9593, or as 112.3 to 100. From the census of 1872, the proportion of males of all ages (including half-castes) to females, is as 125.36 to 100. It must be borne in mind that all these returns for the Sandwich Islands give the proportion of living males to living females, and not of the births; and judging from all civilised countries the proportion of males would have been considerably higher if the numbers had referred to births.[99]

From the several foregoing cases we have some reason to believe that infanticide practised in the manner above explained, tends to make a male-producing race; but I am far from supposing that this practice in the case of man, or some analogous process with other species, has been the sole determining cause of an excess of males. There may be some unknown law leading to this result in decreasing races, which have already become somewhat infertile. Besides the several causes previously alluded to, the greater facility of parturition amongst savages, and the less consequent injury to their male infants, would tend to increase the proportion of live-born males to females. There does not, however, seem to be any necessary connection between savage life and a marked excess of males; that is if we may judge by the character of the scanty offspring of the

[98] This is given in the Rev. H. T. Cheever's 'Life in the Sandwich Islands,' 1851, p. 277.

[99] Dr. Coulter, in describing ('Journal R. Geograph. Soc.,' vol. v. 1835, p. 67) the state of California about the year 1830, says that the natives, reclaimed by the Spanish missionaries, have nearly all perished, or are perishing, although well treated, not driven from their native land, and kept from the use of spirits. He attributes this, in great part, to the undoubted fact that the men greatly exceed the women in number; but he does not know whether this is due to a failure of female offspring, or to more females dying during early youth. The latter alternative, according to all analogy, is very improbable. He adds that "infanticide, properly so called, is not common, though very frequent recourse is had to abortion." If Dr. Coulter is correct about infanticide, this case cannot be advanced in support of Colonel Marshall's view. From the rapid decrease of the reclaimed natives, we may suspect that, as in the cases lately given, their fertility has been diminished from changed habits of life.

I had hoped to gain some light on this subject from the breeding of dogs; inasmuch as in most breeds, with the exception, perhaps, of greyhounds, many more female puppies are destroyed than males, just as with the Toda infants. Mr. Cupples assures me that this is usual with Scotch deerhounds. Unfortunately, I know nothing of the proportion of the sexes in any breed, excepting greyhounds, and there the male births are to the females as 110.1 to 100. Now from enquiries made from many breeders, it seems that the females are in some respects more esteemed, though otherwise troublesome; and it does not appear that the female puppies of the best-bred dogs are systematically destroyed more than the males, though this does sometimes take place to a limited extent. Therefore I am unable to decide whether we can, on the above principles, account for the preponderance of male births in greyhounds. On the other hand, we have seen that with horses, cattle, and sheep, which are too valuable for the young of either sex to be destroyed, if there is any difference, the females are slightly in excess.

lately existing Tasmanians and of the crossed offspring of the Tahitians now inhabiting Norfolk Island.

As the males and females of many animals differ somewhat in habits and are exposed in different degrees to danger, it is probable that in many cases, more of one sex than of the other are habitually destroyed. But as far as I can trace out the complication of causes, an indiscriminate though large destruction of either sex would not tend to modify the sex-producing power of the species. With strictly social animals, such as bees or ants, which produce a vast number of sterile and fertile females in comparison with the males, and to whom this preponderance is of para-mount importance, we can see that those communities would flourish best which contained females having a strong inherited tendency to produce more and more females; and in such cases an unequal sex-producing tendency would be ultimately gained through natural selection. With animals living in herds or troops, in which the males come to the front and defend the herd, as with the bisons of North America and certain ba-boons, it is conceivable that a male-producing tendency might be gained by natural selection; for the individuals of the better defended herds would leave more numerous descendants. In the case of mankind the advantage arising from having a preponderance of men in the tribe is supposed to be one chief cause of the practice of female infanticide.

In no case, as far as we can see, would an inherited tendency to produce both sexes in equal numbers or to produce one sex in excess, be a direct advantage or disadvantage to certain individuals more than to others; for instance, an individual with a tendency to produce more males than females would not succeed better in the battle for life than an individual with an opposite tendency; and therefore a tendency of this kind could not be gained through natural selection. Nevertheless, there are certain animals (for instance, fishes and cirripedes) in which two or more males appear to be necessary for the fertilisation of the female; and the males accordingly largely preponderate, but it is by no means obvious how this male-producing tendency could have been acquired. I formerly thought that when a tendency to produce the two sexes in equal numbers was advantageous to the species, it would follow from natural selection, but I now see that the whole problem is so intricate that it is safer to leave its solution for the future.

7

Reprinted from *The Genetical Theory of Natural Selection,* Dover Publications, Inc., New York, 1958, pp. 158–160

NATURAL SELECTION AND THE SEX RATIO

R. A. Fisher

[*Editors' Note:* In the original, material precedes this excerpt.]

The problem of the influence of Natural Selection on the sex-ratio may be most exactly examined by the aid of the concept of reproductive value developed in Chapter II. As is well known, Darwin expressly reserved this problem for the future as being too intricate to admit of any immediate solution. (*Descent of Man*, p. 399).

In no case, as far as we can see, would an inherited tendency to produce both sexes in equal numbers or to produce one sex in excess, be a direct advantage or disadvantage to certain individuals more than to others; for instance, an individual with a tendency to produce more males than females would not succeed better in the battle for life than an individual with an opposite tendency; and therefore a tendency of this kind could not be gained through natural selection. Nevertheless, there are certain animals (for instance, fishes and cirripedes) in which two or more males appear to be necessary for the fertilization of the female; and the males accordingly largely preponderate, but it is by no means obvious how this male-producing tendency could have been acquired. I formerly thought that when a tendency to produce the two sexes in equal numbers was advantageous to the species, it would follow from natural selection, but I now see that the whole problem is so intricate that it is safer to leave its solution for the future.

In organisms of all kinds the young are launched upon their careers endowed with a certain amount of biological capital derived from their parents. This varies enormously in amount in different species, but, in all, there has been, before the offspring is able to lead an independent existence, a certain expenditure of nutriment in addition, almost universally, to some expenditure of time or activity, which the parents are induced by their instincts to make for the advantage of their young. Let us consider the reproductive value of these offspring at the moment when this parental expenditure on their behalf has just ceased. If we consider the aggregate of an entire generation of such offspring it is clear that the total reproductive value of the males in this group is exactly equal to the total value of all the females, because each sex must supply half the ancestry of all future generations of the species. From this it follows that the sex ratio will so adjust itself, under the influence of Natural Selection, that the total parental expenditure incurred in respect of children of each sex, shall be equal; for if this were not so and the total expenditure incurred in producing males, for instance, were less than the total expenditure incurred in producing females, then since the total reproductive value of the males is equal to that of the females, it would follow that those parents, the innate tendencies of which caused them to produce males in excess, would, for the same expenditure, produce a greater amount of reproductive value; and in consequence would be the progenitors of a larger fraction of future generations than would parents having a congenital bias towards the production of females. Selection would thus raise the sex-ratio until the expenditure upon males became equal to that upon females. If, for example, as in man, the males suffered a heavier mortality during the period of parental expenditure, this would cause them to be more expensive to produce, for, for every hundred males successfully produced expenditure has been incurred, not only for these during their whole period of dependance but for a certain number of others who have perished prematurely before incurring the full complement of expenditure. The average expenditure is therefore greater for each boy reared, but less for each boy born, than it is for girls at the corresponding stages, and we may therefore infer that the condition toward which Natural Selection will tend will be one in which boys are the more numerous at birth, but become less numerous, owing to their higher death-rate, before the end of the period of parental expenditure. The actual sex-ratio in man seems to fulfil these conditions somewhat closely, especially if we make allowance for the large recent diminution in the deaths of infants and children; and since this adjustment is brought about by a somewhat large inequality in the sex ratio at conception, for which no *a priori* reason can be given, it is difficult to avoid the conclusion that the sex-ratio has really been adjusted by these means.

The sex-ratio at the end of the period of expenditure thus depends upon differential mortality during that period, and if there are any such differences, upon the differential demands which the young of such species make during their period of dependency; it will not be influenced by differential mortality during a self-supporting period; the relative numbers of the sexes attaining maturity may thus be influenced without compensation, by differential mortality during the period intervening between the period of dependence and the attainment of maturity. Any great differential mortality in this period will, however, tend to be checked by Natural Selection, owing to the fact that the total reproductive value of either sex, being, during this period, equal to that of the other, whichever is the scarcer, will be the more valuable, and consequently a more intense selection will be exerted in favour of all modifications tending towards its preservation. The numbers attaining sexual maturity may thus become unequal if sexual differentiation in form or habits is for other reasons advantageous, but any great and persistent inequality between the sexes at maturity should be found to be accompanied by sexual differentiations, having a very decided bionomic value.

[*Editors' Note:* Material has been omitted at this point.]

8

Reprinted from *Ann. Human Genetics, London,* **24**(3), 239–244 (1960)

Natural selection and the sex ratio

BY W. F. BODMER AND A. W. F. EDWARDS*

Department of Genetics, University of Cambridge

INTRODUCTION

Darwin (1871) considered whether the sex ratio is subject to Natural Selection. After surveying the available statistics and investigating the selective effects of infanticide at some length he concludes by saying that as far as he can see there is no selective advantage attached to a particular sex ratio, and that 'I now see that the whole problem is so intricate that it is safer to leave its solution to the future'.

There the matter lay until it was taken up by Fisher (1930), who outlined the solution to the problem. His approach depends on essentially economic arguments which must take into account the parental expenditure of effort involved in the rearing of offspring to maturity. This expenditure will be a function of the time and energy which the parents are induced to spend in rearing their offspring, and must therefore depend on the mortality during the period of parental expenditure. Here is a shortened account of Fisher's argument:

Let us consider the reproductive value, or the relative genetic contribution to future generations, of some offspring at the moment when parental expenditure of effort on them has just ceased. It is clear that the total reproductive value of males in a generation of offspring is equal to the total value of the females, because each sex must supply half the ancestry of future generations. From this it follows that the sex ratio will so adjust itself, under the influence of Natural Selection, that the total parental expenditure of effort incurred in respect of children of each sex shall be equal; for if this were not so and the total expenditure incurred in producing males, for instance, were less than that incurred in producing females, then parents genetically inclined to producing males in excess would, for the same expenditure, produce a greater amount of reproductive value, with the result that in future generations more males would be produced. Selection would thus raise the sex ratio until the expenditure upon males became equal to that upon females.

In 1953 Shaw & Mohler tackled the problem, but they consider that 'Fisher's treatment is phrased in non-genetical terms and does not lend itself to further development'. They therefore 'ignore instances involving parental care' and come to the anticipated conclusion that the equilibrium sex ratio at conception should be one-half (we define the sex ratio to be the proportion of males). Since the production of offspring by a sexual organism always involves expenditure by the parents their treatment is necessarily incomplete.

DEVELOPMENT OF FISHER'S THEORY

The purpose of this paper is to put Fisher's theory, and some of the conclusions that may be drawn from it, on an analytic basis. Let us consider first the selective advantage attached to reproduction with a given primary, or conception, sex ratio. Suppose that the mean primary sex ratio in a population is X, and in a given part of the population is x. Let the proportion of males living until the end of the period of parental expenditure be M in the whole population

* Medical Research Council Scholar.

and m in that part of the population producing with sex ratio x, and let the corresponding proportions of females be F and f.

Now consider the reproductive value of an individual from the 'x' part of the population at the moment when parental expenditure on his or her behalf has just ceased. The reproductive value of a male is inversely proportional to the number of males at that stage in the whole population, or proportional to $1/XM$. Similarly, the reproductive value of a female is proportional to $1/(1-X)F$. Thus the average reproductive value of an individual reproducing with sex ratio x is given by the mean of these values weighted by the proportions of the two sexes living at the end of the period of parental expenditure, and is proportional to

$$\frac{\dfrac{xm}{XM} + \dfrac{(1-x)f}{(1-X)F}}{xm + (1-x)f}.$$

If we now write $xm/[xm + (1-x)f] = x'$, the sex ratio at the end of the period of parental expenditure, and $XM/[XM + (1-X)F] = X'$, then the reproductive value is proportional to $x'(1-X') + (1-x')X'$.

We now find the expression for the average parental expenditure required to raise one child to the end of the period of parental expenditure in the 'x' part of the population. Suppose that the expected total expenditure on a child is proportional to the probability of its surviving to the end of the period. This assumes that children dying before the end of the period incur a negligible expenditure compared with those who survive. This is likely in man at least, since the majority of deaths occur early in pregnancy, at a time when the parental expenditure is small. Let the expected expenditure on a male child be to that on a female child as h to $1-h$. Thus the expected expenditure on a male child is mh and on a female $f(1-h)$. The average expenditure per child conceived is therefore $xmh + (1-x)f(1-h)$, and, dividing this by the probability of a child living to the end of the period of parental expenditure, we have the average expenditure required to raise one child to the end of that period:

$$\frac{xmh + (1-x)f(1-h)}{xm + (1-x)f}.$$

Substituting for x' as before gives $x'h + (1-x')(1-h)$.

The reproductive value per unit parental expenditure in the 'x' part of the population is therefore proportional to

$$R = \frac{x'(1-X') + (1-x')X'}{x'h + (1-x')(1-h)},$$

and this expression is a measure of the selective advantage attached to reproduction with particular sex and parental expenditure ratios.

The value of R will be a maximum for variation in x' when

$$\frac{dR}{dx'} = \frac{1 - h - X'}{(x'h + (1-x')(1-h))^2} = 0.$$

Thus when $X' = 1 - h$ the selective advantage is a maximum, and is seen to be independent of x', so that all sex ratios are then equally advantageous. If X' is less than $1-h$ large values of x' are at an advantage and X' moves towards $1-h$ in the next generation; if X' is greater than $1-h$ small values are at an advantage, and X' again moves towards $1-h$. The population is therefore at a stable equilibrium when the mean sex ratio at the end of the period of parental expenditure, X', is $1-h$. We then have $X'h = (1-X')(1-h)$, which gives Fisher's Law, that at equilibrium the total parental expenditure incurred in respect of children of each sex is equal.

It is important to note that this equilibrium refers to the sex ratio X' at the end of the period of parental expenditure, and not to the primary sex ratio or the sex ratio at birth. The selective advantage R also depends only on the sex ratios x' and X', so that allowing for variation in x' takes into account both variation in the primary sex ratio and in the differential mortality during the period of parental dependence. For most species it is likely that the major part of any differences between the two sexes in parental expenditure will depend on this differential mortality and not on any differential demands which the young make on their parents. The quantity h, which is a measure of the latter more restricted part of the differential expenditure, may therefore be expected to remain constant and near to the value one-half. We have thus shown that Natural Selection will tend to maintain the sex ratio at the end of the period of parental expenditure near the value one-half. The numerical equality of the sexes in man in the age group 15–20 years has frequently been commented upon, but in spite of Fisher's theory it has usually been assumed that the optimum sex ratio is one-half at the reproductive age because the chance of encounter between the sexes is then a maximum (see, for example, Crew (1937)). However, it will not be surprising if modern human populations are not in equilibrium for the sex ratio, because prenatal and infant mortalities have been changing very quickly in recent years, and differential mortalities may have been changing more quickly than selection can change them or the primary sex ratio.

PROGRESS OF A POPULATION TOWARDS EQUILIBRIUM

As an example of the implications of the above analysis in the study of changes in the sex ratio, we take the case of a population in which $h = \frac{1}{2}$, and suppose it to be slightly displaced from its equilibrium position, which, since $h = \frac{1}{2}$, will be at $X' = \frac{1}{2}$. Dropping the primes from our notation, let the probability density function of the sex ratio at the end of the period of parental expenditure in generation 0 be $f(x)$, where

$$\int_0^1 f(x)\,dx = 1,$$

and let the distribution have mean X_0 and variance V_0. The selective advantage of a sex ratio x in a population with mean ratio X_0 is proportional to $x(1-X_0)+(1-x)X_0$, since $h = \frac{1}{2}$, or to $X_0 + x(1-2X_0)$.

In deriving the distribution of sex ratios in the next generation we shall, for simplicity, assume that we can multiply the probability density at a given sex ratio by the relevant selective advantage. One genetic system for which this is true is where there is complete genetic determination of the ability of one particular sex to reproduce with a given sex ratio. We may therefore expect that the rate of progress towards equilibrium that is obtained will be an upper limit to the range of possible rates that would be obtained under more realistic, and complex, assumptions.

Performing this multiplication we obtain the probability density function for the next generation:
$$k[X_0 f(x) + x f(x)(1 - 2X_0)],$$

where k is chosen so that the population size remains unity. Integrating the distribution in generation 1 we find
$$1/k = X_0 + (1 - 2X_0) X_0,$$

since
$$\int_0^1 x f(x)\,dx = X_0.$$

90

The mean of generation 1 is therefore

$$X_1 = \frac{X_0 \int_0^1 x f(x)\,dx + (1 - 2X_0) \int_0^1 x^2 f(x)\,dx}{X_0 + (1 - 2X_0) X_0}$$

$$= \frac{X_0^2 + (1 - 2X_0)(V_0 + X_0^2)}{2X_0(1 - X_0)}$$

$$= X_0 + \frac{V_0}{2}\left(\frac{1}{X_0} - \frac{1}{1 - X_0}\right).$$

Thus the change in the mean is

$$X_1 - X_0 = \frac{V_0}{2}\left(\frac{1}{X_0} - \frac{1}{1 - X_0}\right).$$

Similarly, the variance of generation 1 is

$$V_1 = \frac{X_0 \int_0^1 x^2 f(x)\,dx + (1 - 2X_0) \int_0^1 x^3 f(x)\,dx}{X_0 + (1 - 2X_0) X_0} - X_1^2.$$

Now $$\int_0^1 x^3 f(x)\,dx = \int_0^1 (x - X_0)^3 f(x)\,dx + 3X_0 \int_0^1 x^2 f(x)\,dx - 3X_0^2 \int_0^1 x f(x)\,dx + X_0^3$$

$$= 3X_0(V_0 + X_0^2) - 3X_0^3 + X_0^3$$

$$= 3X_0 V_0 + X_0^3,$$

neglecting the third moment about the mean of the original distribution, which in any case is zero if the distribution is symmetrical.

Thus $$V_1 = \frac{X_0(V_0 + X_0^2) + (1 - 2X_0) X_0(3V_0 + X_0^2)}{2X_0(1 - X_0)} - \frac{[X_0^2 + (1 - 2X_0)(V_0 + X_0^2)]^2}{4X_0^2(1 - X_0)^2}$$

$$= V_0 - \left[\frac{V_0}{2}\left(\frac{1}{X_0} - \frac{1}{1 - X_0}\right)\right]^2.$$

We already have $$X_1 = X_0 + \frac{V_0}{2}\left(\frac{1}{X_0} - \frac{1}{1 - X_0}\right)$$

so that $$V_0 - V_1 = (X_0 - X_1)^2,$$

and there is a decrease in the variance from generation to generation.

If the population is near the equilibrium position

$$\frac{V_0}{2}\left(\frac{1}{X_0} - \frac{1}{1 - X_0}\right)$$

is small, and its square may be neglected, in which case $V_1 = V_0$. Thus near the equilibrium we assume that the variance is constant, and the change in the mean from the nth generation to the $(n + 1)$th is given by

$$X_{n+1} - X_n = \frac{V}{2}\left(\frac{1}{X_n} - \frac{1}{1 - X_n}\right).$$

If $X_n - \frac{1}{2}$ is small, as it will be near the equilibrium, a linear approximation to the above equation is

$$X_{n+1} - \frac{1}{2} = (X_n - \frac{1}{2})(1 - 4V),$$

with error of order $(X_n - \frac{1}{2})^3$. Writing $X - \frac{1}{2} = Y$, the difference between the sex ratio and its equilibrium value,

$$Y_n = Y_0(1 - 4V)^n.$$

From these results we can see that *the rate of approach of a population's sex ratio to an equilibrium value of one-half is directly proportional to the genetic variance in sex ratio of that population.* In populations where the equilibrium value is nearly one-half we may expect this result to be at least a good approximation.

DISCUSSION

Knowledge of the variances in sex ratio of populations is very limited. It applies exclusively to the sex ratio at birth, whereas from our analysis we see that knowledge of the variance of the sex ratio at the end of the period of parental expenditure would be more useful. For man Edwards (1958) obtained an estimate of the total variance of the sex ratio at birth of 0·0025, and most of this is probably environmental. An alternative approximate expression for Y_n is

$$Y_n = Y_0 e^{-4Vn}$$

from which we see that the deviation of the sex ratio from one-half will decrease by a factor $e = 2·718$ in $1/4V$ generations. For example, to reduce the sex ratio from 0·5200 to 0·5074 will take this time. If the genetic variance is 0·0025 this is 100 generations, or about 2000 years for man. If, however, the variance is much less, as it may well be, the time is proportionately longer. Thus changes in the human sex ratio due to Natural Selection are almost certainly too slow to be detected over the period for which data are available, although they may not be slow in comparison with many other evolutionary changes.

Since the rate of approach to equilibrium depends on the variance, large variances should be selected for if the ability of a population to approach equilibrium rapidly is advantageous. Now our analysis has only considered intrapopulation selection and is in no way relevant to interpopulation selection. In most sexually reproducing species the reproductive potential of a population will be limited by the number of females it contains. Hence for interpopulation selection an excess of females may be an advantage, although within any population the changes in the sex ratio will be as analysed above. There would thus be no advantage in having a large variance and consequently the ability to approach equilibrium rapidly. This is borne out by the small variances in the sex ratio that have in fact been observed.

It should be possible to mimic the effect of Natural Selection on the sex ratio by artificially changing the differential infant mortality. For example, if in a mammalian population half the males are killed at birth, selection should act so as to bring back the sex ratio at the end of the period of parental expenditure to its original value. The main difficulty in any form of selection for the sex ratio is that the estimate of the probability of a birth being male in a single family, on which selection depends, has a very large variance. A further difficulty, unforeseen by previous experimentalists, is that the type of selection that we have been considering may exert a stronger selection pressure towards equilibrium than the experimental design can exert away from equilibrium. In this respect the above-mentioned experiment, which has been started in this department, is at an advantage. In practice the small amount of variance available will probably only allow changes that are too small to be detected in a reasonable time. It is also likely that in selecting for a change we select for a combination of changed primary sex ratio and infant mortality, which together may have been stabilized by Natural Selection.

It is clear that models for the action of specific genes on the sex ratio must take into account the selective forces which we have analysed. This may call for some modification of the models proposed by Shaw (1958) and of Bennett's (1958) discussion of Wallace's data on the equilibrium of the 'sex ratio' gene in *Drosophila*.

For some time it has been assumed that in order to explain the apparent distribution of the sex ratio with age in human populations it is necessary to postulate that the primary sex ratio is high. We have shown that the prevailing sex ratio has arisen through the interaction of the primary sex ratio and the mortality rates for males and females, and the existing situation in man can be explained equally well by a high primary sex ratio and considerable sex-differential mortality, or by a primary sex ratio near to one-half and little differential mortality. The direct evidence for a high primary sex ratio seems to be somewhat contradictory and has been critically reviewed by McKeown & Lowe (1951). They state that the existing data are inadequate to justify any assumption about its value; that 'at least half of all abortions occur in the first three months' of pregnancy, and that the sex ratio of live foetuses is about one-half at the seventh month. If the primary sex ratio is high there must therefore be considerable differential mortality during early pregnancy, and, since the expenditure on offspring dying during this period is probably small, the difference in expenditure between males and females dying will also be small, and will be even less if the primary sex ratio is near to one-half. It therefore seems likely that the major difference in expenditure between males and females dying during the period of parental expenditure will be incurred in respect of offspring dying just before, and in the year following, birth, when there is considerable differential mortality. Hence this is probably the major contribution to differential expenditure, apart from differences on individual males and females for which we have allowed in the parameter h, which we are neglecting in assuming that the expenditure on an individual conceived is proportional to the probability of its living to the end of the period of parental expenditure. However, it seems unlikely that taking this expenditure into account will alter our basic conclusion that the sex ratio at the end of the period of parental expenditure will be stabilized at a value not far removed from one half.

SUMMARY

We have put Fisher's theory of the control of the sex ratio by Natural Selection on an analytic basis. This has enabled us to derive an expression for the selective advantage attached to reproduction with a given sex ratio, and to show that this depends on the sex ratio at the end of the period in which the offspring incur expenditure by their parents. It is this sex ratio which is probably stabilized near the value one-half by Natural Selection. The rate of approach of a population to its equilibrium sex ratio depends on the available genetic variance in the sex ratio, and since this is probably small, evolutionary changes in the sex ratio of natural populations will almost certainly be too slow to detect.

REFERENCES

BENNETT, J. H. (1958). The existence and stability of selectively balanced polymorphism at a sex-linked locus. *Aust. J. Biol. Sci.* 11, 598–602.

CREW, F. A. E. (1937). The sex ratio. *Amer. Nat.* 71, 529–59.

DARWIN, C. (1871). *The Descent of Man, and Selection in Relation to Sex* (p. 399). London: John Murray.

EDWARDS, A. W. F. (1958). An analysis of Geissler's data on the human sex ratio. *Ann. Hum. Genet., Lond.*, 23, 6–15.

FISHER, R. A. (1930). *The Genetical Theory of Natural Selection*. Oxford University Press.

MCKEOWN, T. & LOWE, C. R. (1951). The sex ratio of still births related to cause and duration of gestation. *Hum. Biol.* 23, 41–60.

SHAW, R. F. (1958). The theoretical genetics of the sex ratio. *Genetics*, 43, 149–63.

SHAW, R. F. & MOHLER, J. D. (1953). The selective significance of the sex ratio. *Amer. Nat.* 87, 337–42.

Editors' Comments
on Papers 9, 10, and 11

9 HALDANE
*A Mathematical Theory of Natural and Artificial Selection:
Part IV*

10 NORTON
Natural Selection and Mendelian Variation

11 FISHER
The Fundamental Theorem of Natural Selection

LIFE HISTORY AND SELECTION: BEGINNINGS

Most early population genetic models dealt with discrete generation populations. However, when individuals differ in their chances of death and reproduction over time, the generations overlap and usually interbreed, and the population then becomes age structured. (The terms "life history," "age structure," and "overlapping generations" are more or less synonymously used in reference to this situation in the literature.) On November 11, 1926, two papers were received by British scientific societies, and the investigation into age-structured populations began. Norton (Paper 10) and Haldane (Paper 9), consulted during the development of their work, and apparently their thinking on this question goes back to 1910 investigations by Norton. No expression of the work seems to exist prior to 1926. Both authors deal with the case in which there is not only an age structure, but the differential mortality and fertility rates are related to the genotypes in the population.

Paper 9 is fourth in a series of nine papers on "A Mathematical Theory of Natural Selection," which summarized Haldane's views on Mendelian population genetics. He begins by proving the stability of Lotka's stable population. Then he considers the rate of spread of a dominant gene in a population in which the bearers of each allele have separate schedules of birth and death by age and sex. Under assumptions of random mating, Haldane derives expressions for the rate of change of gene frequencies, and shows that the rate of change is similar to that in the discrete generation case.

Norton's mathematically complex treatment is closely related to Lotka's concepts of stable population. In the first half of the paper (not reprinted), Norton gives a mathematical demonstration of Lotka's stable age distribution as a special case of a more general result: if the vital rates fluctuate as known functions of time, any two populations subject to them will come to have birth rates that differ according to a fixed proportion. This is a kind of weak ergodicity property (see Lopex, 1961, for more general treatment).

Norton uses the results from the first half of his paper to analyze a composite population of Mendelian types. He assumes that selection operates only through differential mortality, with all genotypes therefore having the same reproductive rates by age and with ages of parents independent of each other. Norton then derives the gene frequencies that would result from a mixed population. These equilibria are expressed in terms of the relations between the real roots of the genotype-specific Lotka renewal equations for the populations. His model is therefore completely Malthusian in that it allows for indefinite growth and expresses results in relative proportions, with absolute numbers free to take on any values.

Norton concludes this treatment with the statement that "subject to the conditions stated, the same results hold as in the case when different generations do not interbreed." Paper 10 has been criticized as being somewhat "contrived" in that the assumptions of equal fertility rates for all genotypes and of independence between parental ages are quite at variance with the facts, especially with human populations (Schull and MacCluer, 1968). However, the assumptions may not make a significant difference in the conclusions reached (Charlesworth, 1970) and the degree of artificiality certainly does not exceed that of a great many population-genetic models when the goal is to predict ultimate results.

Paper 11, another of the early works dealing with genetics, age structure, and population growth, is an excerpt from R. A. Fisher's *The Genetical Theory of Natural Selection.* We have used the 1958 revised edition rather than the original 1930 edition, because Fisher made some changes (of no real historical consequence) and because many of the discussions and critiques of this work refer to the 1958 edition. The reprinted excerpt contains the development of two of Fisher's major contributions to general genetic theory: reproductive value and the fundamental theorem of natural selection; we have included that portion relevant to considerations of age-structured populations.

Beginning with the concept of the life table based on fixed vital rates, Fisher develops an expression for the expected reproductive contribution of an individual in its lifetime. From this, Fisher gives Lotka's equation, and he terms the intrinsic rate of increase the "Malthusian parameter," a term that has become standard among geneticists. He uses this to compute the expected amount of reproduction of an individual age x relative to that of a newborn; this is the *reproductive value*. Fisher discusses the potential action of natural selection in relation to the curve of reproductive value with age, and in so doing he introduces age structure into basic population genetics.

Fisher then develops expressions for the genetic variance used in his *fundamental theorem of natural selection:* "the rate of increase in fitness of any organism at any time is equal to its genetic variance in fitness at that time." Fisher uses the Malthusian parameter as his measure of genetic *fitness;* hence the process of adaptation becomes expressed fundamentally in terms of population growth, which has caused the concept of an ever-growing population to become deeply engraved into population genetics. This theorem is widely cited and discussed, and has caused no end of confusion in population genetics. Some of this confusion is due to Fisher's writing and to typographical (or actual) errors in the equations given, but these are of no moment to our present discussion of demographic effects (the errors are thoroughly reviewed by Price, 1972). A good deal of the confusion is due to the effects of unchecked growth.

In a section following the one reprinted here, Fisher discusses the fact that growth will not continue indefinitely as this relates to the fundamental theorem. His discussion and theorem are anything but clear, as attested by the number of papers written to clarify them. Exegetical approaches to Fisher's book have been made in order to rescue a modern and correct interpretation of the theorem, and are occasionally quite strained in their attempts to show that if Fisher had said what he must have meant his theorem would have rather complete generality (Price and Smith, 1972; and especially Price, 1972). This theorem and the relation between fitness and natural selection are discussed in Kimura (1958), Crow and Kimura (1970), Turner, (1970). For specific aspects of these issues, see Charlesworth (1970), Kempthorne and Pollack (1970), and Wright (1969) on his use of fitness functions. Pollack and Kempthorne (1970; 1971) show that under some circumstances intermating genotypes in age-structured populations have their own Malthusian parameters, and that Fisher's fundamental theorem can hold with regard to these parameters and the variance among them; this was not proved previously. Charlesworth used Kimura's definition of fitness and showed that the fundamental theorem is only approximately true.

In this book we are concerned with the aspects of selection and fitness that relate to demographic structure, and in particular here we are talking about age structure. We cannot argue that any competent geneticist was unaware that population growth must in general be close to zero, and it is true that the standard models of population genetics are Malthusian as a mathematical convenience and with no loss of information as regards most of their evolutionary conclusions. However, the *concept* of Malthusian growth is basic to the context of much of this work; this is due, we believe, to the historical times in which it was done.

9

Reprinted from *Proc. Cambridge Phil. Soc.*, **23**, 607–615 (1927)

A MATHEMATICAL THEORY OF NATURAL AND ARTIFICIAL SELECTION: PART IV

J. B. S. Haldane

[*Received* 11 November, *read* 22 November 1926.]

In such organisms as annual plants, in which successive generations do not overlap, the composition of the $n+1$th generation can be calculated from that of the nth, and the resulting finite difference equation investigated. Where generations overlap we may obtain a similar relation between the compositions of the population at times t and t', but the finite difference equation is now represented by an integral equation. This fact was first pointed out in 1910 by Mr H. T. J. Norton of Trinity College. At a much later date I arrived at the same conclusion, and Mr Norton showed me his results in 1922, stating that he would publish them shortly. He has been prevented from doing so by illness, and, although I believe that all the results here given were reached by me independently, there can be no question that Mr Norton had obtained many of them previously, and had treated the problem rigorously, which I have not done.

The only case considered here is the very simple one in which the intensity of selection is independent of the size of the population. A preliminary lemma will first be discussed.

The growth of a population.

If the death-rate and birth-rate of a population are not functions of its density, its number at any time may be calculated as follows:

Let $N(t)$ be the number at time t. Only the female sex need be considered if the sexes are separate.

$S(x)$ be the probability of an individual surviving to the age x.

$U(t)\,\delta t$ be the number of individuals produced between times t and $t+\delta t$.

$K(x)\,\delta x$ be the probability of an individual between the ages x and $x+\delta x$ producing one (female) offspring. All individuals of this age, both alive and dead, are considered, so that if $P(x)$ be the corresponding function for living individuals only, $K(x) = S(x)\,P(x)$. Then

$$N(t) = \int_0^x U(t-x)\,S(x)\,dx$$
$$U(t) = \int_0^\infty U(t-x)\,K(x)\,dx$$

$$\quad\ldots\ldots\ldots(1.0).$$

Instead of infinity any upper limit exceeding the maximum life of the organism may be taken. Equation (1·0) has been considered by Herglotz*. Let $U(t) = ce^{zt}$. Then

$$\int_0^\infty e^{-zx} K(x)\,dx = 1, \text{ or } \int_0^a e^{-zx} K(x) = 1 \ldots\ldots\ldots(1\cdot1),$$

where a is sufficiently large. Since $K(x)$ is always real and zero or positive, the above integral is a monotone function of z when z is real, and can have any real positive value. Hence it has one and only one real root for z, say α_0. The complex roots clearly occur in pairs $\alpha_r \pm \iota\beta_r$. Then

$$\int_0^\infty e^{-\alpha_r x} K(x)\,dx > \int_0^\infty e^{-(\alpha_r + \iota\beta_r)x} K(x)\,dx$$
$$= 1.$$

Therefore $\qquad\qquad\qquad \alpha_r < \alpha_0.$

If any two functions of t are solutions of (1·0) so is their sum. Therefore

$$\left.\begin{aligned} U(t) &= \alpha_0 e^{\alpha_0 t} + \sum_{r=1}^{\infty} a_r e^{\alpha_r t} \cos \beta_r (t - b_r) \\ N(t) &= c_0 e^{\alpha_0 t} + \sum_{r=1}^{\infty} c_r e^{\alpha_r t} \cos \beta_r (t - d_r) \end{aligned}\right\} \quad \ldots\ldots(1\cdot2).$$

In general there will be an infinite number of terms. The values of a_r and b_r, and hence of c_r and d_r, depend on the initial conditions. Where multiple roots occur there will be periodic terms including powers of t as factors. Since $\alpha_r < \alpha_0$, all the periodic terms become negligible compared with the first after the lapse of a sufficient time. That is to say, oscillations of the population about an exponential function of the time are either damped or at least increase less rapidly than the population itself. In particular, if $\int_0^\infty K(x)\,dx = 1$, so that the population is in equilibrium, oscillations are damped and the equilibrium is stable. We are therefore justified in neglecting periodic terms in the solution of equations which only differ by small terms from (1·0), and which occur in the subsequent analysis. It is proposed to discuss the stability of the equilibrium when $S(x)$ and $K(x)$ depend on the number of the population in a subsequent paper.

Incidentally, if $C(x)\,\delta x$ be the probability of any member of the population being between the ages x and $x + \delta x$, then

$$C(x) = \frac{U(t-x)\,S(x)}{N(t)}.$$

* Herglotz, *Math. Ann.* 65, p. 87.

Hence, when oscillations have died down,

$$C(x) = \frac{e^{-a_0 x} S(x)}{\int_0^\infty e^{-a_0 x} S(x)\, dx}.$$

This constitutes a new proof of Lotka's[*] theorem on the stability of the normal age distribution.

Selection of an autosomal factor.

Consider a population consisting, at time t, of $D(t)$ female zygotes possessing a dominant factor A, $R(t)$ female recessives. The sex-ratio at birth is taken as fixed.

Let $F(t)\,\delta t$ be the number of fertile A ova produced between times t and $t + \delta t$.

$f(t)\,\delta t$ be the number of fertile a ova produced between times t and $t + \delta t$.

$M(t)\,\delta t$ be the number of A spermatozoa produced between times t and $t + \delta t$.

$m(t)\,\delta t$ be the number of a spermatozoa produced between times t and $t + \delta t$.

$S(x)$ be the probability of a female dominant reaching the age x.

$s(x)$ be the probability of a female recessive reaching the age x.

$K(x)\,\delta x$ be the probability of a female dominant (alive or dead, as above) producing a female offspring between the ages x and $x + \delta x$.

$[K(x) - k(x)]\,\delta x$ be the same probability for a female recessive.

$L(x)\,\delta x$ be the same probability for a male dominant.

$[L(x) - l(x)]\,\delta x$ be the same probability for a male recessive.

$$S = \int_0^\infty S(x)\, dx, \quad s = \int_0^\infty s(x)\, dx.$$

$$K = \int_0^\infty K(x)\, dx, \quad K' = \int_0^\infty x K(x)\, dx, \quad k = \int_0^x k(x)\, dx.$$

$$L = \int_0^\infty L(x)\, dx, \quad L' = \int_0^x x L(x)\, dx, \quad l = \int_0^x l(x)\, dx.$$

In general the functions $S(x)$, $s(x)$, $K(x)$, etc., will not be functions of age alone, but of $D(t)$, $R(t)$, etc. We make the assumption however that selection and population growth are proceeding so slowly that $k(x)$ and $l(x)$, and $K - 1$ are small, and $S(x)$, etc., do not vary appreciably in the course of a generation.

[*] Lotka, *Proc. Nat. Ac. Sci.* 8, p. 339, 1922.

If mating be at random, the rates of production of the three female phenotypes at time t are

$$AA, \frac{F(t) M(t)}{M(t) + m(t)}; \quad Aa, \frac{F(t) m(t) + f(t) M(t)}{M(t) + m(t)}; \quad aa, \frac{f(t) m(t)}{M(t) + m(t)}.$$

The group aged x at time t was hatched or born at time $t - x$. Therefore

$$
\left.
\begin{aligned}
F(t) &= \int_0^\infty \frac{2F(t-x) M(t-x) + F(t-x) m(t-x) + f(t-x) M(t-x)}{M(t-x) + m(t-x)} K(x)\, dx \\
f(t) &= \tfrac{1}{2} \int_0^\infty \frac{F(t-x) m(t-x) + f(t-x) M(t-x) + 2f(t-x) m(t-x)}{M(t-x) + m(t-x)} K(x)\, dx \\
&\qquad - \int_0^\infty \frac{f(t-x) m(t-x) k(x)\, dx}{M(t-x) + m(t-x)} \\
M(t) &= \tfrac{1}{2} \int_0^\infty \frac{2F(t-x) M(t-x) + F(t-x) m(t-x) + f(t-x) M(t-x)}{M(t-x) + m(t-x)} L(x)\, dx \\
m(t) &= \tfrac{1}{2} \int_0^\infty \frac{F(t-x) m(t-x) + f(t-x) M(t-x) + 2f(t-x) m(t-x)}{M(t-x) + m(t-x)} L(x)\, dx \\
&\qquad - \int_0^\infty \frac{f(t-x) m(t-x) l(x)\, dx}{M(t-x) + m(t-x)}
\end{aligned}
\right\} \dots (2\cdot0).
$$

Since selection and population growth are slow, we may put $F(t-x) = F(t) - xF'(t)$, etc., $M(t) = \lambda F(t)$, $m(t) = \lambda f(t)$, all to the first order of small quantities. Hence, to this degree of approximation,

$$
\begin{aligned}
F(t) &= \tfrac{1}{2} K F(t) + \frac{\tfrac{1}{2} K [F(t) + f(t)] M(t)}{M(t) + m(t)} - \tfrac{1}{2} K' F'(t) - \frac{\tfrac{1}{2} K' [F'(t) + f'(t)] M(t)}{M(t) + m(t)} \\
&\qquad - \frac{\tfrac{1}{2} K' [F(t) + f(t)] M'(t) m(t) - M(t) m'(t)}{[M(t) + m(t)]^2} \\
&= \tfrac{1}{2} K F(t) + \frac{\tfrac{1}{2} K [F(t) + f(t)] M(t)}{M(t) + m(t)} - K' F'(t).
\end{aligned}
$$

Similarly

$$f(t) = \tfrac{1}{2} K f(t) + \frac{\tfrac{1}{2} K [F(t) + f(t)] m(t)}{M(t) + m(t)} - K' f(t) - \frac{k [f(t)]^2}{F(t) + f(t)}$$

$$M(t) = \tfrac{1}{2} L F(t) + \frac{\tfrac{1}{2} L [F(t) + f(t)] M(t)}{M(t) + m(t)} - L' F'(t)$$

$$m(t) = \tfrac{1}{2} L f(t) + \frac{\tfrac{1}{2} L [F(t) + f(t)] m(t)}{M(t) + m(t)} - L' f'(t) - \frac{l [f(t)]^2}{F(t) + f(t)},$$

all approximately. Therefore

$$
\begin{aligned}
\frac{M(t)}{m(t)} &= \frac{\left[2LF(t) + Lf(t) - 2L'F(t)\right]\frac{M(t)}{m(t)} + LF(t) - 2L'F'(t)}{\left[LF(t) - 2L'f'(t) - \frac{2l\{f(t)\}^2}{F(t)+f(t)}\right]\frac{M(t)}{m(t)} + LF(t) + 2Lf(t) - 2L'f'(t) - \frac{2l[f(t)]^2}{F(t)+f(t)}} \\
&= \frac{L[2F(t) + f(t)]\frac{M(t)}{m(t)} + LF(t) - \dfrac{2L'F'(t)\,[F(t)+f(t)]}{f(t)}}{Lf(t)\frac{M(t)}{m(t)} + LF(t) + 2Lf(t) - \dfrac{2L'f'(t)\,[F(t)+f(t)]}{f(t)} - 2lf(t)}
\end{aligned}
$$

approximately,

$$= \frac{F(t)}{f(t)} + \frac{2L'[F(t)f'(t) - F'(t)f(t)]}{L[F(t)+f(t)]} + \frac{lF(t)[f(t)]^2}{L[F(t)+f(t)]^2},$$

approximately, by solving the quadratic. Therefore

$$(K-1)F(t) - K'F'(t) + \frac{L'[F(t)f'(t) - F'(t)f(t)]}{L[F(t)+f(t)]} + \frac{lF(t)[f(t)]^2}{L[F(t)+f(t)]^2} = 0$$

$$(K-1)f(t) - K'f'(t) - \frac{L'[F(t)f'(t) - F'(t)f(t)]}{L[F(t)+f(t)]} - \frac{lF(t)[f(t)]^2}{L[F(t)+f(t)]^2} - \frac{k[f(t)]^2}{F(t)+f(t)} = 0$$

$$\dotfill (2\cdot1).$$

If
$$u(t) = F(t)/f(t),$$

$$\frac{d}{dt}u(t) = \frac{(l+kL)u(t)}{(L'+K'L)[1+u(t)]} \dotfill (2\cdot2).$$

This is equivalent to equation (2·1) of Part I*, $\dfrac{l+kL}{L'+K'L}$ being
the coefficient of selection. In general this quantity is not independent of t, hence the equation cannot be integrated, but if its upper and lower limits are known, the march of the composition of the population can be roughly calculated from equation (2·3) of Part I. If, however, the population is very nearly in equilibrium, and either dominants or recessives are very rare, more accurate results are possible.

When recessives are rare, $F(x)$ is large and equal to a constant F, $D(x)$ being also large and equal to a constant $N \cdot K = 1$, and $F'(t)$ is negligible, while $f(t)$ is small. Therefore

$$K'f'(t) + \frac{L'}{L}f'(t) + \left(\frac{l+kL}{L}\right)\frac{[f(t)]^2}{F} = 0,$$

and
$$f(t) = \frac{(L'+K'L)F}{(l+kL)(t-t_0)},$$

where t_0 is an integration constant. But

$$N = D(t) = \int_0^\infty F(t-x)S(x)\,dx = FS,$$

$$R(t) = \int_0^\infty \frac{[f(t-x)]^2 s(x)\,dx}{F(t-x)+f(t-x)} = \frac{(L'+K'L)^2 Fs}{(l+kL)^2(t-t_0)^2}.$$

$$R(t) = \frac{(L'+K'L)^2 sN}{(l+kL)^2 S(t-t_0)^2} \dotfill (2\cdot3).$$

Hence selection proceeds at the same rate as when generations are separate, with a selection coefficient equal to

$$\left(\frac{l+kL}{L'+K'L}\right)\left(\frac{S}{s}\right)^{\frac{1}{2}}.$$

* Haldane, *Trans. Camb. Phil. Soc.* **23**, p. 19, 1924.

When dominants are rare, $f(x)$ is large and equal to a constant f, $K = 1 + k$, and $F(x)$ is small. Therefore

$$\left(k + \frac{l}{L}\right) F(t) + \left(K' + \frac{L'}{L}\right) F'(t) = 0,$$

$$F(t) = e^{\frac{(l + kL)(c - t)}{L' + K'L}},$$

where c is a constant of integration. Therefore

$$D(t) = \int_0^\infty 2F(t - x) S(x) dx = 2SF(t)$$

$$= e^{\frac{(l + kL)(t_0 - t)}{L' + K'L}} \quad\quad\dots\dots\dots\dots\dots\dots\dots(2\cdot4),$$

where t_0 is an arbitrary constant. Selection therefore proceeds as when generations are separate, but with a selection coefficient

$$\frac{l + kL}{L' + K'L}.$$

When the death rates and fertility rates are the same in the two sexes, or in a hermaphrodite species, we have in general,

$$\frac{d}{dt} u(t) = \frac{ku(t)}{K'[1 + u(t)]} \dots\dots\dots\dots\dots(2\cdot5),$$

when recessives are rare,

$$R(t) = \frac{K'^2 sN}{k^2 S(t - t_0)^2} \quad\quad\dots\dots\dots\dots\dots(2\cdot6),$$

when dominants are rare,

$$D(t) = e^{\frac{k(t_0 - t)}{K'}} \quad\quad\dots\dots\dots\dots\dots(2\cdot7).$$

Now k has approximately the same meaning as when generations are separate, provided $K = 1$. Hence the two cases become comparable if we choose our unit of time, or "generation," so as to make K' or $\int_0^\infty x K(x) dx = 1$, as in the calculations of Dublin and Lotka[*] on the rate of increase of a population. In each case, if functions analogous to $S(x)$, $s(x)$ are known for the males, their numbers can be calculated.

[*] Dublin and Lotka, *Journ. Amer. Stat. Assoc.* 1925, p. 306.

Selection of a sex-linked factor.

Here, using the same notation as above, we find

$$(K-1)\,F(t) - K'F'(t) + \frac{L'\,[F(t)\,f'(t) - F'(t)\,f(t)] + lF(t)\,f(t)}{2L\,[F(t)+f(t)]} = 0$$

$$(K-1)\,f(t) - K'f'(t) - \frac{L'\,[F(t)\,f'(t) - F'(t)\,f(t)] + lF(t)\,f(t)}{2L\,[F(t)+f(t)]} - \frac{k\,[f(t)]^2}{F(t)+f(t)} = 0$$

$$\left.\right\}\dots(3\cdot1).$$

In general, if $\qquad u = F(t)/f(t)$,

$$(L' + 2K'L)\,\frac{d}{dt}\,u(t) = lu(t) + \frac{2kLu(t)}{1+u(t)} \quad\dots\dots(3\cdot2),$$

an equation analogous to $(2\cdot0)$ of Part III*.

When recessives are rare,

$$(L' + 2K'L)\,f'(t) + lf(t) + \frac{2kL\,[f(t)]^2}{F} = 0,$$

where F is the birth-rate of dominants. Three cases occur:

(*a*) If kL is negligible compared with lN, which will be the case if selection is of the same order of intensity in the two sexes, or more intense among males, then

$$f(t) = e^{\frac{l(t_0 - t)}{L' + 2K'L}}\dots\dots\dots\dots\dots(3\cdot31).$$

The number of recessive males is proportional to $f(t)$, of recessive females to its square.

(*b*) If kL is of the same order of magnitude as lF, then

$$\frac{2kLf(t)}{2kLf(t) + lF} = e^{\frac{l(t_0 - t)}{L' + 2k'L}} \quad\dots\dots\dots\dots(3\cdot32).$$

Hence if V, v are quantities corresponding to S, s for the male sex, the proportion of recessive males is

$$\frac{lv}{2kLV\left(e^{\frac{l(t - t_0)}{L' + 2K'L}} - 1\right)},$$

of recessive females

$$\frac{l^2 s}{4k^2 L^2 S\left(e^{\frac{l(t - t_0)}{L' + 2K'L}} - 1\right)^2}.$$

* Haldane, *Proc. Camb. Phil. Soc.* 23, p. 363, 1926.

(c) If kL is much larger than lF, i.e. selection is confined to females, then

$$f(t) = \frac{(2K'L + L')F}{2kL(t - t_0)} \quad \dots\dots\dots\dots(3\cdot33).$$

Hence the proportion of recessive males is

$$\frac{(L' + 2K'L)v}{2kLV(t - t_0)},$$

of recessive females

$$\frac{(L' + 2K'L)^2 s}{4k^2 L^2 S(t - t_0)^2}.$$

When dominants are rare,

$$F(t) = e^{\frac{(l + 2kL)(t - t_0)}{L' + 2K'L}} \quad \dots\dots\dots\dots(3\cdot4).$$

The number of male dominants is proportional to $F(t)$, that of females being double that of males.

When the intensity of selection is equal in both sexes, these equations simplify to

$$\frac{d}{dt}u(t) = \frac{ku(t)[3 + u(t)]}{3K'[1 + u(t)]} \quad \dots\dots\dots(3\cdot5),$$

$$f(t) = e^{\frac{k(t_0 - t)}{3K'}} \quad \dots\dots\dots\dots\dots(3\cdot6),$$

$$F(t) = e^{\frac{k(t_0 - t)}{K'}} \quad \dots\dots\dots\dots\dots(3\cdot7),$$

analogous to equation (7·2) of Part I.

DISCUSSION.

The most satisfactory table of $K(x)$ known to me is that given by Dublin and Lotka[*] for certain American women. Here the population is growing, and $K = 1\cdot17$, while $\frac{K'}{K}$, the length of a "generation," is 28·45 years. No satisfactory values of k are known in the present state of genetics, though the data on mice discussed in Part I suggest that here $k = \cdot04$ approximately. In man mating is highly assortative for age, and the above formulae cannot be applied. Moreover, a change in the coefficient of correlation between the ages of spouses would undoubtedly affect the values of $K(x)$, etc., if other conditions remained equal. Thus old men would beget more children if they were more likely to have young

[*] Dublin and Lotka, *loc. cit.*

105

wives. It is thus impossible to calculate the effect of this correlation on selection. But a consideration of the extreme case when the age of the wife fixes that of the husband makes it clear that selection must follow equations of the type here arrived at, with changes in the parameters only.

SUMMARY.

Expressions are found for the progress of slow selection in a Mendelian population where generations overlap. The changes are very similar to those which occur when generations are separate.

10

Reprinted from *Proc. London Math. Soc.*, **28**(1639), 1–4, 25–45 (1928)

NATURAL SELECTION AND MENDELIAN VARIATION

By H. T. J. Norton.

[Received and read 11 November, 1926.]

Part I.

THE object of the present paper is to take the simplest known law of heredity, the Mendelian law, and to see how far it is possible to work out the effects of selection on a variety, inherited by this law, without making specializing assumptions. The conclusions, however, which are here reached are subject to the assumption that the ages of parents vary independently : that is to say, that in regard to the children born at any one moment, the age distribution in the fathers of those whose mothers are of age t is independent of t, and similarly with the age distribution in the mothers of those whose fathers are of age t. This, and the assumption that the birth-rates and death-rates in each type are functions merely of age and that the birth-rate is the same in all three types, are the main limitations. The equations involved are integral equations, and the generality of the results is a consequence of the well known fact that most properties of these equations are analogous to those of linear equations. The Mendelian equations are not linear, but the essential point is the analogue of the property of convergence possessed by the convergents to a continued fraction. A similar (but wider) generalization to difference equations of the third and higher order has been published by Perron*. The effect of selection, when generations do not overlap,

---

* *Math. Annalen*, 64 (1907), 1–76.

has been discussed by Haldane*, both in the case of the simple Mendelian law and of more complicated laws. So far as I can see, all the main results, which are true when generations do not interbreed, are also true when they do, and it is not necessary to assume any particular form for the function which connects the birth- and death-rates with age. In the following paper some preliminary questions are discussed in 1–10, and the Mendelian equations are stated and considered in 11–17.

It will be supposed that the population forms a community. By a community is meant an endogamous population, which consists at any time of the descendants of the original group which formed the population at the beginning of the period considered. The inhabitants of an oceanic island would form a community if there were no emigration nor immigration. The main result of §§ 1–9 is an asymptotic formula for the number of births in a community in terms of the instantaneous birth-rate and death-rate. These are the quantities that measure the chance that a member of a group, born at a certain time, should die at a certain age, or should become a parent at a certain age; the definitions are as follows : the instantaneous birth-rate and death-rate+ at age t for births at time x are $v(t, x)$ and $\mu(t, x)$ when $v(t, x)dt$ is the ratio of the number of members of the group, that were born between x and $x+dx$ and become parents between the ages of t and $t+dt$, to the number that were born between x and $x+dx$ and are alive at age t; and when $\mu(t,x)dt$ is the ratio of the number, that were born between x and $x+dx$ and die at ages between t and $t+dt$, to the number that were born between x and $x+dx$ and survive to age t.

The only type of community considered here in detail is one in which births occur at all times in the year. But it will make the argument clearer to consider briefly the communities in which births occur only at breeding seasons which can be treated as momentary. In this case, u_n, the number of births in the n-th season, is given by

$$u_n = K(n, n-1) u_{n-1} + K(n, n-2) u_{n-2} + \ldots + K(n, n-r) u_{n-r}, \quad (1.1)$$

where $K(n, s)$ is the chance that an individual, born in the s-th season, should produce a child in the n-th, and r is the number of seasons in which he can produce children. $K(n, s)$ is easily expressible in terms of the instantaneous birth- and death-rates. (1.1) is a linear difference equa-

---

* J. B. S. Haldane, *Trans. Camb. Phil. Soc.*, 23 (1924), 19-41.

† *Encyclopédie des Sciences mathématiques*, tome 1, vol. 4, fasc. 3, p. 472. Only the definition of the instantaneous death-rate is given here, but that of the birth-rate is an obvious extension.

tion, and the general solution is expressible as a linear combination of at most r particular solutions. Suppose $r \geqslant s \geqslant 1$, and let $-2k(r, s)$ be the number of descendants, born in the n-th season, of a couple who were born in the s-th; the factor -2 is to keep the notation the same as that used in §§ 4–8. Then the general solution of (1.1) is of the form

$$u_n = -\sum_{s=1}^{r} C_s k(n, s), \tag{1.2}$$

where C_1, C_2, ... are constants, depending on the initial age-distribution in the community. The actual value of C_s is

$$u_s - \sum_{t=1}^{s-1} K(s, t)\, u(t).$$

If, now, the eventual increase of the community is independent of its initial composition, u_n must be of the form $Q_n A_n$, where Q_n depends merely on the K's and A_n is a function depending upon the particular community, and tending to a constant positive limit as n tends to infinity. The necessary condition for this is that

$$\lim_{n \to \infty} \frac{k(n, s)}{k(n, 1)} \quad (s = 1, 2, ..., r)$$

should exist and be positive. Now $k(n, s)$, considered as a function of the K's, belongs to a well known type. In the particular case $r = 2$,

$$-k(2, 1) = K(2, 1); \quad -k(3, 1) = K(3, 2)\,K(2, 1) + K(3, 1),$$

$$-k(2, 2) = 1; \quad -k(3, 2) = K(3, 2), \text{ etc.}$$

Thus $-k(n, 1)$ is the denominator and $-k(n, 2)$ is the numerator of the n-th convergent to the continued fraction

$$\frac{1}{K(2, 1)+} \frac{K(3, 1)}{K(3, 2)+} \frac{K(4, 2)}{K(1, 3)+ \ldots}.$$

For larger values of r, the functions $k(n, s)$ have similar properties, and have been considered by various writers. Thus it appears that, in the case $r = 2$, the necessary and sufficient condition that the eventual increase of the community should not depend on its initial composition is that the continued fraction should be convergent. Suppose births to occur continually. Then the main result of §§ 2–8 is to show that two communities with the same birth- and death-rates are asymptotically equivalent if condition (5.1) is satisfied. If we call the chance that an individual, born at x, should produce a child at the age t, "the survival ratio at time x for age t'', this condition can be stated in the form that there is a certain age, the same for all values of x, at which the survival ratio is not zero and does not tend to zero as x increases. The general line of argument in §§ 4–7 is the same as that outlined above; an exact equation (4.4) is first found, and an asymptotic formula is deduced from it by an argument of which the first part is essentially the same as that used by Perron to show that, in the case of Jacobi's algorithm,

$$\frac{k(n,\, s)}{k(n,\, 0)}$$

tends to a limit.

[*Editors' Note:* Material has been omitted at this point.]

10. I propose in the next part to apply the preceding argument to the selection of varieties inherited by the Mendelian law, and I think it may be convenient to consider briefly a simpler case first. Suppose, then, that a community—for instance, one of annual flowers—has definite breeding seasons, which can be treated as momentary, and that no individual breeds in any season but that next after his birth; and let a simple pair of Mendelian factors be inherited in the community. If this is so, all the members of the community belong to one or another of three classes, the two pure types and the hybrid type, and there are rules assigning the probable types of the children, born of any couple. These are that all the children of two members of the same pure type belong to that type, and all the children of two members of different pure types are hybrids; that of the children of hybrids, half are hybrids and a-quarter are of each pure type, and of the children of a hybrid and a member of a pure type half are of that pure type, half hybrids. This being so, suppose that in the n-th season, u_n, w_n, $2v_n$ are the number born, the last being the number of hybrids. A proportion of each type will live and have children in the next breeding season; and if these proportions are supposed given, it is possible to calculate, on the assumption of random inter-breeding, the values of u_{n+1}, w_{n+1}, $2v_{n+1}$. The formulae to which this leads are

$$u_{n+1} + v_{n+1} = a_n u_n + b_n v_n,$$

$$v_{n+1} + w_{n+1} = b_n v_n + c_n w_n,$$

$$u_{n+1} w_{n+1} = v_{n+1}^2,$$

a_n, b_n, c_n being the proportion of each type that survive to have children. If we write $u_n : 2v_n : w_n = 1 : 2p_n : p_n^2$, this gives the formula

$$p_{n+1} = \frac{c_n p_n^2 + b_n p_n}{b_n p_n + a_n}.$$

If p_n tends to infinity, as n tends to infinity, the community will eventually consist of members of the w-type alone; if p_n tends to zero, of the u-type alone; and in all other cases it will contain all three indefinitely. There is one case of particular importance, in which the hybrids are indistinguishable from one of the pure types; when this is so, the factor is said to exhibit dominance, and the pure type, which the hybrids resemble, is called the dominant type, and the other the recessive. Suppose, now, that in the difference equation just written c_n/a_n, b_n/a_n are constants independent of n. Then it is easy to show that

$$\text{if } \frac{c_n}{a_n} \geqslant \frac{b_n}{a_n} > 1, \quad \text{or} \quad \frac{c_n}{a_n} > \frac{b_n}{a_n} = 1, \quad p_n \to \infty .$$

The case $c_n/a_n = b_n/a_n$ would include all cases of complete dominance in which the dominant is selected; the case $b_n/a_n = 1$ all cases in which the recessive is selected. Thus, whenever the hybrids are neither more advantageous nor less advantageous than both the pure types, the selected pure type eventually monopolizes the community. There are certain other results which follow as to the rate at which changes occur, but I need not mention them now. My main object is to generalize these results to the kind of community considered in the last part.

Part II.

11. I consider in the following sections the application of the preceding argument to the case of Mendelian varieties: it will be found that, in rough outline, the results follow easily from the equations of § 7. Suppose that, during a certain period P, the members of the community belong to the three zygotic types of a simple pair of Mendelian allelomorphic factors which are not correlated with sex or fertility: thus each type consists of those members of the community who possess a certain peculiarity of appearance or structure which is inherited according to the law stated in the next section; and every member of the community belongs to one type or other. If the death-rates are not the same in all three types, the distribution of the community between the types will vary with the time, and my object is to connect the changes in numbers of the types with the magnitudes of their advantages. It will be supposed that the death-rate and birth-rate in each type remain constant, for a given age, throughout the period P; that there is no difference in any type between the male and female death-rates, and that the ages of parents vary independently. By the last statement it is meant that the age distribution in the fathers of children whose mothers were aged t at the children's birth is independent of t, and so is the age distribution in the mothers of children whose fathers were of age t. The last two assumptions do not correspond to the facts, and are introduced to simplify the argument. Perhaps it may be said that in so far as the presence of a death-rate, varying with sex, or of a correlation between the ages of

parents affects the proportions of the types, selection is no longer ordinary selection, but is partly sex-limited or sexual.

12. Let x be any moment in P, and, in the short interval x to $x+dx$, let $u(x)dx$ and $w(x)dx$ be the numbers of the pure types born, and $2v(x)dx$ be the number of hybrids born. At a subsequent moment y, a certain proportion of the individuals born between x and $x+dx$ will still survive; let the proportion of the three types who still survive be respectively $K_1(y, x)$, $K_3(y, x)$, $K_2(y, x)$. Of those that do survive, a certain proportion will become parents in the short interval between y and $y+dy$. Suppose, now, that $\pi_y(t, \theta)\, dt\, d\theta\, dy$ is the total number of couples of individuals of respective ages between t and $t+dt$ and between θ and $\theta+d\theta$, who produce a child in the interval between y and $y+dy$. Then, it being supposed that there is no sexual selection of the typical characteristics, and no variation in fertility between the three types, the total number of couples, consisting of a member of the u-type, whose age is between t and $t+dt$, and a member of the v-type, whose age is between θ and $\theta+d\theta$, who produce a child between y and $y+dy$, is

$$\frac{2K_1(y, y-t)\, K_2(y, y-\theta)\, u(y-t)\, v(y-\theta)\, \pi_y(t, \theta)\, dt\, d\theta}{N(t)\, N(\theta)}\, dy,$$

where $N(t)dt$ has been written for the total number of persons between the ages t and $t+dt$, and is equal to

$$[u(y-t)\, K_1(y, y-t)+2v(y-t)\, K_2(y, y-t)+w(y-t)\, K_3(y, y-t)]\, dt.$$

The total number of children, born to members of the u-type and v-type in the interval y to $y+dy$ is, therefore,

$$dy \int_{\lambda_1}^{\lambda_2} \int_{\lambda_1}^{\lambda_2} \frac{2\pi_y(t, \theta)}{N(t)\, N(\theta)}\, K_1(y, y-t)\, K_2(y, y-\theta)\, u(y-t)\, v(y-\theta)\, dt\, d\theta.$$

Since it is obvious that

$$\pi_y(t, \theta) = \pi_y(\theta, t),$$

this can also be written

$$dy \int_{\lambda_1}^{\lambda_2} \int_{\lambda_1}^{\lambda_2} \frac{\pi_y(t, \theta)}{N(t)\, N(\theta)}\, [K_1(y, y-t)\, K_2(y, y-\theta)\, u(y-t)\, v(y-\theta)$$
$$+ K_1(y, y-\theta)\, K_2(y, y-t)\, u(y-\theta)\, v(y-t)]\, dt\, d\theta.$$

Similarly the number of children born to two u's is

$$\tfrac{1}{2}dy \int_{\lambda_1}^{\lambda_2} \int_{\lambda_1}^{\lambda_2} \frac{\pi_y(t, \theta)}{N(t)\, N(\theta)}\, K_1(y, y-t)\, K_1(y, y-\theta)\, u(y-t)\, u(y-\theta)\, dt\, d\theta.$$

By the definition of Mendelian allelomorphic factors, if u is a **pure** type, all the offspring of u-u matches, half those of u-v matches, and a quarter those of v-v matches are u's, and none of the offspring of other combinations. Adding the three numbers together, we get

$$\tfrac{1}{2}dy \int_{\lambda_1}^{\lambda_2} \int_{\lambda_1}^{\lambda_2} \frac{\pi_y(t,\,\theta)}{N(t)\,N(\theta)} \left[K_1(y,\,y-t)\,u(y-t)+K_2(y,\,y-t)\,v(y-t)\right]$$
$$\times \left[K_1(y,\,y-\theta)\,u(y-\theta)+K_2(y,\,y-\theta)\,v(y-\theta)\right]dt\,d\theta,$$

and this must be equal to $u(y)dy$. Similarly, if v is a hybrid type, half the offspring of u-v matches, v-v matches, and v-w matches, and all the offspring of u-w matches are hybrids. We thus get

$$2v(y) = \int_{\lambda_1}^{\lambda_2} \int_{\lambda_1}^{\lambda_2} \frac{\pi_y(t,\,\theta)}{N(t)\,N(\theta)} \left[K_1(y,\,y-t)\,u(y-t)+K_2(y,\,y-t)\,v(y-t)\right]$$
$$\times \left[K_2(y,\,y-\theta)\,v(y-\theta)+K_3(y,\,y-\theta)\,w(y-\theta)\right]dt\,d\theta.$$

And similarly,

$$w(y) = \tfrac{1}{2} \int_{\lambda_1}^{\lambda_2} \int_{\lambda_1}^{\lambda_2} \frac{\pi_y(t,\,\theta)}{N(t)\,N(\theta)} \left[K_3(y,\,y-t)\,w(y-t)+K_2(y,\,y-t)\,v(t)\right]$$
$$\times \left[K_2(y,\,y-\theta)\,v(y-\theta)+K_3(y,\,y-\theta)\,w(y-\theta)\right]dt\,d\theta.$$

If we add half the second equation to the first equation and then to the third,

$$u(y)+v(y) = \tfrac{1}{2} \int_{\lambda_1}^{\lambda_2} \left[K_1(y,\,y-t)\,u(y-t)+K_2(y,\,y-t)\,v(y-t)\right] \int_{\lambda_1}^{\lambda_2} \frac{\pi_y(t,\,\theta)}{N(t)} \,d\theta\,dt,$$

$$v(y)+w(y) = \tfrac{1}{2} \int_{\lambda_1}^{\lambda_2} \left[K_2(y,\,y-t)\,v(y-t)+K_3(y,\,y-t)\,w(y-t)\right] \int_{\lambda_1}^{\lambda_2} \frac{\pi_y(t,\,\theta)}{N(t)} \,d\theta\,dt.$$

The quantity $\int_{\lambda_1}^{\lambda_2} \frac{\pi_y(t,\,\theta)}{N(t)}\,d\theta$ is what was called, in the preceding part, the instantaneous birth-rate in the community, $v(t,\,y-t)$. If now we introduce the assumption that the ages of parents vary independently,

$$\pi_y(t,\,\theta) = \pi_y(t)\,\pi_y(\theta),$$

and we have, in the first place, the equation

$$v^2(y) = u(y)\,w(y), \tag{12.1}$$

and, secondly, we can consider $v(t,\,y-t)$ as independent of the age distribution in the community, and equal to $v(t)$. If the ages of parents do not vary independently, $v(t,\,y-t)$ is necessarily a function of the age distribution. Assuming, further, that the death-rates in the three types are constant, and writing

$$\tfrac{1}{2}K_1(y, y-t)\,\nu(t) = K'(t), \quad \tfrac{1}{2}K_2(y, y-t)\,\nu(t) = K''(t),$$

$$\tfrac{1}{2}K_3(y, y-t)\,\nu(t) = K'''(t), \quad (12.11)$$

we have the equations

$$u(y)+v(y) = \int_{\lambda_1}^{\lambda_2} K'(t)\,u(y-t)\,dt + \int_{\lambda_1}^{\lambda_2} K''(t)\,v(y-t)\,dt, \quad (12.2)$$

$$(y \geqslant \lambda_2)$$

$$v(y)+w(y) = \int_{\lambda_1}^{\lambda_2} K'''(t)\,w(y-t)\,dt + \int_{\lambda_1}^{\lambda_2} K''(t)\,v(y-t)\,dt, \quad (12.3)$$

$$u(y)+v(y) = f_1(y) + \int_{\lambda_1}^{y} K'(t)\,u(y-t)\,dt + \int_{\lambda_1}^{y} K''(t)\,v(y-t)\,dt, \quad (12.21)$$

$$(\lambda_2 \geqslant y \geqslant 0)$$

$$v(y)+w(y) = f_2(y) + \int_{\lambda_1}^{y} K''(t)\,v(y-t)\,dt + \int_{\lambda_1}^{y} K'''(t)\,w(y-t)\,dt, \quad (12.31)$$

where, in the last two equations, the integrals vanish if $\lambda_1 \geqslant y$, and $f_1(y)$, $f_2(y)$ are given functions of their arguments.

13. With regard to the functions $K'(t)$, etc., it will be supposed that they are continuous and never negative if $\lambda_2 \geqslant t \geqslant \lambda_1$, and that numbers a, β, ϵ exist such that $\epsilon > 0$, $\lambda_2 \geqslant a > \beta \geqslant \lambda_1$, and if $a \geqslant t \geqslant \beta$, $K'(t)$, $K''(t)$, $K'''(t) > \epsilon$.

It will be supposed, as in the last part, that a, β, ϵ have been chosen once for all such that these conditions are satisfied; any function $K(y, x)$, which is continuous and never negative in the strip

$$0 \leqslant x_0 \leqslant x, \quad x \leqslant y \leqslant x+\lambda,$$

and is such that, if $a \geqslant y-x \geqslant \beta$, $K(y, x) > \epsilon$, will be said to satisfy conditions A; and η will be used for a constant, which depends only on a, β, ϵ, λ_2, and M, the upper bound of $K'(t)$, $K''(t)$, $K'''(t)$. It will be convenient to define $K'(t)$, $K''(t)$, $K'''(t)$ as zero, if t is negative or is greater than λ_2. There is no objection to regarding the functions so defined as continuous.

The first question that arises is as to the existence of a solution (12.1.2); but, since $\lambda_1 > 0$, its existence is obvious. If $f_1(y)$, $f_2(y)$ are arbitrary continuous functions and are never negative, the right-hand sides of (12.21.31) are given functions for $0 \leqslant y \leqslant \lambda_1$, and hence, by (12.1), $u(y)$, $v(y)$, and $w(y)$ are given in that interval. If these are substituted in the right-hand sides of (12.21.31), the values of $u(y)$, $v(y)$, and $w(y)$ are found, with the help of (12.1), for

$2\lambda_1 \geqslant y \geqslant \lambda_1$. This process can be repeated indefinitely. The functions so obtained are finite and never negative, provided that $f_1(y)$, $f_2(y)$ are finite, and are continuous, except for a possible abrupt jump at $y = \lambda_2$.

Further, since the K's satisfy conditions A, (12.2) gives

$$u(y) + v(y) \geqslant \epsilon \int_a^\beta [u(y-t) + v(y-t)] \, dt.$$

It follows* from this that, from a certain point η on,

$$u(y) + v(y) > 0 ;$$

similarly with $v(y) + w(y)$, and hence with $u(y)$, $v(y)$, and $w(y)$. Since the initial point is arbitrary, it may be supposed that

$$u(y), \quad v(y), \quad w(y) > 0 \quad (y \geqslant 0).$$

On the conditions assumed, the three functions of z :

$$\int_0^\lambda e^{-zt} K'(t) \, dt - 1, \quad \int_0^\lambda e^{-zt} K''(t) \, dt - 1, \quad \int_0^\lambda e^{-zt} K'''(t) \, dt - 1,$$

have each one, and only one, real zero. Let these real zeros be ρ, σ, τ; it may be supposed, without loss of generality, that $\rho \leqslant \tau$. Consider, now, (12.2); it may be written in the form

$$z_1(y) = \int_{y-\lambda}^y Q_1(y, x) z_1(x) \, dx \quad (y > \lambda), \qquad (14.1)$$

where

$$Q_1(y, x) = \frac{K'(y-x) u(x) + K''(y-x) v(x)}{u(x) + v(x)}.$$

Thus $Q(y, x)$ satisfies conditions A, since $K'(y-x)$, $K''(y-x)$ do so, and $u(x)$ and $v(x)$ are positive and continuous. Hence, by § 7, if $q_1(y, x)$ is the function reciprocal to $Q_1(y, x)$, $\lim\limits_{y \to \infty} \dfrac{q_1(y, x)}{q_1(y, 0)}$ exists, and, if we write it equal to $L_1(x)$,

$$L_1(x) = \int_x^{x+\lambda} L_1(y) Q_1(y, x) \, dy,$$

and, by (7.9), (14.1),

$$\eta > \frac{z_1(x) L_1(x)}{\displaystyle\int_0^\lambda L_1(t) f_1(t) \, dt} > \eta'.$$

* Cf. the proof of (5.2).

116

Let p be a positive quantity and ω_p the real zero of

$$\int_0^\lambda e^{-zt}\left[\frac{K'(t)+pK''(t)}{1+p}\right]dt-1,$$

and suppose that $\sigma > \rho$. For any fixed value of p, the function of z just written decreases, as z increases from ρ to σ, and it is positive if $z = \rho$, negative if $z = \sigma$; therefore $\sigma > \omega_p > \rho$. Also, if $p' > p$,

$$\int_0^\lambda e^{-\omega_p t}\left[\frac{K'(t)+p'K''(t)}{1+p'}\right]dt-1$$

$$= \int_0^\lambda e^{-\omega_p t}\left[\frac{K'(t)+p'K''(t)}{1+p'}\right]dt - \int_0^\lambda e^{-\omega_p t}\left[\frac{K'(t)+pK''(t)}{1+p}\right]dt$$

$$= \frac{(p-p')}{(1+p)(1+p')}\int_0^\lambda e^{-\omega_p t}[K'(t)-K''(t)]\,dt$$

$$= \frac{(p-p')}{(1+p)(1+p')}\left[\int_0^\lambda \{e^{-\omega_p t}-e^{-\rho t}\}K'(t)\,dt+\int_0^\lambda \{e^{-\sigma t}-e^{-\omega_p t}\}K''(t)\,dt > 0.$$

Hence, if $p' > p$, $\omega_{p'} > \omega_p$.

Take any interval $x_1 \geqslant x \geqslant x_0$, and let p' be the maximum, p the minimum value of $v(x)/u(x)$ in it; p' and p are positive and finite. Then

$$L_1(x)\,e^{\omega_{p'}x} = \int_x^{x+\lambda} L_1(y)\,e^{\omega_{p'}y}\,Q_1(y, x)\,e^{-\omega_{p'}(y-x)}\,dy,$$

where

$$\int_x^{x+\lambda} e^{-\omega_{p'}(y-x)}Q_1(y, x)\,dy$$

$$= \frac{u(x)}{u(x)+v(x)}\int_0^\lambda e^{-\omega_{p'}t}K'(t)\,dt+\frac{v(x)}{u(x)+v(x)}\int_0^\lambda e^{-\omega_{p'}t}K''(t)\,dt \leqslant 1.$$

Hence, if M_n be the greatest value of $L_1(x)e^{\omega_{p'}x}$ in the interval

$$x_0+(n-1)\lambda \leqslant x \leqslant x_0+n\lambda,$$

$$M_n \leqslant M_{n+1} \qquad\qquad (14.21)$$

Further, by (7.3), $\eta > -L_1(x)\,q_1(x, 0) > \eta'$ $(x > \eta)$,

and, by (5.4), $\qquad \eta > \dfrac{q_1(x', 0)}{q_1(x, 0)} > \eta'$ $(x+\lambda \geqslant x' \geqslant x > \eta)$.

Therefore if N_n is the least value of $L_1(x)e^{\omega_{p'}x}$ in

$$x_0+(n-1)\lambda \leqslant x \leqslant x_0+n\lambda$$

$$\eta > \frac{M_n}{N_n} > \eta'. \qquad\qquad (14.22)$$

It follows from (14 . 21 . 22) that

$$L_1(x) > \eta e^{-\omega_{p'}(x-x_0)} L_1(x_0) \quad (x_1 \geqslant x \geqslant x_0).$$

A similar argument holds for ω_p, so that

$$L_1(x_0) \, \eta' e^{-\omega_p(x-x_0)} > L_1(x) > \eta e^{-\omega_{p'}(x-x_0)} L_1(x_0)$$

$$(\sigma > \rho \, ; \; x_1 \geqslant x \geqslant x_0). \quad (14 . 3)$$

If $\rho = \sigma$, then,* by § 7,

$$L_1(x) = e^{-\rho x} \quad (\sigma = \rho). \quad (14 . 31)$$

Again, if ω_p' is the real zero of

$$\int_0^\lambda e^{-zt} \left[\frac{K''(t) + p K'''(t)}{1+p} \right] dt - 1,$$

and $L_2(x)$ the function, analogous to $L_1(x)$, for the equation (12 . 3), then, since $w(x)/v(x) = v(x)/u(x)$,

$$L_2(x_0) \, \eta' e^{-\omega'(x-x_0)} > L_2(x) > \eta e^{-\omega_{p'}'(x-x_0)} L_2(x_0)$$

$$(\tau > \sigma \, ; \; x_1 \geqslant x \geqslant x_0), \quad (14 . 32)$$

$$L_2(x) = e^{-\tau x} \quad (\tau = \sigma). \quad (14 . 33)$$

In addition to these inequalities we shall require the following formula :

If ω is any constant and

$$U(y) = \int_{y-\lambda}^y K(y, x) \, U(x) \, dx \quad (y > \lambda),$$

where $K(y, x)$ is finite and its discontinuities, if it has any, are regularly distributed, then

$$\int_t^{t+\lambda} e^{-\omega y} \int_{y-\lambda}^t K(y, x) \, U(x) \, dx \, dy$$

$$= \int_\lambda^{2\lambda} e^{-\omega y} \left[U(y) - \int_\lambda^y K(y, x) \, U(x) \, dx \right] dy$$

$$- \int_\lambda^t e^{-\omega x} U(x) \left[1 - \int_0^\lambda e^{-\omega\theta} K(x+\theta, x) \, d\theta \right] dx \quad (t \geqslant \lambda).$$

$$(14 . 4)$$

To prove this, write the left-hand side equal to $F(t)$, and differentiate

---

* For, $e^{-\rho r} = \int_x^{x+\lambda} e^{-\rho y} Q_1(y, x) \, dy$, and $e^{-\rho\lambda} = 1$; therefore, by § 7, $\lim \dfrac{q_1(y, x)}{q_1(y, 0)} = e^{-\rho x}$.

with respect to t,

$$\frac{\partial F}{\partial t} = -e^{-\omega t} \int_{t-\lambda}^{t} K(t, x)\, U(x)\, dx + U(t) \int_{t}^{t+\lambda} K(y, t)\, e^{-\omega y}\, dy$$

$$= -U(t)\, e^{-\omega t} \left[1 - \int_{0}^{\lambda} e^{-\omega \theta} K(t+\theta, t)\, d\theta \right].$$

Hence, $\quad F(t) = F(\lambda) - \int_{\lambda}^{t} U(x)\, e^{-\omega x} \left[1 - \int_{0}^{\lambda} e^{-\omega \theta} K(\theta + x, x)\, d\theta \right] dx.$

Since $\qquad F(\lambda) = \int_{\lambda}^{2\lambda} e^{-\omega y} \left[U(y) - \int_{\lambda}^{y} U(x)\, K(y, x)\, dx \right] dy,$

this proves the result.

In particular, if we write $U = z_1$, $K = Q_1$, we have

$$\int_{t}^{t+\lambda} e^{-\omega y} \int_{y-\lambda}^{t} \left[K'(y-x)\, u(x) + K''(y-x)\, v(x) \right] dx\, dy$$

$$= \int_{\lambda}^{2\lambda} e^{-\omega y} \left[u(y) + v(y) - \int_{\lambda}^{y} \left[u(\theta)\, K'(y-\theta) + v(\theta)\, K''(y-\theta) \right] d\theta \right] dy$$

$$- \left[1 - \int_{0}^{\lambda} e^{-\omega \theta} K'(\theta)\, d\theta \right] \int_{\lambda}^{t} e^{-\omega \theta} u(\theta)\, d\theta$$

$$- \left[1 - \int_{0}^{\lambda} e^{-\omega \theta} K''(\theta)\, d\theta \right] \int_{\lambda}^{t} e^{-\omega \theta} v(\theta)\, d\theta. \qquad (14.5)$$

15. Consider, now, the cases $\rho \leqslant \sigma \leqslant \tau$.

(1) $\rho = \sigma = \tau$. This is the case of no selection. By $(14.33.31)$ $L_1(x) = L_2(x) = e^{-\rho x}$. By (7.9),

$$\eta > \frac{z_1(x)\, L_1(x)}{\displaystyle\int_{0}^{\lambda} L_1(t)\, f_1(t)\, dt} > \eta' \quad (x > \eta''), \qquad (15.01)$$

with a similar inequality for $z_2(x)$. Hence

$$\eta > \frac{z_1(x)\, e^{-\rho x}}{\displaystyle\int_{0}^{\lambda} e^{-\rho t} f_1(t)\, dt} > \eta' \; ; \quad \eta > \frac{z_2(x)\, e^{-\rho x}}{\displaystyle\int_{1}^{\lambda} e^{-\rho t} f_2(t)\, dt} > \eta \quad \begin{matrix} (x > \eta''), \\[4pt] (\rho = \sigma = \tau). \end{matrix}$$

$$(15.1)$$

Thus $z_1(x)e^{-\rho x}$ and $z_2(x)e^{-\rho x}$ oscillate between fixed limits.

(2) $\rho < \sigma = \tau$. This includes all cases of selected dominants. By (14.33), (7.9),

$$\eta > \frac{z_2(x)\, e^{-\tau x}}{\displaystyle\int_{0}^{\lambda} e^{-\tau t} f_2(t)\, dt} > \eta' \quad (x > \eta''). \qquad (15.21)$$

Put $\omega = \tau$ in (14.5) :

$$\int_t^{t+\lambda} e^{-\tau y} \int_{y-\lambda}^t \left[K'(y-x)\, u(x) + K''(y-x)\, v(x) \right] dx\, dy$$

$$= \int_\lambda^{2\lambda} e^{-\tau y} \left[u(y) + v(y) - \int_\lambda^y \{ u(x)\, K'(y-x) + v(x)\, K''(y-x) \}\, dx \right] dy$$

$$- \left[1 - \int_0^\lambda e^{-\tau \theta} K'(\theta)\, d\theta \right] \int_\lambda^t e^{-\tau x} u(x)\, dx \quad (t > \lambda). \quad (15.22)$$

Since the term on the left is never negative, and the first term on the right is independent of t, and $1 > \int_0^\lambda e^{-\tau t} K'(t)\, dt$,

$$\int_\lambda^t e^{-\tau x} u(x)\, dx$$

must be less than a constant for all values of t ; and, therefore, since $u(x)$ is positive,

$$\int_\lambda^\infty e^{-\tau x} u(x)\, dx$$

must be convergent.   Hence

$$\underline{\lim}\, e^{-\tau x} u(x) = 0.$$

Multiplying (15.21) by $u(x) e^{-\tau x}$, we have

$$\eta u(x) e^{-\tau x} > [e^{-2\tau x} u(x)\, w(x)]/C_2 = e^{-2\tau x} v^2(x)/C_2,$$

where, here and in future,

$$C_2 = \int_0^\lambda e^{-\tau t} f_2(t)\, dt, \quad C_1 = \int_0^\lambda e^{-\rho t} f_1(t)\, dt.$$

Hence

$$\underline{\lim}\, v(x)\, e^{-\tau x} = 0, \quad \underline{\lim}\, z_1(x)\, e^{-\tau x} = 0, \quad \overline{\lim}\, L_1(x)\, e^{\tau x} = \infty.$$

But

$$L_1(x)\, e^{\tau x} = \int_x^{x+\lambda} L_1(y)\, e^{\tau y}\, Q_1(y,\, x)\, e^{-\tau(y-x)}\, dy,$$

and

$$\int_x^{x+\lambda} e^{-\tau(y-x)}\, Q_1(y,\, x)\, dy \leqslant 1.$$

Thus, as before, if M_n, N_n are the greatest and least values of $L_1(x) e^{\tau x}$ in the interval $(n-1)\lambda \leqslant x \leqslant n\lambda$,

$$M_n \leqslant M_{n+1}, \quad \eta > \frac{M_n}{N_n} > \eta' ;$$

so that, since $\overline{\lim} M_n = \infty$,

$$M_n, \; N_n \to \infty; \quad L_1(x) \, e^{\tau x} \to \infty; \quad z_1(x) \, e^{-\tau x} \to 0.$$

It follows, from this and (15.21), that

$$\eta > \frac{w(x) \, e^{-\tau x}}{C_2} > \eta'. \quad v(x) \, e^{-\tau x} \to 0, \quad u(x) \, e^{-\tau x} \to 0 \quad (\rho < \sigma = \tau). \quad (15.23)$$

Further, by (14.3), (15.01), (15.21), in any interval $x_1 \geqslant x \geqslant x_0$, in which the maximum and minimum values of $v(x)/u(x)$ are p' and p,

$$\frac{v(x_0)}{w(x_0)} \, \eta e^{-(\tau - \omega_{p'})(x - x_0)} > \frac{v(x)}{w(x)} > \eta' e^{-(\tau - \omega_p)(x - x_0)} \, \frac{v(x_0)}{w(x_0)}.$$

$\tau - \omega_p$ tends to zero as p tends to infinity, so that, for every positive δ,

$$e^{\delta x} v(x) / w(x) \to \infty. \quad (15.24)$$

Further, since the left-hand side of (15.22) tends to zero,

$$\left[1 - \int_0^\lambda e^{-\tau t} K'(t) \, dt \right] \int_\lambda^\infty e^{-\tau x} u(x) \, dx$$

$$= \int_\lambda^{2\lambda} e^{-\tau y} \left[u(y) + v(y) - \int_\lambda^y \{ K'(y - x) \, u(x) + K''(y - x) \, v(x) \} \, dx \right] dy$$

$$(\rho < \sigma = \tau). \quad (15.25)$$

(3) $\rho = \sigma < \tau$. This includes all cases of selected recessives. By (14.31), (7.9),

$$\eta > \frac{[u(x) + v(x)] \, e^{-\rho x}}{C_1} > \eta' \quad (x > \eta''). \quad (15.3)$$

Further, $\qquad L_2(x) \, e^{\rho x} = \int_x^{x+\lambda} L_2(y) \, e^{\rho y} \, Q_2(y, x) \, e^{-\rho(y-x)} \, dy,$

and $\qquad\qquad \int_x^{x+\lambda} e^{-\rho(y-x)} Q_2(y, x) \, dy \geqslant 1.$

Hence, if M_n', N_n' are the greatest and the least values of $L_2(x) e^{\rho x}$ in the interval $(n-1)\lambda \leqslant x \leqslant n\lambda$,

$$N_n' \geqslant N_{n+1}', \quad \eta > \frac{M_n'}{N_n'} > \eta' \quad (n > \eta).$$

Thus $M_n' \leqslant \eta N_{n_0}'$, if $n > n_0$, and

$$\frac{\overline{\lim} \, z_2(x) \, e^{-\rho x}}{\int_0^\lambda L_2(t) \, f_2(t) \, dt} \geqslant \eta \, \overline{\lim} \left[\frac{1}{L_2(x) \, e^{\rho x}} \right] \geqslant \frac{\eta}{N_{n_0}'}.$$

Hence, by (15.3), $\varliminf \dfrac{v(x)}{u(x)} \geqslant \dfrac{C_4}{C_1} \dfrac{\eta}{N'_{n_0}}$ (C_4 = constant),

and it follows, since $v(x)/u(x)$ is positive, that, in (14.32), we may put $r > p_0$, where p_0 is independent of x_1 so that

$$L_2(x)\, e^{+\rho(x-x_0)} \leqslant \eta e^{-(\omega'_{p_0}-\rho)(x-x_0)} L_2(x_0) \quad (x > x_0);$$

which tends to zero as x tends to infinity.   Thus

$$w(x)\, e^{-\rho x} \to \infty\ ; \quad u(x)\, e^{-\rho x} \to 0\ ; \quad \eta > \dfrac{v(x)\, e^{-\rho x}}{C_1} > \eta' \quad (\rho = \sigma < \tau). \quad (15.31)$$

Again, in any interval $x_1 \geqslant x \geqslant x_0$, in which the greatest and least values of $v(x)/u(x)$ are p' and p,

$$\dfrac{v(x_0)}{w(x_0)}\, \eta e^{-(\omega'_p - \rho)(x-x_0)} > \dfrac{v(x)}{w(x)} > \eta' e^{-(\omega'_{p'} - \rho)(x-x_0)}\, \dfrac{v(x_0)}{w(x_0)}.$$

But $\omega'_{p'} - \rho$ tends to zero as p' tends to zero; and it follows that, given any interval $x_0 + m\lambda \geqslant x \geqslant x_0 + (m-1)\lambda$, there is no positive δ, such that

$$\dfrac{v(x)}{w(x)} < e^{-\delta(x-x_0)}\, \dfrac{v(x_0)}{w(x_0)} \quad (\rho = \sigma < \tau), \quad (15.32)$$

for all values of $v(x_0)/w(x_0)$.

(4) $\rho < \sigma < \tau$.   This includes those cases of imperfect dominance in which the advantages of the hybrids are intermediate between those of the pure forms.   As before, if M_n, N_n are the greatest and least values of $L_1(x)\, e^{\sigma x}$ in $n\lambda \geqslant x \geqslant (n-1)\lambda$, and M'_n, N'_n are the greatest and least values of $L_2(x)\, e^{\sigma x}$ in the same interval,

$$M_n \leqslant M_{n+1}, \quad \eta > \dfrac{M_n}{N_n} > \eta',$$

$$N'_{n+1} \leqslant N_n, \quad \eta > \dfrac{M'_n}{N'_n} > \eta' \quad (x > \eta).$$

Thus $\eta > \dfrac{M_n}{N'_n} \cdot \dfrac{M'_n}{N_n} > \eta',$

or, since $\dfrac{M_n}{N'_n} > \eta\, \dfrac{z_2(x)}{z_1(x)}\, \dfrac{C'_2}{C'_4},$

$$\dfrac{M'_n}{N_n} > \eta\, \dfrac{z_1(x)}{z_2(x)}\, \dfrac{C_4}{C'_2},$$

where C_4', C_2' are constants, if p_n' is the greatest, and p_n the least value of $z_2(x)/z_1(x)$ in $n\lambda \geqslant x \geqslant (n-1)\lambda$,

$$\eta > \frac{p_n'}{p_n} > \eta' \,;$$

hence, for any fixed m and x_0,

$$\eta^m > \frac{p'}{p} > \eta' \quad (x_0+m\lambda \geqslant x \geqslant x_0),$$

where p', p are the maximum and minimum of $z_2(x)/z_1(x)$ in this interval. Further,

$$\int_0^\lambda e^{-\omega_{p'}t} K'(t)\,dt + p' \int_0^\lambda e^{-\omega_{p'}t} K''(t)\,dt = 1+p' = 1+p' \int_0^\lambda e^{-\sigma t} K''(t)\,dt\,;$$

so that $\quad \displaystyle\int_0^\lambda \left[e^{-\omega_{p'}t}-e^{-\sigma t}\right] K''(t)\,dt = +\frac{1}{p'}\left[1-\int_0^\lambda e^{-\omega_{p'}t} K'(t)\,dt\right].$

Similarly, $\quad \displaystyle\int_0^\lambda \left[e^{-\sigma t}-e^{-\omega_p' t}\right] K''(t) = +p\left[\int_0^\lambda e^{-\omega_p' t} K'''(t)\,dt-1\right],$

and therefore

$$\int_0^\lambda \left[e^{-\omega_{p'}t}-e^{-\omega_p' t}\right] K''(t)\,dt = \frac{1}{p'}\left[1-\int_0^\lambda e^{-\omega_{p'}t} K'(t)\,dt\right]$$
$$+p\left[\int_0^\lambda e^{-\omega_p' t} K'''(t)\,dt-1\right].$$

Each term on the right-hand side is positive; as p tends to infinity, the second term is greater than $\epsilon_1 > 0$, and, as p' tends to zero, the first term is greater than $\epsilon_2 > 0$. Hence, if p', p vary in the triangle

$$\infty \geqslant p \geqslant 0, \quad p' \geqslant \eta p,$$

where η is a fixed positive number, the right-hand side must have a positive minimum, that is, if

$$\frac{p'}{p} > \eta, \quad \int_0^\lambda \left[e^{-\omega_{p'}t}-e^{-\omega_p' t}\right] K''(t)\,dt > \epsilon',$$

and, therefore, $\omega_p' - \omega_{p'} > \epsilon''$.

Now, by (14.3), (14.32),

$$L_2(x) < \eta e^{-\omega_p'(x-x_0)} L_2(x_0),$$

$$L_1(x) > \eta e^{-\omega_{p'}(x-x_0)} L_1(x_0),$$

so that $\qquad \displaystyle\frac{L_2(x)}{L_1(x)} < \eta e^{-\epsilon(x-x_0)} \frac{L_2(x_0)}{L_1(x_0)}.$

Since this holds for all values of x if $\delta > 0$,

$$e^{(\epsilon-\delta)x}\,\frac{L_2(x)}{L_1(x)} \to 0\;;$$

and, hence, for a positive δ,

$$e^{\delta x}\,\frac{v(x)}{w(x)} \to 0 \quad (\rho < \sigma < \tau). \tag{15.41}$$

This is to be contrasted with (15.24) and (15.32). The behaviour of $u(x)$, $v(x)$, $w(x)$ at infinity follows more easily from the lemma in the next section.

There are two cases remaining, $\sigma > \tau$, ρ and $\sigma < \tau$, ρ; I shall only consider them very briefly, but I am anxious not to omit them, for the sake of a general conclusion. If $\sigma > \tau$, ρ or $\sigma < \tau$, ρ there exist a real ω and a positive l such that

$$1+l = \int_0^\lambda e^{-\omega t}\,K'(t)\,dt + l\int_0^\lambda e^{-\omega t}\,K''(t)\,dt,$$

$$1+l = \int_0^\lambda e^{-\omega t}\,K''(t)\,dt + l\int_0^\lambda e^{-\omega t}\,K'''(t)\,dt.$$

This follows on eliminating l, when we get the equation

$$\left[1-\int_0^\lambda e^{-\omega t}\,K''(t)\,dt\right]^2 = \left[1-\int_0^\lambda e^{-\omega t}\,K'(t)\,dt\right]\left[1-\int_0^\lambda e^{-\omega t}\,K'''(t)\,dt\right].$$

This has a root between σ and τ, if $\sigma > \tau$, and one between σ and ρ, if $\sigma < \rho$; and the corresponding values of l are positive. If $\sigma > \tau$, (14.3), (14.32) lead to the inequality :

$$\eta p(x_0)\,e^{(\omega'_\rho-\omega_p)(x-x_0)} > p(x) > \eta p(x_0)\,e^{(\omega'_{p'}-\omega_{p'})(x-x_0)}$$

$$(x > x_0,\; \sigma > \tau \geqq \rho),$$

where $\omega'_\pi - \omega_\pi \gtreqless 0$ as $\pi \lesseqgtr l$. Hence there is a contradiction if either $\overline{\lim}\, p(x) < l$ or $\underline{\lim}\, p(x) > l$, so that

$$\overline{\lim}\, p(x) \geqslant l \geqslant \underline{\lim}\, p(x) \quad (\sigma > \tau \geqslant \rho).$$

(2) If $\sigma < \rho$, these inequalities become

$$\eta p(x_0)\,e^{(\omega'_{p'}-\omega_{p'})(x-x_0)} > p(x) > \eta p(x_0)\,e^{(\omega'_p-\omega_p)(x-x_0)},$$

where $\omega'_\pi - \omega_\pi \gtreqless 0$ as $\pi \gtreqless l$. Hence, if $p(x_0)$ is small enough, $p(x)$ tends to zero, and, if $p(x_0)$ is large enough, to ∞.

16. We can round off some of these results by means of the following lemma :

If $K'(t)$, $K''(t)$ *satisfy conditions* A; *if* $f(t)$ *is* **continuous** *for*

$0 \leqslant t \leqslant \lambda$, *and* $v(x)$, $u(x)$ *are never negative and continuous, and if*

$$u(x)+v(x) = \int_0^\lambda K'(t)\,u(x-t)\,dt + \int_0^\lambda K''(t)\,v(x-t)\,dt \quad (x > \lambda),$$

$$= f(x) + \int_0^x K'(t)\,u(x-t)\,dt + \int_0^x K''(t)\,v(x-t)\,dt$$

$$(\lambda \geqslant x \geqslant 0),$$

then (1), *if* $0 = \sigma > \rho$,

$$\int_0^\infty u(t)\,dt \leqslant \left[\int_0^\lambda f(t)\,dt\right] \Big/ \left[1 - \int_0^\lambda K'(t)\,dt\right],$$

$$u(x)+v(x) \to \left[\int_0^\lambda f(t)\,dt - \left\{1 - \int_0^\lambda K'(t)\,dt\right\} \int_0^\infty u(t)\,dt\right] \Big/ \int_0^\lambda tK''(t)\,dt,$$

and (2), *if* $0 = \sigma = \rho$, *and* $u(x) \to 0$,

$$v(x) \to \left[\int_0^\lambda f(t)\,dt\right] \Big/ \int_0^\lambda tK''(t)\,dt.$$

This can be proved by writing the equation in the form

$$u(x)+v(x) = \int_0^\lambda K''(t)\,[u(x-t)+v(x-t)]\,dt + \phi(x),$$

and applying (4.4), (9.2), (14.4).

If $k''(x)$ is the reciprocal function to $K''(x)$, we have, on multiplying (16.1) by $k(y-x)$ and integrating,

$$u(y)+v(y) = -\int_0^\lambda k''(y-t)\,f(t)\,dt + \int_0^y u(\theta)\,R(y-\theta)\,d\theta$$

$$-\int_0^y k''(y-\theta)\int_0^\theta u(x)\,R(\theta-x)\,dx\,d\theta,$$

$$(y > \lambda).$$

where $\quad R(y-\theta) = K'(y-\theta) - K''(y-\theta)$, if $\quad \lambda \geqslant y-\theta \geqslant 0$,

$$= 0, \quad \text{otherwise.}$$

Take any positive number A, greater than λ. Then we can write the preceding equation

$$u(y)+v(y) = -\int_0^\lambda k''(y-t)\,f(t)\,dt$$

$$+ \int_{y-A}^y u(\theta)\left[R(y-\theta) - \int_\theta^y k''(y-t)\,R(t-\theta)\,dt\right]d\theta$$

$$- \int_0^{y-A} u(\theta)\int_\theta^{\theta+\lambda} k''(y-t)\,R(t-\theta)\,dt\,d\theta. \tag{1}$$

125

If $\sigma = 0$, by (9.2),

$$-k''(y) = \left[\int_0^\lambda tK''(t)\,dt\right]^{-1}[1-\psi(y)\,e^{-\eta y}] \quad \text{where} \quad |\psi(y)| < \eta' \quad (y > \eta').$$

Substitute this in the right-hand side of (1):

$$-\int_0^\lambda k''(y-t)\,f(t)\,dt = \frac{\displaystyle\int_0^\lambda f(t)\,dt}{\displaystyle\int_0^\lambda tK''(t)\,dt} + \psi_2(y)\int_0^\lambda e^{-\eta(y-t)} f(t)\,dt,$$

where $$|\psi_2(y)| < \eta;$$

so that, as $y \to \infty$,

$$-\int_0^\lambda k''(y-t)f(t)\,dt \to \left(\int_0^\lambda f(t)\,dt\right)\Big/\int_0^\lambda tK''(t)\,dt,$$

$$\int_0^{y-A} u(\theta)\int_\theta^{\theta+\lambda} k''(y-t)\,R(t-\theta)\,dt\,d\theta = \frac{\left[\displaystyle\int_0^\lambda K'(t)\,dt-1\right]}{\displaystyle\int_0^\lambda tK''(t)\,dt}\int_0^{y-A} u(\theta)\,d\theta + I_1,$$

where $$I_1 = \left[\int_0^{y-A} u(\theta)\int_\theta^{\theta+\lambda} e^{-\eta(y-t)}\psi(y-t)\,R(t-\theta)\,dt\,d\theta\right]\Big/\int_0^\lambda tK''(t)\,dt$$

$$\leqslant \eta e^{-\eta A}\int_0^{y-A} u(\theta)\,e^{-\eta(y-A-\theta)}\,d\theta;$$

finally

$$\left|\int_{y-A}^y u(\theta)\left[R(y-\theta) - \int_\theta^y k''(y-t)\,R(t-\theta)\,dt\right]d\theta\right| \leqslant \eta\int_{y-A}^y u(\theta)\,d\theta.$$

Hence

$$u(y)+v(y) = \frac{\displaystyle\int_0^\lambda f(t)\,dt}{\displaystyle\int_0^\lambda tK''(t)\,dt} + \frac{\displaystyle\int_0^\lambda K'(t)\,dt-1}{\displaystyle\int_0^\lambda tK''(t)\,dt}\int_0^{y-A} u(\theta)\,d\theta + I_1$$

$$+\eta\int_{y-A}^y u(\theta)\,d\theta + \psi(y), \quad (16.1)$$

where $\psi(y) \to 0$, as $y \to \infty$.
    By (14.5), if $0 = \sigma > \rho$,

$$\int_0^\infty u(\theta)\,d\theta$$

is convergent; hence, for every A,

$$\lim_{y \to \infty} \int_{y-A}^{y} u(\theta)\, d\theta = 0,$$

and, as y tends to infinity,

$$\lim I_1 \leqslant \eta e^{-\eta A} \int_0^\infty u(\theta)\, d\theta < Ce^{-\eta A} \quad (C = \text{constant}).$$

If, therefore, in equation (16.1), we first let y tend to infinity and then A,

$$\lim_{y \to \infty} [u(y)+v(y)] = \left[\int_0^\lambda f(t)\, dt - \left\{1 - \int_0^\lambda K'(t)\, dt\right\} \int_0^\infty u(\theta)\, d\theta\right] \Big/ \int_0^\lambda tK''(t)\, dt.$$

If $0 = \rho = \sigma$, and $u(x) \to 0$, then since, for every positive δ,

$$\int_0^y u(\theta)\, e^{-\delta(y-\theta)}\, d\theta \to 0 \quad \text{as} \quad y \to \infty,$$

provided that $\qquad\qquad u(\theta) \to 0,$

it follows that, for every positive A, as y tends to infinity,

$$I_1 e^{\eta A} \to 0.$$

And hence, letting y tend to infinity, in (16.1),

$$\lim_{y \to \infty} [u(y)+v(y)] = \int_0^\lambda f(t)\, dt \Big/ \int_0^\lambda tK''(t)\, dt.$$

This proves the result.

If now we apply this result to cases (2), (3), (4), it leads to the conclusion that, in all cases,

$$e^{-\sigma y}u(y) \to 0,$$

$$e^{-\sigma y}v(y) \to \text{constant} \quad (\rho < \sigma \leqslant \tau,\ \rho = \sigma < \tau),$$

$$e^{-\tau y}w(y) \to \text{constant},$$

and there is no difficulty in proving that the only case in which the constants can be zero is $\rho < \sigma = \tau$, when

$$e^{-\sigma y}v(y) \to 0,$$

$$e^{-\tau y}w(y) \to \text{constant} > 0.$$

For instance, suppose that $\rho = \sigma < \tau$. In applying the lemma it is arbitrary what is considered the initial point. Thus for any x_0, if

$$A_1 = \int_0^\lambda te^{-\sigma t} K''(t) \, dt, \qquad A_2 = \int_0^\lambda te^{-\tau t} K'''(t) \, dt,$$

$$1 - A_3 = \int_0^\lambda e^{-\tau t} K''(t) \, dt,$$

then $e^{-\sigma x} v(x) \to \dfrac{1}{A_1} \displaystyle\int_\lambda^{2\lambda} e^{-\sigma y} \int_{y-\lambda}^\lambda \left[K'(y-\theta) \, u(\theta) + K''(y-\theta) \, v(\theta) \right] d\theta \, dy$

$$= \frac{1}{A_1} \int_{x_0}^{x_0+\lambda} e^{-\sigma y} \int_{y-\lambda}^{x_0} \left[K'(y-\theta) \, u(\theta) + K''(y-\theta) \, v(\theta) \right] d\theta \, dy = C, \text{ say;}$$

$$e^{-\tau x} w(x) \to \frac{1}{A_2} \left[\int_{x_0}^{x_0+\lambda} e^{-\tau y} \int_{y-\lambda}^{x_0} \left[K''(y-\theta) \, v(\theta) + K'''(y-\theta) \, w(\theta) \right] d\theta \, dy \right.$$

$$\left. - A_3 \int_{x_0}^\infty e^{-\tau y} v(y) \, dy \right].$$

Given any ϵ, we can choose x_0 so large that

$$\left| \frac{e^{-\sigma x} v(x)}{C} - 1 \right| < \epsilon \quad (x > x_0),$$

and, hence, so large that

$$\frac{C(1+\epsilon) e^{-(\tau-\sigma)x}}{\tau-\sigma} > \int_x^\infty e^{-\tau y} v(y) \, dy > \frac{C(1-\epsilon) e^{-(\tau-\sigma)x}}{\tau-\sigma} \quad (x > x_0).$$

If this is substituted in the expression for $e^{-\tau x} w(x)$,

$$\lim_{x \to \infty} e^{-\tau x} w(x) = \frac{1}{A_2} \int_{x_0}^{x_0+\lambda} e^{-\tau y} \int_{y-\lambda}^{x_0} \left\{ \left[K'''(y-\theta) \, w(\theta) + K''(y-\theta) \, v(\theta) \right] \right.$$

$$\left. - \phi \left[K''(y-\theta) \, v(\theta) + K'(y-\theta) \, u(\theta) \right] \right\} d\theta \, dy,$$

where $\qquad |\phi| <$ a constant, independent of x_0.

But the integral on the right is necessarily positive if x_0 is large enough, since

$$\frac{v(\theta)}{w(\theta)}, \quad \frac{u(\theta)}{w(\theta)} \to 0, \text{ as } \theta \to \infty.$$

Hence $\qquad \lim e^{-\tau x} w(x) > 0.$

The other case can be treated in the same way.

I think that equation (15.25) is worth noticing, because it has a simple interpretation in the present application. Since the initial point is arbitrary, it can be written, if $\rho = \sigma < \tau$,

$$\int_x^\infty e^{-\tau t} u(t)\, dt$$

$$= \frac{\int_x^{x+\lambda} e^{-\tau t} \left[u(t) + v(t) - \int_x^t \{ K'(t-\theta)\, u(\theta) + K''(t-\theta)\, v(\theta) \}\, d\theta \right] dt}{1 - \int_0^\lambda e^{-\tau t} K'(t)\, dt}.$$

Suppose that $\tau = 0$.

The integral on the left-hand side is the total number of recessives which are born after a fixed date, when the survival ratio of the dominants corresponds to that in a stationary population; the integral on the right depends on the number of recessives and hybrids in the community in the first generation after that date; so that, in the case of the selected dominant, there is a simple relation connecting the two. Similarly, in the case of the selected recessive, we have

$$\int_{-\infty}^x e^{-\rho t} w(t)\, dt$$

$$= \frac{\int_x^{x+\lambda} e^{-\rho t} \left[v(t) + w(t) - \int_x^t [K''(t-\theta)\, v(\theta) + K'''(t-\theta)\, w(\theta)]\, d\theta \right] dt}{\int_0^\lambda e^{-\rho t} K'''(t)\, dt - 1},$$

where the left-hand side is the total number of recessives born before a fixed date, on the assumption that the survival ratio of the dominant corresponds to that in a stationary population. Similarly, again, in the simple community mentioned in § 10, we have, if we make a slight change in the notation, and write p_n for the number of dominants born in the n-th breeding season, $2q_n$ for the number of hybrids, and r_n for the number of recessives, and suppose their survival ratios to be, respectively, a, b, c, then, if the dominant is selected,

$$\sum_{s=n}^\infty r_s a^{-s} = a^{-n} [q_n + r_n] \Big/ \left(1 - \frac{c}{a} \right) \quad (c < b = a),$$

and if the recessive is selected,

$$\sum_{s=-\infty}^{n-1} r_s a^{-s} = a^{-n} [q_n + r_n] \Big/ \left(\frac{c}{a} - 1 \right) \quad (c > b = a).$$

Here, again, the left-hand sides are, respectively, the number of recessives born after a fixed date, and the number born before a fixed date, on the assumption that the survival ratio of the dominants corresponds to that in a stationary population.

In conclusion, it may be mentioned that (15 . 25) makes clear the sense in which selection is capable of establishing a dominant variety. This formula can be written

$$\int_\lambda^\infty e^{-\sigma t}\, u(t)\, dt = A\Big/\Big[1-\int_0^\lambda e^{-\sigma\tau}\, K'(t)\, dt\Big],$$

where A is a constant depending on the initial condition in the community. This is to be compared with the case in which the pure types do not interbreed, when

$$\int_\lambda^\infty e^{-\sigma t}\, u(t)\, dt = A'\Big/\Big[1-\int_0^\lambda e^{-\sigma t}\, K'(t)\, dt\Big],$$

where A' is another constant. If $\sigma = 0$, the dominants are stationary, the recessives decrease to zero, and the left-hand sides of (15 . 251 . 252) represent the total number of recessives born after a fixed date. The formulae are of the same type. The difference between the two cases lies in the time taken before the last recessive is born, which may be one generation in the second case, while it is indefinitely long in the other.

The main results of the previous part can be summarized as follows. If the ages of parents vary independently, and the instantaneous birth-rate is the same in all three types and the same for a given age at all moments, and if the death-rates are the same for a given type and age at all moments, the advantages possessed by the three types can be measured by the real zeros of the functions

$$\int_0^\lambda e^{-zt}\, K'(t)\, dt - 1, \quad \int_0^\lambda e^{-zt}\, K''(t)\, dt - 1, \quad \int_0^\lambda e^{-zt}\, K'''(t)\, dt - 1,$$

where $K'(t)$, $K''(t)$, $K'''(t)$ are the functions defined by (12 . 11). If ρ, σ, τ are the three respective zeros, and $u(x)$, $v(x)$, $w(x)$ the density of births in the types at moment x, then

$$\frac{w(x)}{v(x)}, \quad \frac{v(x)}{u(x)} \to \infty ;$$

and
$$\int_0^\infty e^{-\tau x}\, u(x)\, dx$$

is less than a constant depending on the initial state of the community; the integral last written represents the total number of the unselected pure type born after a fixed date, if the birth- and death-rates in the other pure type correspond to those in a stationary population. If $\rho < \sigma < \tau$, there exists a positive δ, depending only on the K's, such that

$$\frac{w(x)}{v(x)}\, e^{-\delta x} \to \infty ;$$

if $\rho < \sigma = \tau$, then in any interval $x_1 \geqslant x \geqslant x_0$ in which $w(x)/v(x) < p'$, there exists a positive δ, such that

$$e^{-\delta x} \frac{w(x)}{v(x)} > e^{-\delta x_0} \frac{w(x_0)}{v(x_0)},$$

but eventually, for every δ,

$$e^{-\delta x} \frac{w(x)}{v(x)} \to 0 ;$$

if $\rho = \sigma < \tau$, then in any interval in which $w(x)/v(x) > p > 0$ there is a positive δ such that

$$e^{-\delta x} \frac{w(x)}{v(x)} > e^{-\delta x_0} \frac{w(x_0)}{v(x_0)},$$

but there is no δ such that this holds for every p; if $\rho = \sigma = \tau$, $w(x)/v(x)$ oscillates between finite positive limits. These results can all be summarized in the statement that, subject to the conditions stated, the same results hold as in the case when different generations do not interbreed.

The preceding results are very incomplete. But I think that enough has been proved to make clear that, if the advantages possessed by a type are measured by the constant of the geometric ratio in which a community would increase, if it had the same birth-rate and death-rate as the type, and if the other conditions are fulfilled, then the changes in composition of the community depend essentially on the relative magnitude of the advantages of the types ; and that, in particular, if the hybrids are neither the most nor the least advantageous type, then the pure type which has the greater advantages increases indefinitely in proportion to the others.

11

Reprinted from *The Genetical Theory of Natural Selection*, Dover Publications, Inc., New York, 1958, pp. 22–30, 37–41

THE FUNDAMENTAL THEOREM OF NATURAL SELECTION

R. A. Fisher

The life table and the table of reproduction. The Malthusian parameter of population increase. Reproductive value. The genetic element in variance. Natural Selection. The nature of adaptation. Deterioration of the environment. Changes in population. Summary.

One has, however, no business to feel so much surprise at one's ignorance, when one knows how impossible it is without statistics to conjecture the duration of life and percentage of deaths to births in mankind. DARWIN, 1845. (*Life and Letters*, ii, 33.)

In the first place it is said—and I take this point first, because the imputation is too frequently admitted by Physiologists themselves—that Biology differs from the Physico-chemical and Mathematical sciences in being 'inexact'. HUXLEY, 1854.

The life table

IN order to obtain a distinct idea of the application of Natural Selection to all stages in the life-history of an organism, use may be made of the ideas developed in the actuarial study of human mortality. These ideas are not in themselves very recondite, but being associated with the laborious computations and the technical notation employed in the practical business of life insurance, are not so familiar as they might be to the majority of biologists. The textbooks on the subject, moreover, are devoted to the chances of death, and to monetary calculations dependent on these chances, whereas in biological problems at least equal care and precision of ideas is requisite with respect to reproduction, and especially to the combined action of these two agencies in controlling the increase or decrease of the population.

The object of the present chapter is to combine certain ideas derivable from a consideration of the rates of death and reproduction of a population of organisms, with the concepts of the factorial scheme of inheritance, so as to state the principle of Natural Selection in the form of a rigorous mathematical theorem, by which the rate of improvement of any species of organisms in relation to its environment is determined by its present condition.

The fundamental apparatus of the actuary's craft is what is known as a life table. This shows, for each year of age, of the population considered, the proportion of persons born alive who live to attain that age. For example, a life table may show that the proportion of persons living to the age of 20 is 88 per cent., while only 80 per cent. reach the age of 40. It will be easily inferred that 12 per cent. of those born alive die in the first 20 years of life, and 8 per cent. in the second 20 years. The life table is thus equivalent to a statement of the frequency distribution of the age of death in the population concerned. The amount by which each entry is less than the preceding entry represents the number of deaths between these limits of age, and this divided by the number living at the earlier age gives for these the probability of death within a specified time. Since the probability of death changes continuously throughout life, the death rate at a given age can only be measured consistently by taking the age interval to be infinitesimal. Consequently if l_x is the number living to age x, the death rate at age x is given by:

$$\mu_x = -\frac{1}{l_x}\frac{d}{dx}l_x = -\frac{d}{dx}(\log l_x),$$

the logarithm being taken, as in most mathematical representations, to be on the Natural or Naperian system. The life table thus contains a statement of the death rates at all ages, and conversely can be constructed from a knowledge of the course taken by the death rate throughout life. This in fact is the ordinary means of constructing the life tables in practical use.

It will not be necessary to discuss the technical procedure employed in the construction of life tables, the various conventions employed in this form of statement, nor the difficulties which arise in the interpretation of the observational data available in practice for this purpose. It will be sufficient to state only one point. As in all other experimental determinations of theoretical values, the accuracy attainable in practice is limited by the extent of the observations; the result derived from any finite number of observations will be liable to an error of random sampling, but this fact does not, in any degree, render such concepts as death rates or expectations of life obscure or inexact. These are statements of probabilities, averages &c., pertaining to the hypothetical population sampled, and depend only upon its nature and circumstances. The inexactitude of our methods of measurement has no more reason in statistics than it has in physics to dim our conception of that which we measure. These conceptions would be equally clear if we were stating the chances of death of a single individual of unique genetic constitution, or of one exposed to an altogether transient and exceptional environment.

The table of reproduction

The life table, although itself a very comprehensive statement, is still inadequate to express fully the relation between an organism and its environment; it concerns itself only with the chances or frequency of death, and not at all with reproduction. To repair this deficiency it is necessary to introduce a second table giving rates of reproduction in a manner analogous to the rates of death at each age. Just as a person alive at the beginning of any infinitesimal age interval dx has a chance of dying within that interval measured by $\mu_x dx$, so the chance of reproducing within this interval will be represented by $b_x dx$, in which b_x may be called the rate of reproduction at age x. Again, just as the chance of a person chosen at birth dying within a specified interval of age dx is $l_x \mu_x dx$, so the chance of such a person living to reproduce in that interval will be $l_x b_x dx$.

Owing to bisexual reproduction a convention must be introduced into the measurement of b_x, for each living offspring will be credited to both parents, and it will seem proper to credit each with one half in respect of each offspring produced. This convention will evidently be appropriate for those genes which are not sex-linked (autosomal genes) for with these the chance of entering into the composition of each offspring is known to be one half. In the case of sex-linked genes those of the heterogametic parent will be perpetuated or not according as the offspring is male or female. These sexes, it is true, will not be produced in exactly equal numbers, but since both must co-operate in each act of sexual reproduction, it is clear that the different frequencies at birth must ultimately be compensated by sexual differences in the rates of death and reproduction, with the result that the same convention appears in this case to be equally appropriate.

A similar convention, appropriate in the sense of bringing the formal symbolism of the mathematics into harmony with the biological facts, may be used with respect to the period of gestation. For it will happen occasionally that a child is born after the death of its father. The children born to fathers aged x should in fact be credited to males aged three-quarters of a year younger. Such corrections are not a necessity to an exact mathematical representation of the facts, but are a manifest convenience in simplifying the form of expression; thus with mankind we naturally think of the stage in the life-history as measured in years from birth. With other organisms the variable x which with man represents this age, may in some cases be more conveniently used to indicate rather the stage in the life history irrespective of chronological age, merely to give greater vividness to the meaning of the symbolism, but without altering the content of the symbolical statements.

The Malthusian parameter of population increase

If we combine the two tables giving the rates of death and repro-duction, we may, still speaking in terms of human populations, at once calculate the expectation of offspring of the newly-born child. For the expectation of offspring in each element of age dx is $l_x b_x dx$, and the sum of these elements over the whole of life will be the total expectation of offspring. In mathematical terms this is

$$\int_0^\infty l_x b_x dx,$$

where the integral is extended from zero, at birth, to infinity, to cover every possible age at which reproduction might conceivably take place. If at any age reproduction ceases absolutely, b_x will thereafter be zero and so give a terminating integral under the same form.

The expectation of offspring determines whether in the population concerned the reproductive rates are more or less than sufficient to balance the existing death rates. If its value is less than unity the reproductive rates are insufficient to maintain a stationary popula-tion, in the sense that any population which constantly maintained the death and reproduction rates in question would, apart from temporary fluctuations, certainly ultimately decline in numbers at a calculable rate. Equally, if it is greater than unity, the population biologically speaking is more than holding its own, although the actual number of heads to be counted may be temporarily decreasing.

This consequence will appear most clearly in its quantitative aspect if we note that corresponding to any system of rates of death and reproduction, there is only one possible constitution of the population in respect of age, which will remain unchanged under the action of this system. For if the age distribution remains unchanged the relative rate of increase or decrease of numbers at all ages must be the same; let us represent the relative rate of increase by m; which will also represent a decrease if m is negative. Then, owing to the constant rates of reproduction, the rate at which births are occurring at any epoch will increase proportionately to e^{mt}. At any particular epoch, for which we may take $t=0$, the rate at which births were occurring x years ago will be proportional to e^{-mx}, and this is the rate at which births were occurring at the time persons now of age x were being born. The number of persons in the infinitesimal age interval dx will therefore be proportional to $e^{-mx}l_x dx$, for of those born only the fraction l_x survive to this age. The age distribution is therefore determinate if the number m is uniquely determined. But knowing the numbers living at each age, and the reproductive rates at each age, the rate at which births are now occurring can be calculated, and this can be equated to the known rate of births appropriate to $t = 0$. In fact, the contribution to the total rate, of persons in the age interval dx, must be $e^{-mx}l_x b_x dx$ and the aggregate for all ages must be

$$\int_0^\infty e^{-mx}l_x b_x dx,$$

which, when equated to unity, supplies an equation for m, of which one and only one real solution exists. Since e^{-mx} is less than unity for all values of x, if m is positive, and is greater than unity for all values of x, if m is negative, it is evident that the value of m, which reduces the integral above expressed to unity, must be positive if the expectation of offspring exceeds unity, and must be negative if it falls short of unity.

The number m which satisfies this equation is thus implicit in any given system of rates of death and reproduction, and measures the relative rate of increase or decrease of a population when in the steady state appropriate to any such system. In view of the emphasis laid by Malthus upon the 'law of geometric increase' m may appropriately be termed the Malthusian parameter of population increase. It evidently supplies in its negative values an equally good measure of population decrease, and so covers cases to which, in respect of mankind, Malthus paid too little attention.

In view of the close analogy between the growth of a population supposed to follow the law of geometric increase, and the growth of capital invested at compound interest, it is worth noting that if we regard the birth of a child as the loaning to him of a life, and the birth of his offspring as a subsequent repayment of the debt, the method by which m is calculated shows that it is equivalent to answering the question—At what rate of interest are the repayments the just equivalent of the loan ? For the unit investment has an expectation of a return $l_x b_x dx$ in the time interval dx, and the present value of this repayment, if m is the rate of interest, is $e^{-mx}l_x b_x dx$; consequently the Malthusian parameter of population increase is the rate of interest at which the present value of the births of offspring to be expected is equal to unity at the date of birth of their parent. The actual values of the parameter of population increase, even in sparsely populated dominions, do not, however, seem to approach in magnitude the rates of interest earned by money, and negative rates of interest are, I suppose, unknown to commerce.

Reproductive value

The analogy with money does, however, make clear the argument for another simple application of the combined death and reproduction rates. We may ask, not only about the newly born, but about persons of any chosen age, what is the present value of their future offspring; and if present value is calculated at the rate determined as before, the question has the definite meaning—To what extent will persons of this age, on the average, contribute to the ancestry of future generations ? The question is one of some interest, since the direct action of Natural Selection must be proportional to this contribution. There will also, no doubt, be indirect effects in cases in which an animal favours or impedes the survival or reproduction of its relatives; as a suckling mother assists the survival of her child, as in mankind a mother past bearing may greatly promote the reproduction of her children, as a foetus and in less measure a sucking child inhibits conception, and most strikingly of all as in the services of neuter insects to their queen. Nevertheless such indirect effects will in very many cases be unimportant compared to the effects of personal reproduction, and by the analogy of compound interest the present value of the future offspring of persons aged x is easily seen to be given by the equation

$$v_x/v_0 = \frac{e^{mx}}{l_x} \int_x^\infty e^{-mt}l_t b_t dt.$$

Each age group may in this way be assigned its appropriate

reproductive value. Fig. 2 shows the reproductive value of women
according to age as calculated from the rates of death and reproduction current in the Commonwealth of Australia about 1911. The
Malthusian parameter was at that time positive, and as judged from

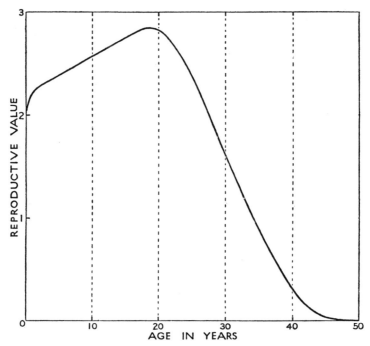

Fig. 2. Reproductive value of Australian women.
    The reproductive value for female persons calculated from the birth- and death-
rates current in the Commonwealth of Australia about 1911. The Malthusian
parameter is +0·01231 per annum.

female rates was nearly equivalent to 1¼ per cent. compound interest;
the rate would be lower for the men, and for both sexes taken together,
owing to the excess of men in immigration. The reproductive value,
which of course is not to be confused with the reproductive rate,
reaches its maximum at about 18½, in spite of the delay in reproduction caused by civilized marriage customs; indeed it would have
been as early as 16, were it not that a positive rate of interest gives
higher value to the immediate prospect of progeny of an older woman,
compared to the more remote children of a young girl. If this is the

case among a people by no means precocious in reproduction, it would be surprising if, in a state of society entailing marriage at or soon after puberty, the age of maximum reproductive value should fall at any later age than twelve. In the Australian data, the value at birth is lower, partly by reason of the effect of an increasing population in setting a lower value upon remote children and partly because of the risk of death before the reproductive age is reached. The value shown is probably correct, apart from changes in the rate since 1911, for such a purpose as assessing how far it is worth while to give assistance to immigrants in respect of infants (though of course, it takes no account of the factor of eugenic quality), for such infants will usually emigrate with their parents; but it is overvalued from the point of view of Natural Selection to a considerable extent, owing to the capacity of the parents to replace a baby lost during lactation. The reproductive value of an older woman on the contrary is undervalued in so far as her relations profit by her earnings or domestic assistance, and this to a greater extent from the point of view of the Commonwealth, than from that of Natural Selection. It is probably not without significance in this connexion that the death rate in Man takes a course generally inverse to the curve of reproductive value. The minimum of the death rate curve is at twelve, certainly not far from the primitive maximum of the reproductive value; it rises more steeply for infants, and less steeply for the elderly than the curve of reproductive value falls, points which qualitatively we should anticipate, if the incidence of natural death had been to a large extent moulded by the effects of differential survival.

A property that well illustrates the significance of the method of valuation, by which, instead of counting all individuals as of equal value in respect of future population, persons of each age are assigned an appropriate value v_x, is that, whatever may be the age constitution of a population, its total reproductive value will increase or decrease according to the correct Malthusian rate m, whereas counting all heads as equal this is only true in the theoretical case in which the population is in its steady state. For suppose the number of persons in the age interval dx is $n_x dx$; the value of each element of the population will be $n_x v_x dx$; in respect of each such group there will be a gain in value by reproduction at the rate of $n_x b_x v_o dx$, a loss by death of $n_x \mu_x v_x dx$, and a loss by depreciation of $-n_x dv_x$, or in all

$$n_x \{ (b_x v_o - \mu_x v_x)\, dx + dv_x \},$$

but by differentiating the equation by which v_x is defined, it appears that

$$\frac{1}{v_x}\frac{dv_x}{dx} + \frac{1}{l_x}\frac{dl_x}{dx} - m = \frac{-l_x b_x e^{-mx}}{\frac{v_x}{v_o}l_x e^{-mx}} = -\frac{b_x v_o}{v_x},$$

or that

$$dv_x - \mu_x v_x dx + b_x v_o dx = mv_x dx.$$

Consequently the rate of increase in the total value of the population is m times its actual total value, irrespective of its constitution in respect of age. A comparison of the total values of the population at two census epochs thus shows, after allowance for migration, the genuine biological increase or decrease of the population, which may be entirely obscured or reversed by the crude comparison of the number of heads. The population of Great Britain, for example, must have commenced to decrease biologically at some date obscured by the war, between 1911 and 1921, but the census of 1921 showed a nominal increase of some millions, and that of 1931 will, doubtless in less degree, certainly indicate a further spurious period of increase, due to the accumulation of persons at ages at which their reproductive value is negligible.

[*Editors' Note:* Material has been omitted at this point.]

Natural Selection

Any group of individuals selected as bearers of a particular gene, and consequently the genes themselves, will have rates of increase which may differ from the average. The excess over the average of any such selected group will be represented by a, and similarly the average effect upon m of introducing the gene in question will be represented by α. Since m measures fitness by the objective fact of representation in future generations, the quantity

$$\Sigma'(2pa\alpha)$$

will represent the contribution of each factor to the genetic variance in fitness. The total genetic variance in fitness being the sum of these contributions, which is necessarily positive, or, in the limiting case, zero. Moreover, any increase dp in the frequency of the chosen gene will be accompanied by an increase $2\alpha\,dp$ in the average fitness of the species, where α may, of course, be negative. But the definition of a requires that

$$\frac{d}{dt} \log p = a$$

or $$dp = (pa)dt$$

hence $$(2\alpha)dp = (2pa\alpha)dt$$

which must represent the rate of increase of the average fitness due to the change in progress in frequency of this one gene. Summing for all allelomorphic genes, we have

$$dt\Sigma'(2pa\alpha)$$

and taking all factors into consideration, the total increase in fitness is

$$\Sigma\alpha\,dp = dt\Sigma\Sigma'(2pa\alpha) = W\,dt\;.$$

If therefore the time element dt is positive, the total change of fitness $W\,dt$ is also positive, and indeed the rate of increase in fitness due to all changes in gene ratio is exactly equal to the genetic variance of fitness W which the population exhibits. We may consequently state the fundamental theorem of Natural Selection in the form:

The rate of increase in fitness of any organism at any time is equal to its genetic variance in fitness at that time.

141

The rigour of the demonstration requires that the terms employed should be used strictly as defined; the ease of its interpretation may be increased by appropriate conventions of measurement. For example, the frequencies p should strictly be evaluated at any instant by the enumeration, not necessarily of the census population, but of all individuals having reproductive value, weighted according to the reproductive value of each.

Since the theorem is exact only for idealized populations, in which fortuitous fluctuations in genetic composition have been excluded, it is important to obtain an estimate of the magnitude of the effect of these fluctuations, or in other words to obtain a standard error appropriate to the calculated, or expected, rate of increase in fitness. It will be sufficient for this purpose to consider the special case of a population mating and reproducing at random. It is easy to see that if such chance fluctuations cause a difference δp between the actual value of p obtained in any generation and that expected, the variance of δp will be

$$\frac{pq}{2n},$$

where n represents the number breeding in each generation, and $2n$ therefore is the number of genes in the n individuals which live to replace them. The variance of the increase in fitness, $\Sigma 2\alpha dp$, due to this cause, will therefore be

$$\frac{1}{2n}(2pq\alpha^2).$$

Now, with random mating, the chance fluctuation in the different gene ratios will be independent, and the values of a and α are no longer distinct, it follows that, on this condition, the rate of increase of fitness, when measured over one generation, will have a standard error due to random survival equal to

$$\frac{1}{T}\sqrt{\frac{W}{2n}}$$

where T is the time of a generation. It will usually be convenient for each organism to measure time in generations, and if this is done it will be apparent from the large factor $2n$ in the denominator, that the random fluctuations in W, even measured over only a single generation, may be expected to be very small compared to the average rate of progress. The regularity of the latter is in fact guaranteed by the same circumstance which makes a statistical assemblage of particles, such as a bubble of gas obey, without appreciable deviation, the laws of gases. A visible bubble will indeed contain several billions of molecules, and this would be a comparatively large number for an organic population, but the principle ensuring regularity is the same. Interpreted exactly, the formula shows that it is only when the rate of progress, W, when time is measured in generations, is itself so small as to be comparable to $1/n$, that the rate of progress achieved in successive generations is made to be irregular. Even if an equipoise of this order of exactitude, between the rates of death and reproduction of different genotypes, were established, it would be only the rate of progress for spans of a single generation that would be shown to be irregular, and the deviations from regularity over a span of 10,000 generations would be just a hundredfold less.

It will be noticed that the fundamental theorem proved above bears some remarkable resemblances to the second law of thermodynamics. Both are properties of populations, or aggregates, true irrespective of the nature of the units which compose them; both are statistical laws; each requires the constant increase of a measurable quantity, in the one case the entropy of a physical system and in the other the fitness, measured by m, of a biological population. As in the physical world we can conceive of theoretical systems in which dissipative forces are wholly absent, and in which the entropy consequently remains constant, so we can conceive, though we need not expect to find, biological populations in which the genetic variance is absolutely zero, and in which fitness does not increase. Professor Eddington has recently remarked that 'The law that entropy always increases—the second law of thermodynamics—holds, I think, the supreme position among the laws of nature'. It is not a little instructive that so similar a law should hold the supreme position among the biological sciences. While it is possible that both may ultimately be absorbed by some more general principle, for the present we should note that the laws as they stand present profound differences— (1) The systems considered in thermodynamics are permanent; species on the contrary are liable to extinction, although biological improvement must be expected to occur up to the end of their existence. (2) Fitness, although measured by a uniform method, is qualitatively different for every different organism, whereas entropy, like temperature, is taken to have the same meaning for all physical systems. (3) Fitness may be increased or decreased by changes in the environment, without reacting quantitatively upon that environment. (4) Entropy changes are exceptional in the physical world in being irreversible, while irreversible evolutionary changes form no exception among biological phenomena. Finally, (5) entropy changes lead to a progressive disorganization of the physical world, at least from the human standpoint of the utilization of energy, while evolutionary changes are generally recognized as producing progressively higher organization in the organic world.

The statement of the principle of Natural Selection in the form of a theorem determining the rate of progress of a species in fitness to survive (this term being used for a well-defined statistical attribute of the population), together with the relation between this rate of progress and its standard error, puts us in a position to judge of the validity of the objection which has been made, that the principle of Natural Selection depends on a succession of favourable chances. The objection is more in the nature of an innuendo than of a criticism, for it depends for its force upon the ambiguity of the word chance, in its popular uses. The income derived from a Casino by its proprietor may, in one sense, be said to depend upon a succession of favourable chances, although the phrase contains a suggestion of improbability more appropriate to the hopes of the patrons of his establishment. It is easy without any very profound logical analysis to perceive the difference between a succession of favourable deviations from the laws of chance, and on the other hand, the continuous and cumulative action of these laws. It is on the latter that the principle of Natural Selection relies.

In addition to the genetic variance of any measurable character there exists, as has been seen, a second element comprised in the total genotypic variance, due to the heterozygote being in general not equal to the mean of the two corresponding homozygotes. This component, ascribable to dominance, is also in a sense capable of exerting evolutionary effects, not through any direct effect on the gene ratios, but through its possible influence on the breeding system. For if, in general, heterozygotes were favoured as compared with homozygotes, it is evident that the offspring of outcrosses would be at an advantage compared with those of matings between relatives, or of self-fertilization, and any heritable tendencies favouring such matings might come to be eliminated, with consequent increase in the proportion of heterozygotes.

This indirect and conditional factor in selection seems to have been able to produce effects of considerable importance, such as the separation of the sexes, self-sterility in many plants, and flowers made attractive by colour, scent and nectar.   A first step to the understanding of these effects of dominance has been made in Chapter III, but the author would emphasize that in his opinion no satisfactory selective model has been set up competent even to derive a distylic species like the primrose from a monostylic species of the same genus.   Possibly, therefore, the course of evolutionary change has been complex and circuitous.

Such effects ascribable to the dominance component of the genotypic variation are not in reality additional to the evolutionary changes accounted for by the fundamental theorem; for in that theorem they are credited to the gene-substitutions needed, for example, to develop bigger or brighter flowers; although the selective advantage conferred by these may be wholly due to dominance deviations in fitness recognizable in numerous other factors.

[*Editors' Note:* Material has been omitted at this point.]

Editors' Comments
on Papers 12 Through 15

12   **ROUGHGARDEN**
     Density-Dependent Natural Selection

13   **CHRISTIAN**
     Social Subordination, Population Density, and Mammalian Evolution

14   **ANDERSON and KING**
     Age-Specific Selection

15   **CHARLESWORTH and GIESEL**
     Selection in Population with Overlapping Generations: II. Relations Between Gene Frequency and Demographic Variables

LIFE HISTORY AND ECOLOGICAL GROWTH

After Fisher's work in 1930, few attempts were made to relate age structure to population biology until about 1948, when Birch published an analysis of the meaning of Lotka's intrinsic growth rate as it applied (in particular) to insects. In 1954, L. C. Cole published an influential paper analyzing the way in which various possible reproductive strategies affected population growth. The study of the relationship between growth and age structure goes back to 1907, but, as Cole notes, this has been insufficiently considered by ecologists. By allowing all types to have equal total reproductive fitness, Cole showed that, under certain conditions of birth and death schedule by age ("life histories"), different growth rates would result. For example, Cole investigates the way in which evolution from semelparity (reproducing only once in life) to iteroparity (repeated reproduction) could occur. He relates the litter size, type of reproduction, and life history to the resulting Malthusian parameter. Cole looks at whole species, rather than competition within a population (although the results would probably be the same if intertype breeding were random or at least followed a known pattern), and he equates success with more rapid growth (he does acknowledge ultimate density effects).

Cole was followed by 1965 by Lewontin, who expanded his work and made it more general. By assuming a simple form for the reproductive function (life history), Lewontin shows how differing patterns result in different rates of growth. He looked for types of reproductive strategy that would give advantage to populations colonizing a new area and hence free to grow. Lewontin recognized that growth would eventually have to stop as the new environment is settled. Lewontin also deals with aspects of interdemic selection, that is, selection among noninterbreeding demes for colonizing ability, seen as small changes in their life histories that would give them higher growth rates.

Other studies followed Lewontin's in looking at various aspects of selection for the scheduling of birth and death rates as they affected a population structured along the lines specified by Lotka (e.g., Meats, 1971; Emlen, 1970; Goodman, 1971; Demetrius, 1969). Meanwhile, other papers were dealing with density and population growth from the "ecological" viewpoint.

DENSITY AND NATURAL SELECTION

In his discussion of the fundamental theorem, Fisher discusses the effects of a deteriorating environment on the growth rate; this topic has been mentioned by various authors, notable examples of which would be Kimura (1958), Wright (1960), and Haldane (1953). However, the most significant and substantial paper to treat population density as a basic factor in natural selection was by MacArthur (1962). Since it deals with the fundamental theorem, and we do not wish to overweight this volume with discussions of that idea, we have not included MacArthur's paper here; his ideas are fully developed in many other readily available sources (MacArthur and Wilson, 1967; MacArthur, 1972; Emlen, 1973; Slobodkin, 1961; see also Hairston et al., 1970, and Pianka, 1972, for discussions of this topic).

Concepts of density-dependent growth were common in the ecological literature as they related to competition between populations or size changes within single populations, following Volterra and Lotka, but only beginning in 1971 does the problem of density-dependent natural selection seem to have been given sufficiently serious attention, after MacArthur's lead. In that year, at least four papers, three with nearly identical titles, dealt with this topic (by Roughgarden, Charlesworth, Anderson, and Nei). Paper 12 is representative of these approaches (Nei actually deals primarily with another topic). These papers expand on MacArthur's concepts or r and K selection. Paper 12 demonstrates that, if genotype-specific reproduction and growth are affected by population density, then differing density effects place different relative advantages of an r strategist, one that does competitively better at low population densities when growth potential is important, and a K strategist, which does better under crowded conditions. Much of the formulation of Paper 12 can be traced back to the original concepts of Volterra.

As MacArthur (1972) notes, if density effects on selection are such that the same genotype has a fitness advantage at all densities, then standard models of population genetics will suffice to explain the dynamics of the population. Furthermore, if the population spends most of its time at or near carrying capacity, this probably will be true. But if there are subpopulations colonizing marginal or new territories, or if there are rapid fluctuations in carrying capacity (e.g., a harsh seasonal environment), then the gene frequencies in a population or species can fluctuate considerably, with r or K strategists being favored depending on the conditions.

Concepts of growth, age structure, and density are explicitly treated in another series of recent papers, which we shall treat momentarily. However, the regulation of size of animal populations, competition between populations, and the nature of colonizing populations all involve the social behavior of animals. This has become one of the most active areas of present biology, and since many of these topics involve demographic considerations, it is worthwhile to look at them briefly.

SOCIAL ASPECTS

Competition between groups has been a subject of conversation, discussion, and scientific examination at least since the Social Darwinism of the nineteenth century. Certainly, Volterra's approach to population ecology and the use of Lotka's population concepts just discussed are interdemic models. When the demes in question are of the same species, we can consider that their genetic interaction is of demographic import.

A traditional approach to interdemic, or "group," selection would find one of several competing demes possessing some genetic advantage that would be reflected in a higher Malthusian parameter. The genes from this more rapidly growing deme would eventually swamp the gene pool of the species. When the successful group grows more rapidly owing to the higher reproductivity of individuals bearing the advantageous genes, standard genetic theory applies. However, there are apparent cases in nature where individuals seem to deny their own survival or reproduction in a way that allows the group as a whole to "survive" better. Various forms of altruism and self-sacrifice have been cited; the classic case is that of sterile castes in ants and bees.

The natural behavior of many animal species exhibits characteristics less dramatic than sterile castes but which still cause problems for Malthusian models of genetic competition, especially as they apply to demic competition. Darwin observed long ago that species do not grow at the maximum rate that their reproductivity could produce. Ethologists have recently developed much evidence that this is often largely due to the inhibition of reproduction rather than to overreproduction and massive mortality. Questions of self-regulation of animal numbers are of great importance (for a good discussion, see McLaren, 1971).

Paper 13 deals with the evolutionary genetic effects of several social aspects of animal populations and the way in which genetic conservatism is exercised by a population on its home range. Those who are excluded from this area by *social* means are the potential colonizers of new territories should they become available; they are also the "social mortality" that so intrigued Wynne-Edwards, for *animals excluded from a home territory do not fight to the death to prevent that ostracization.*

The social behavior of animals relates directly to mating patterns, vital rates, and territory. In short, almost everything demographic has a social connection; therefore, it would seem that any realistic appraisal of the evolution of things other than simple traits must increasingly become concerned with the effect of social behavior. Eventually, we must come to understand the evolution of social behavior, for social behavior accounts for at least as much effect on vital rates as do "natural" causes. Recent quantitative treatments of this topic may be found in van Valen (1971), Boorman and Levitt (1973), and Maynard-Smith (1972). These issues will come to dominate investigations of the major trends in animal evolution.

LIFE HISTORY, GROWTH, DENSITY: SYNTHESIS

We have seen the birth of age-specific selection models in the work of Haldane, Norton, and Fisher, who used the Malthusian treatment of population begun by Lotka. Also, we have traced the "ecological" concepts of population from the pioneering work of Lotka and Volterra. This work prompted later investigations of population growth rate that were ecological and "selective" in a general sense: they were interdemic models of competition which examined those reproductive schedules, that, given the total output, would maximize population growth.

These works which owe their intellectual heritage to Lotka and Volterra and to such workers as Gause, who used their mathematics, began to delve into the question of age distribution, but they did so in a clearly Malthusian context. We have mentioned that MacArthur and other ecologists were concerned with the effects of population densities and carrying capacities. Papers on these subjects appeared in the literature throughout the 1960s, but the year 1970 seems to have marked the beginning of studies that synthesize the Malthusian models on the one hand and the "logistic" models on the other.

Whereas the "demographic" approach treated the age structure of a population as an internal property of the aggregate (and few papers looked at this subject at all), the "ecological" approach treated populations as undifferentiated aggregates with various powers to grow or compete with other groups and an environment. Uniting these two approaches was conceptually simple: the competing populations of Volterra are now the genotypes within a population, and each is assigned its own set of vital rates that depend on population density (or other factors in the environment). The familiar logistic relationship to the environment is the most frequently used, but other forms of feedback also are applied.

We have somewhat exaggerated the difference between the "demographic" and "ecological" approaches in segregating papers by this criterion. However, although various authors have discussed the combined effects of regulated populations with age structure, we believe that only after about 1970 did the genetic structure of ecologically regulated populations with overlapping, intermarrying generations become a major concern of a substantial group of authors. No longer is the goal the prediction of ultimate or equilibrium gene frequencies or population sizes, which could be done with previous methods. Now the goals are to understand the way in which environmental effects on age-specific birth and death rates produce the specific characteristics of populations. The more or less independent approaches of previous decades have been explicitly combined in a serious way only in very recent years.

Anderson and King (Paper 14) and King and Anderson (1971) provide a matrix projection, or simulation, of an age-specific selection regime in a density-limited environment. These papers are a reflection, in our opinion, of a turning point in thinking about the relationships between growth, age structure, density, and genetics. Paper 14 deals with a Malthusian model of age-specific selection in which mortality and fertility rates for various genotypes vary. A Leslie matrix approach is used to project gene frequencies; because it can handle more factors than simpler approaches, it seems fairly certain that computer simulation will be a major tool in future research along these lines. The vital rates are then effectively damped by a standard logistic formula, so that each genotype has its own reproductive relationship to a particular carrying capacity and to the present population size. This is discussed in the context of r and K selection for the various genotypes. Placed in different environments, or at different times during the filling up of a territory, the genetic structure of the population will differ.

Paper 15 is the second in a continuing series of studies of the genetics of populations with overlapping generations, the first of which appeared in 1970 (Charlesworth, 1970). It represents a modern synthesis of the Volterran and Malthusian approaches regarding age-structured populations and is related directly to what we are calling demographic genetics. The first paper in the series set the historical and fundamental groundwork for the rest of the series. In it, Charlesworth builds on the work of Haldane, Norton, Fisher, and Kimura, and examines aspects of Fisher's fundamental theorem in relation to age-structured populations.

In a sexually mating, age-structured population, mating occurs between individuals who were born at different times. On the assumptions that genotype-specific fertility rates by age were the same for each genotype and that mating between individuals was independent with respect to their ages, Charlesworth shows that the fitness of each genotype will not in general be constant over time, and that Hardy–Weinberg proportions will not necessarily obtain even if there is random mating (demonstrating a point noted by others; e.g., see Turner, 1970). Charlesworth uses a definition of fitness derived from Kimura that differs from

Fisher's somewhat; he concludes that Fisher's fundamental theorem is not exactly true, and that deviation from it will be greatest if either selection is strong or the annual number of births fluctuates rapidly.

Paper 15 is largely devoted to demonstrating that changes in vital rates, "which may be nonspecific with respect to genotype," can produce significant changes in the genotype frequencies that exist, owing to the fact that each genotype has its own particular schedule of birth and death rates by age and will lead to a different equilibrium frequency for a different overall population growth rate. In growing populations, those genotypes with high reproduction early in life will proliferate relative to those reproducing later in life; in diminishing populations, the reverse will be true. Changes in genotype frequency observed in nature may thus not necessarily be related to specific genetic selection, but only to overall changes in the population growth.

Later papers in this series further develop these ideas and augment them by consideration of density dependence and other factors (Charlesworth, 1972, 1973, 1974; Charlesworth and Giesel, 1972b).

12

Copyright © 1971 by the Ecological Society of America

Reprinted from *Ecology*, **52**(3), 453–468 (1971)

DENSITY-DEPENDENT NATURAL SELECTION[1]

Jonathan Roughgarden

Department of Biology, University of Massachusetts, Boston 02116
Department of Biology, Harvard University, Cambridge, Massachusetts 02138

Abstract. Density-dependent selective values illustrate the evolutionary effect of population-regulating processes that diminish an individual's probability of survival with increased crowding. The selective values, assumed to decrease as a linear function of density, lead in a mild environment to the evolution of phenotypes having a high carrying capacity, *K*, at the expense of a low intrinsic rate of increase, *r*. A graphical technique shows that selection causes evolution of phenotypes having a high *r* at the expense of a low *K* in harsh seasonal environments.

A mathematical technique developed for analyzing evolution in coarse-grained seasonal environments reveals genetic mechanisms, including ones with full dominance, with which a moderately harsh seasonal environment causes stable polymorphism between high-*r* and high-*K* genes.

The energy balance equation demonstrates the role of high-*r* and high-*K* phenotypes in the population's energy flow. A high-*r* phenotype makes a large expected contribution to the population's productivity under conditions of negligible crowding, and a high-*K* phenotype has, for a given contribution to the population's productivity under uncrowded conditions, a low sensitivity to having that contribution diminished by crowding.

Ecologists (e.g., Birch 1960) have recognized that natural processes that cause population regulation also effect, by way of natural selection, the qualitative properties of the organisms in the regulated populations. Many of these regulatory processes dimish an individual's probability of survival with increasing population density, while others diminish an individual's expected fertility. This paper examines the evolutionary effect of population-regulating processes that diminish the individual's probability of survival. Clearly, for any natural selection to occur at all, the processes must diminish the probabilities of survival to a different extent for various genotypes present in the population.

Early theoretical work making reference to density effects in evolution was done by MacArthur (1962) and by MacArthur and Wilson (1967). These authors form a distinction between two kinds of population genetics, one in which fitness is defined as *r*, the intrinsic rate of increase, and one in which fitness is defined as *K*, the carrying capacity. They assert that the population genetics in which fitness is the intrinsic rate of increase applies to expanding populations under conditions of negligible crowding where population density has no influence on birth and death rates; "*r*-selection" is said to be occurring in such a situation. The population genetics in which fitness is the carrying capacity applies to crowded populations where "*K*-selection" is said to be occurring. Also, MacArthur and Wilson write "where climates are rigorously seasonal and winter survivors recolonize each spring, we expect *r*-selection . . . ; where climates are uniformly benign, *K*-selection . . . ." In addition, they interpret the result of *r*-selection to be the evolution of phenotypes having high pro-

ductivity, and of *K*-selection, phenotypes having high efficiency.

This present paper differs from and extends the earlier work of MacArthur in certain ways. First, the fitness is throughout measured by the selective value introduced by Sewall Wright. The analysis shows *r*-selection to occur in situations where phenotypes possessing high *r* while sacrificing *K* have the highest selective value, and *K*-selection to occur in situations where phenotypes possessing high *K* while sacrificing *r* have the highest selective value. Evolution in every case favors the genes producing phenotypes with the highest selective value, and hence one need not think there exist two (or more) kinds of population genetics each with its own definition of fitness. Second, this paper is primarily concerned with evolution in environments of intermediate harshness. It establishes the environmental conditions leading to stable coexistence between high-*r* and high-*K* genes, and it determines what happens in the region of environmental harshness marking the transition between *r*-selection and *K*-selection. Third, the study shows that *K*-selection results in phenotypes characterized not by a high efficiency, but by a low sensitivity to having their productivity diminished by crowding; however, the result of *r*-selection is indeed phenotypes with high productivity.

SIMULTANEOUS EQUATIONS FOR ΔN AND Δp

In this section we derive simultaneous equations to predict both the change in population size and gene pool composition at one locus from measurements of a single set of parameters. The equations are essentially those of Sewall Wright, and the reader familiar with his work may skip to the next section. We assume the population under study has a life

[1] Received May 14, 1970; accepted December 13, 1970.

153

history clearly divided into discrete generations. During any generation the population consists initially of zygotes, then of immature individuals, and finally of adults. Near the end of their lives the adults interbreed and the resultant crop of zygotes initiates the next generation. We chose to census the population at the time of interbreeding, as it seems more interesting to talk of the properties of adults, but another possible choice is to census the zygotes. The population is potentially polymorphic for a trait determined at one autosomal locus with two alleles, A and B. When the interbreeding occurs the mating is random with respect to the locus of interest.

The set of basic symbols is defined below:

N^i = total number of adults at time of interbreeding for ith generation

N^i_{AA} = number of adults with AA genotype at time of interbreeding for ith generation

N^i_{AB} = (similar to above)

N^i_{BB} = (similar to above)

$2m$ = average number of zygotes resulting from a mating—assumed to be independent of the parents' genotype at the locus of interest

l_{AA} = average fraction of zygotes with AA genotype surviving to take part in breeding

l_{AB} = (similar to above)

l_{BB} = (similar to above)

Note especially that the superscript i denotes the generation. It is *not* an exponent.

Since m, the average individual fertility, is independent of the genotype, selection occurs through differential survival only. Both m and the genotypic specific survival fractions, l_{AA}, etc., will in general depend on population size and gene pool composition; they are not constants. Since there are only three genotypes at the locus of interest,

$$N^i = N^i_{AA} + N^i_{AB} + N^i_{BB} \qquad (1)$$

The gene pool composition is described by the frequency of the A allele, p, or equivalently of the B allele, q. By definition,

$$p^i = \frac{2N^i_{AA} + N^i_{AB}}{2(N^i_{AA} + N^i_{AB} + N^i_{BB})}$$

$$q^i = \frac{2N^i_{BB} + N^i_{AB}}{2(N^i_{AA} + N^i_{AB} + N^i_{BB})} \qquad (2)$$

The frequencies of the various genotypes are defined as:

$$D^i = \frac{N^i_{AA}}{N^i}, \quad H^i = \frac{N^i_{AB}}{N^i}, \quad R^i = \frac{N^i_{BB}}{N^i} \qquad (3)$$

Clearly,

$$p^i + q^i = 1, \quad D^i + H^i + R^i = 1 \qquad (4)$$

$$p^i = D^i + (\tfrac{1}{2})H^i, \quad q^i = R^i + (\tfrac{1}{2})H^i$$

Table 1 presents the steps in deriving the numbers of each genotype present in the $(i + 1)$th generation from the population size and genotypic frequencies in the ith generation. The first column tabulates the number of matings between all possible pairs of parental genotypes, assuming random mating. The first entry in this column is formed, for example, as follows: the probability that any arbitrary mating in the ith generation is between an AA parent and another AA parent, where the parents are chosen at random, is $(D^i)^2$. Since there are N^i organisms, and each mates on the average once, there are $N^i/2$ total matings. Then the expected number of matings between two AA parents is $\frac{1}{2}(D^i)^2 N^i$. A similar argument is made for the other entries in the column. The progression through the remaining columns is self-explanatory.

One will observe from the last column in Table 1 that the parental fertility and the various offspring survival fractions appear together as a product. This fact allows definition for each genotype, say AA, of a number designated by $W_{AA}(N^i, p^i)$ and called the selective value of the phenotype expressed by the AA genotype in the ith generation. The selective values are defined according to

$$W_{AA}(N^i, p^i) = l_{AA}(N^i, p^i)m(N^i, p^i)$$
$$W_{AB}(N^i, p^i) = l_{AB}(N^i, p^i)m(N^i, p^i) \qquad (5)$$
$$W_{BB}(N^i, p^i) = l_{BB}(N^i, p^i)m(N^i, p^i)$$

The selective value is a very useful measure of the fitness of the phenotype expressed by a given genotype. (The selective value in this paper is the "absolute selective value" of Wright 1959, p. 112). The selective value is empirically determined, as the definition indicates, from the product of the average parental fertility and an average offspring survival fraction. Statistical analysis of data for any particular situation will usually show both the parental fertility and offspring survival fractions, and therefore the selective value, to be functions of the parental population size and gene pool composition. The entire range of selective values for various parental population sizes and gene pool compositions is given by a function of two variables: $W_{AA}(N,p)$, which, for application to any generation, say j, is evaluated at $N = N^j$ and $p = p^j$. Birch (1954) presents evidence for genotypes in *Drosophila* where the selective values are functions of density.

Using the last column of Table 1 and the definition of the gene frequency, p, from (2), we obtain

$$p^{i+1} = \frac{p^i W_{AA}(N^i, p^i) + q^i W_{AB}(N^i, p^i)}{\overline{W}(N^i, p^i)} p^i \qquad (6)$$

where

$$\overline{W}(N^i, p^i) = (p^i)^2 W_{AA}(N^i, p^i) + 2p^i q^i W_{AB}(N^i, p^i) + (q^i)^2 W_{BB}(N^i, p^i) \qquad (7)$$

TABLE 1. Derivation of the numbers of breeding adults of each genotype in the $(i + 1)$th generation from the genotypic proportions and population size in the ith generation

Parental genotypes	Average number of matings between given parental genotypes in ith generation	Average number of zygotes from matings of given parent combinations	Number of zygotes of various genotypes according to Mendelian ratios		
			AA	AB	BB
AA × AA	$(\frac{1}{2})(D^i)^2 N^i$	$(D^i)^2 N^i m$	$(D^i)^2 N^i m$		
AA × AB	$(\frac{1}{2}) 2 D^i H^i N^i$	$2 D^i H^i N^i m$	$D^i H^i N^i m$	$D^i H^i N^i m$	
AB × AB	$(\frac{1}{2})(H^i)^2 N^i$	$(H^i)^2 N^i m$	$(\frac{1}{4})(H^i)^2 N^i m$	$(\frac{1}{2})(H^i)^2 N^i m$	$(\frac{1}{4})(H^i)^2 N^i m$
AA × BB	$(\frac{1}{2}) 2 D^i R^i N^i$	$2 D^i R^i N^i m$		$2 D^i R^i N^i m$	
AB × BB	$(\frac{1}{2}) 2 H^i R^i N^i$	$2 H^i R^i N^i m$		$H^i R^i N^i m$	$H^i R^i N^i m$
BB × BB	$(\frac{1}{2})(R^i)^2 N^i$	$(R^i)^2 N^i m$			$(R^i)^2 N^i m$

Offspring genotypes	Total number of zygotes	Number of breeding adults in $(i + 1)$th generation
AA	$N^i m [(D^i)^2 + D^i H^i + (\frac{1}{4})(H^i)^2] =$ $N^i m [D^i + (\frac{1}{2}) H^i]^2 = N^i m (p^i)^2$	$l_{AA} m (p^i)^2 N^i$
AB	$N^i m [D^i H^i + (\frac{1}{2})(H^i)^2 + 2 D^i R^i + H^i R^i] =$ $N^i m 2 [D^i + (\frac{1}{2}) H^i][R^i + (\frac{1}{2}) H^i] = N^i m 2 p^i q^i$	$l_{AB} m 2 p^i q^i N^i$
BB	$N^i m [(R^i)^2 + H^i R^i + (\frac{1}{4})(H^i)^2] =$ $N^i m [R^i + (\frac{1}{2}) H^i]^2 = N^i m (q^i)^2$	$l_{BB} m (q^i)^2 N^i$

Similarly, using the last column of Table 1 and equation (1) we obtain:

$$N^{i+1} = \overline{W}(N^i, p^i) N^i \tag{8}$$

Equations (6) and (8) can be rewritten as

$$\Delta p^i = p^{i+1} - p^i =$$

$$\frac{p^i W_{AA}(N^i, p^i) + q^i W_{AB}(N^i, p^i) - \overline{W}(N^i, p^i)}{\overline{W}(N^i, p^i)} p^i \tag{9}$$

$$\Delta N^i = N^{i+1} - N^i = [\overline{W}(N^i, p^i) - 1] N^i \tag{10}$$

Equations (9) and (10) are the simultaneous equations for the change in population size and gene frequency over a generation. An extremely important property of these equations is that measurements of a single set of parameters, the selective value functions, $W_{AA}(N,p)$, $W_{AB}(N,p)$ and $W_{BB}(N,p)$, enable prediction of both ΔN and Δp.

One of the interesting uses for the equations predicting gene frequency and population size is in determining trajectories on the phase plane. The phase plane is a graph whose horizontal axis is scaled for p and vertical axis for N. A trajectory indicates the entire change in N and p over many consecutive generations starting from some initial values, N^0 and p^0. Clearly, an entire trajectory is produced by successive applications of equations (6) and (8) starting with N^0 and p^0.

THE MODEL

In the last section the selective values, e.g., $W_{AA}(N,p)$, were regarded as being density and frequency dependent without any additional assumption specifying the functional form of that dependence. To assign some particular function as a selective

value is to invoke a particular model of density- and frequency-dependent selective pressures. In this section the selective values will express what seems the simplest model of density-dependent selection. The selective values are assumed to drop off linearly with population size, as in the equation $W(N) = a - bN$. But instead of writing the parameters as a and b, we write $(r + 1)$ and (r/K), for reasons that will be clear shortly. Thus the selective values are:

$$W_{AA}(N) = (r_A + 1) - (r_A/K_A)N$$
$$W_{AB}(N) = (r_H + 1) - (r_H/K_H)N \tag{11}$$
$$W_{BB}(N) = (r_B + 1) - (r_B/K_B)N$$
$$r_A > r_B, \quad K_B > K_A$$

Refer in Fig. 1 to the graphs of $W_{AA}(N)$ and $W_{BB}(N)$. At low population sizes the AA homozy-

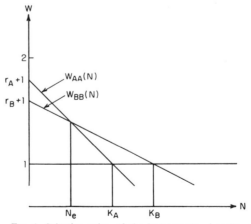

FIG. 1. Selective value of the homozygote phenotypes as a function of population density.

gotes are more fit while at high population sizes the BB homozygotes are more fit. The intercepts are labeled in terms of r and K. A is the high-r allele, and B the high-K allele. N_e designates the population size at which the selective values of both homozygotes are equal.

The selective values determine simultaneously both gene pool evolution, Δp, and population growth, ΔN. The meaning of the parameters, r and K, in the homozygote selective value functions, $W_{AA}(N)$ and $W_{BB}(N)$, can be deduced by considering the population growth entailed by these functions if the locus is fixed for either allele. Suppose the locus is fixed for the A allele. Then substituting $p = 1$, and the selective value functions, (11), into the equation for ΔN, (10), yields

$$\Delta N = r_A N \frac{(K_A - N)}{K_A} \qquad (12)$$

Equation (12) describes logistic growth with r_A as the intrinsic rate of increase, and K_A the carrying capacity. A similar argument exists for r_B and K_B. Fig. 2 illustrates growth curves for populations consisting entirely of AA homozygotes or BB homozygotes. As illustrated, the growth curves predicted by the logistic difference equation lead to an asymptotically stable equilibrium at $N = K$ provided that r is between 0 and 1, and hence we will restrict r to within (0,1). (See Smith 1968, p. 26.) A selective value thus has two complementary interpretations. The first, of special relevance to evolutionary biology, is as a measure of the fitness of a phenotype; the second, of special relevance to population dynamics, is as a determinant of the growth characteristics of a population containing that phenotype.

The existence of these two complementary interpretations to the selective value is of practical importance because the number of methods for measuring selective values is increased. The original method of measuring a selective value function, say $W_{AA}(N)$, following the definition of the selective value in equations (5), consists of determining the

product of the individual fertility and survival fraction, (ml_{AA}), at various densities. The best fit of this data to a line with negative slope yields an equation in N whose intercept at $N = 0$ is $(r_A + 1)$ and which equals one at $N = K_A$. An alternative method now possible is to isolate a true breeding strain with the phenotype of the AA homozygotes, and to observe the population growth curve. The best fit of the population growth data to a logistic equation provides the values for r_A and K_A.

Two complementary interpretations also exist for the heterozygote selective value. However, the method of measuring the selective value from observations on population growth may need to be somewhat contrived. One must either obtain a true breeding strain with the phenotype of the AB heterozygotes—a somewhat unlikely possibility unless perhaps convergent phenotypes are drawn from populations of different evolutionary stock—or, using actual AB heterozygotes, one must observe population growth under conditions such that he substitutes offspring of AB genotype for any offspring produced of other genotypes before differential survival is significant.

The evolutionary result of density-dependent natural selection in any environment is contingent on the fitness of the heterozygotes relative to the two homozygotes. The set of possible relations between the fitness of the heterozygote and fitness of the homozygotes generates a set of different cases of density-dependent selection, many requiring separate analysis. All the cases are listed in Table 2. The situation described by most of the cases is self-explanatory except for cases 11, 12, and 13, which all involve a heterozygote intermediate in both r and K. Referring to Fig. 3, it is clear that $W_{AB}(N)$ may pass above, through, or below the point where $W_{AA}(N)$ intersects $W_{BB}(N)$. If $W_{AB}(N)$ passes above this point of intersection, then there is an interval of population sizes in which the heterozygote

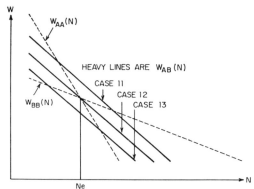

FIG. 3. Selective value of the heterozygote phenotype as a function of population density for the three cases where the heterozygote is intermediate in both r and K.

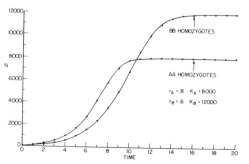

FIG. 2. Population growth for populations consisting entirely of AA or BB homozygotes.

TABLE 2. Possible cases of heterozygote fitness

Case	Properties of heterozygote	Requirement on r_H and K_H	
1	Same as High-r homozygote	$K_H = K_A,$	$r_H = r_A$
2	Same as High-K homozygote	$K_H = K_B,$	$r_H = r_B$
3	Superior K, Superior r	$K_H > K_B,$	$r_H > r_A$
4	" Intermediate r	"	$r_B < r_H < r_A$
5	" Inferior r	"	$r_H < r_B$
6	Inferior K, Superior r	$K_H < K_A,$	$r_H > r_A$
7	" Intermediate r	"	$r_B < r_H < r_A$
8	" Inferior r	"	$r_H < r_B$
9	Intermediate K, Superior r	$K_A < K_H < K_B,$	$r_H > r_A$
10	" Inferior r		$r_H < r_B$
	Intermediate K, Intermediate r		$r_B < r_H < r_A$
11	Heterozygote superior at some densities	$W_{AB}(N_e) > W_{AA}(N_e)$	
12	Heterozygote intermediate at all densities	$W_{AB}(N_e) = W_{AA}(N_e)$	
13	Heterozygote inferior at some densities	$W_{AB}(N_e) < W_{AA}(N_e)$	

is superior to both homozygotes. If $W_{AB}(N)$ passes below this point of intersection, then there is an interval where the heterozygote is inferior. Only if $W_{AB}(N)$ actually passes through the point is the heterozygote intermediate in fitness at all population sizes (except N_e).

The evolutionary outcome of density-dependent natural selection is analyzed first in a constant environment. A constant environment need not be a theoretical abstraction. A natural enviroment is understood as constant, with respect to a particular species, if in that environment consecutive generations of the species face the same kinds of selective pressures. Since selective pressures are specified with the selective functions, a constant environment can be defined as one where the selective value functions for consecutive generations have the same functional form and whose parameters keep the same numerical values. For the model of density-dependent selective pressures, the selective values through consecutive generations must remain of the form $W(N) = (r + 1) - (r/K)N$, and the numerical values of r and K must remain the same. Thus a natural environment having much spatial and temporal heterogeneity may very well be constant for some species simply because species there receive the same kinds of selective pressures in successive generations.

Full dominance, cases 1 and 2

The case where the high-r allele, A, is fully dominant is presented in Fig. 4; where the high-K allele, B, is fully dominant in Fig. 5. The family of trajectories for both cases include certain common features:

(1) All trajectories of the population converge eventually to an asymptotically stable equilibrium point with the population size equaling the larger of the carrying capacities and with the gene pool fixed for the allele yielding the larger carrying capacity.

(2) During the early generations of any trajectory, the population always shows an increase in the *proportion* of high-r phenotypes so long as the population size is less than N_e. N_e is the population size at which the selective values of the high-r and high-K phenotypes are equal. The *proportion* of high-r phenotypes decreases whenever the population size is greater than N_e.

(3) During the early phases of any trajectory the population growth is rapid with both phenotypes increasing in *number*. A later phase is entered when the population size exceeds the smaller of the two carrying capacities, K_A. During the late phase the *number* of high-r phenotypes decreases because the selective value of that phenotype is then less than unity. The population does not grow rapidly and evolution can be interpreted as a substitution of high-K phenotypes for high-r phenotypes in a population of approximately constant size.

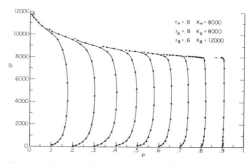

FIG. 4. Trajectories assuming the high-r allele, A, is dominant. N is the population size. The abscissa, p, is the gene frequency of A. Populations founded with 100 individuals containing various initial gene frequencies progress after each generation to the next higher dot on their trajectories; they all eventually converge at $N = 12,000$ and $p = 0$.

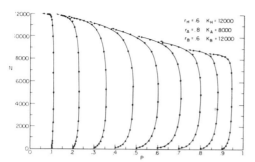

FIG. 5. Trajectories assuming the high-K allele, B, is dominant.

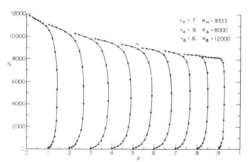

FIG. 6. Trajectories assuming the heterozygote is intermediate in fitness at all population sizes.

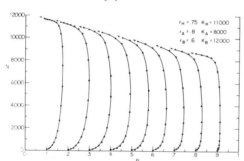

FIG. 7. Trajectories assuming the heterozygote is superior in fitness at medium population sizes but is intermediate in fitness at high and low population sizes.

The family of trajectories for both cases shows the following differences:

(1) The plateau across the phase plane between the levels of the two carrying capacities, K_B and K_A, is at its highest elevation for the high-K allele dominant case. This result occurs because the recessive high-r allele, when present at lower frequencies, is very likely to be found in a heterozygote where it is not expressed. Consequently, the population size is large since most of the individuals possess phenotypes permitting a large carrying capacity.

(2) The number of generations required to achieve

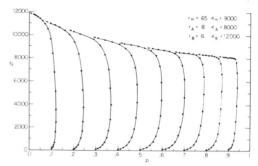

FIG. 8. Trajectories assuming the heterozygote is inferior in fitness at medium population sizes but is intermediate in fitness at high and low population sizes.

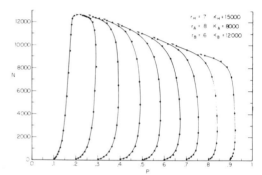

FIG. 9. Trajectories assuming the heterozygote has the largest K and is thus superior in fitness to both homozygotes at high population sizes. The trajectories converge at a stable polymorphism.

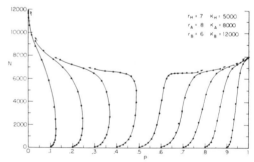

FIG. 10. Trajectories assuming the heterozygote has the lowest K and is thus inferior in fitness to both homozygotes at high population sizes. The eventual outcome of selection depends on the composition and size of the founding population.

fixation of the high-K allele, starting from intermediate gene frequencies, is highest for the high-K allele dominant case. This result is another instance of the generalization that more time is required to select against a recessive allele than a dominant allele because the presence of a recessive allele is masked in heterozygotes.

Intermediate heterozygote, cases 11, 12, and 13

The case where the heterozygote is intermediate at all population densities is presented in Fig. 6. The entire pattern of the trajectories is intermediate between the two extremes of dominance. The plateau is not at as high an elevation as in the high-K dominant case but is higher than in the high-r dominant case. The time required for a given degree of high-K allele fixation is similarly intermediate. It is important to note that the values of r_H and K_H necessary to produce this fully intermediate case are not themselves intermediate. For example, the figures presented all use $r_A = .8$, $r_B = .6$, $K_A = 8,000$, $K_B = 12,000$, but a heterozygote with $r_H = .7$ and $K_H = 10,000$ would be superior over a definite interval of population sizes, and hence would not be an instance of this fully intermediate case.

The case where the heterozygote is superior for an interval of population densities is shown in Fig. 7. The pattern differs from Fig. 6 at intermediate densities in that the trajectories lean more toward the center of the graph. For example, the segment of each trajectory between N equaling 3,000 to 7,500 and originating from p equal to .1, .2, or .3 inclines more steeply to the right in Fig. 7 than in Fig. 6, while a similar segment of each trajectory originating from p equal to .7, .8, or .9 inclines more steeply to the left. This difference does not affect the outcome of selection in a constant environment but will prove important when considering a seasonal environment later.

The case where the heterozygote is inferior over some population densities appears in Fig. 8. The pattern differs from Fig. 6 in that at intermediate densities the trajectories lean away from the center.

Heterozygote with intermediate K, cases 9 and 10

The cases where the heterozygote has an intermediate K lead to essentially the same patterns as illustrated in Fig. 4 through 8. The major differences are that with a superior r, the trajectories at low densities lean toward the center of the graph, while with an inferior r, the trajectories at low densities lean away from the center.

Heterozygote with superior K, cases 3, 4, and 5

The pattern of trajectories produced with a superior heterozygote K is shown in Fig. 9. The pattern is different from the previous. The trajectories converge at some non-zero gene frequency maintaining a stable polymorphism. The population size at the stable equilibrium point is less than K_H, the carrying capacity of a true breeding strain with the heterozygote phenotype, and greater than K_B, the larger of the homozygote carrying capacities. The equilibrium values of N and p are obtained by letting both $\Delta N = 0$ and $\Delta p = 0$ in equations (9) and (10) and

solving for the resultant N and p. The pattern leading to a stable polymorphism occurs regardless of the relative magnitude of r_H, ($r < 1$). In spite of the special pattern of trajectories associated with a superior K heterozygote, the result of natural selection is, as before, the maintenance of the phenotype with the highest K in as high a proportion as possible in population.

Heterozygote with inferior K, cases 6, 7, and 8

The pattern of trajectories for a heterozygote with an inferior K is presented in Fig. 10. This pattern also is quite different from any of the patterns previously discussed. The trajectories lead to fixation of either the high-r or high-K allele depending upon the initial gene frequency. This pattern occurs regardless of the relative magnitude of r_H, ($r < 1$).

GRAPHICAL ANALYSIS OF SEASONAL ENVIRONMENT

Many, perhaps most, species of organisms in the world live in an environment with intense seasonal variability. Do the conclusions derived in a constant environment still apply for a seasonal environment? Records of environmental variation obtained with thermometers, etc., are not enough to entail that the previous constant environment analysis is inappropriate for any given species. The necessary and sufficient condition that the constant environment treatment be inappropriate is that members of consecutive generations in the given species face selective pressures specified by different equations, or by the same equations but with changed parameter values. Two examples: in one generation the selective values show negative linear-density dependence and in the next generation follow some other equation; in one generation the r's and K's take one set of values and in the next generation an alternate set of values. Clearly this condition disallowing a constant environment analysis bears more heavily in some groups of organisms than in others. One can imagine the same habitat being a constant environment to a vertebrate while being a seasonal environment to an insect.

We will assume, for simplicity, a particular kind of seasonal environment. A very distinct growing season and season of dormancy occur in this environment. Throughout the growing season the density-dependent selective values obey equations (11) and the population changes just as if it were in a constant environment. However, during the dormancy season the population does not function well, with the result that at the end of each dormancy season the population is left with only N^0 individuals possessing the gene frequency which occurred at the end of the preceding growing season. Thus, each growing season starts with a founding population of N^0 individuals with the gene pool left from the end of the preceding growing season. Considerable attention will be focused on N^0 which, for a growing season of some given number of generations, serves as a good measure of the harshness of the environment. The seasonal environment described above lends itself very readily to analysis by a graphical technique.

Full dominance

Consider first the case where the high-r allele is dominant. The trajectories appear in Fig. 4. One must specify two parameters of the seasonal environment in advance: N^0, the size of the founding population at the beginning of each growing season, and L, the length of the growing season in generations. Initially, let L be two generations. Then consider a harsh environment that causes N^0 to be 2,500. Draw a horizontal line across the phase plane at the level of N equal to 2,500. Next estimate the vertical distance the population will have moved after two generations starting at N equal to 2,500. This new vertical distance is about at N equal to 5,500. Draw a horizontal line at this level. One can now estimate the trajectory of the population through successive growing seasons. Suppose the population starts out originally at N equal to 2,500 and p equal to .1; then at the end of the first season N is about 5,500 and p is about .12. The beginning of the next growing season is entered with p equal to .12 and N equal to 2,500, and so on. The result after successive seasons is that p gets progressively larger until eventually the high-r allele is fixed. The same result will clearly occur if the environment is harsher making N^0 lower, assuming L is still two generations.

Now consider a milder environment with N^0 equal to 3,900 but with L still two generations. For this environment successive growing seasons lead eventually to fixation of the high-K allele. Clearly the same result will occur in milder environments with growing seasons of two generations in length. Finally, consider an environment with N^0 equal to 2,500 again but with L set at four generations. For this environment evolution in successive growing seasons leads to fixation of the high-K allele.

Thus, the outcome of natural selection in a harsh seasonal environment may be entirely opposite to that in the constant environment because the high-r allele can be fixed instead of the high-K allele. However, the long-term intensity of natural selection in favor of the high-r allele is diminished as the dormancy season becomes milder and/or as the growing season becomes longer. Indeed, if the dormancy season is mild enough or the growing season long enough, the high-K allele will be fixed instead of the high-r allele. These results are also true for the case where the high-K allele is dominant, as presented in Fig. 5. The limitation of the graphical technique is clearly one of accuracy. It is impossible to determine graphically the degree of harshness marking the transition between selection for the high-r allele instead of the high-K allele. This question and others of a quantitative sort about the seasonal environment are deferred until the next section.

Intermediate heterozygote

The case where the heterozygote shows superiority through an interval of intermediate densities has some interesting properties in a seasonal environment. By applying the graphical method to Fig. 7, one can verify that with reasonably short growing seasons, about two to four generations, three results can occur depending upon the harshness of the environment. If the environment is very harsh, with N^0 in the range of 1,500 and below, then the high-r allele is either fixed, or exists in very high frequency. If the environment is very mild, with N^0 in the range of 6,000 and above, then the high-K allele is either fixed or exists in very high frequency. But if the environment is of intermediate harshness with N^0 in the range of 1,500 to 6,000, then a significant stable polymorphism results. A more accurate analysis is presented in the next section.

The case where the heterozygote shows inferiority through an interval of intermediate densities can also show three results with reasonably short growing seasons depending upon the harshness of the environment. A very harsh environment results in fixation of the high-r allele, a mild environment in fixation of the high-K allele, while an environment of intermediate harshness results either in fixation of the high-r allele or the high-K allele, depending on the original gene frequency in the population.

The case where the heterozygote is of intermediate fitness at all densities shows the same results as the cases with full dominance previously discussed.

Additional cases

The additional cases involve a heterozygote superior in r or inferior in r. As might be expected, in a harsh environment with a short growing season, evolution with a heterozygote superior in r leads to a stable polymorphism, while with a heterozygote inferior in r, evolution leads either to fixation of the high-r or of the high-K allele, depending upon the original gene pool frequency. These evolutionary results for a harsh environment with a short growing season are independent of the value of the heterozygote K.

Let us agree to call selection which results in evolution of high-K phenotypes at the expense of a low r as "K-selection," and that which results in evolution of high-r phenotypes at the expense of a low K as "r-selection." The model, as discussed so far, clearly indicates that K-selection occurs in constant and mild seasonal environments and r-selection in harsh seasonal environments with short growing seasons. However, K-selection and r-selection do not exhaust the possible outcomes, for we have already seen one case, with a phenotypically intermediate heterozygote, where density-dependent selection in a seasonal environment of medium harshness results in polymorphism, and more such cases are discussed in the next section.

Another kind of environment leading to r-selection, discussed by Gadgil and Bossert (1970), presumes some continuously operating cause of mortality extrinsic to the population itself. This mortality is assumed to maintain the population density at a low enough level so that the selective value of the high-K phenotype is still less than that of the high-r phenotype. Such an environment leads to the fixation of the high-r gene. This environment would be appropriate, for example, to weeds in a field where mowing regularly holds the plant density to a low value.

MacArthur and Wilson (1967) proposed that there should exist a latitudinal gradient in r-selection and K-selection; that proportionally more species in tropical latitudes face K-selection instead of r-selection than do species of temperate latitudes. Scientists working in the tropics often doubt the truth of this hypothesis because the evident seasonal variability in tropical habitats appears to disqualify these habitats from being the uniformly benign environments where K-selection was supposed to occur. But, as we have seen, seasonal variability, per se, does not rule out K-selection. It is the magnitude of harshness that is critical. And it seems true that the harshness of the seasonal variability in most (not all) tropical habitats is not as large as in comparable temperate habitats. So the hypothesis that there is a latitudinal gradient in r-selection and K-selection should be taken seriously. However, the gradient should only occur in certain groups, groups to whom the seasonal variability appears coarse grained. Long-lived vertebrates are probably facing K-selection all over the world, but it may be otherwise with some insects and other invertebrates with short life span.

The hypothesis of a latitudinal gradient in r-selection and K-selection is testable. We know, for example, that growth of genetically homogeneous populations of *Drosophila melanogaster* follows the logistic curve (Buzzati-Traverso 1955, p. 178). It would be interesting to see whether the r's of different *Drosophila* populations are positively correlated with latitude and the K's negatively correlated.

MATHEMATICAL ANALYSIS OF SEASONAL ENVIRONMENT

We now turn to the mathematical analysis of evolution in seasonal environments.

The distinguishing feature of the seasonal environment is that consecutive generations are expected to face different regimes of selective pressures. The eventual outcome of evolution can only be anticipated by knowing the long-term fitness of the various phenotypes. The long-term fitness must refer to the performance of a phenotype over the number, L, of generations required for the entire expected regime of selective pressures to be realized. This number, L, is called the period of the environment, and the seasonal environment is seen as a continuing repetition of a sequence of L regimes of selective pressures. One approach to the analysis of the seasonal environment is to construct a measure of the long-term fitness from the set of single-generation selective values representing each of the expected regimes of natural selection.

Suppose initially that the period of the environment is two generations. The two regimes of selective pressures that occur are represented by two sets of selective value functions, e.g., $W_{AA,0}(N, p)$ and $W_{AA,1}(N, p)$. The numerical subscript refers to a position in the sequence of occurrence of selective regimes; thus within any period $W_{AA,0}(N, p)$ acts for the initial generation and $W_{AA,1}(N, p)$ for the next. Let p^0 be the frequency of the A allele at the beginning of a period, p^1 be the frequency in the middle, and p^2 the final frequency. (Remember, superscripts in italics are not exponents; they serve to identify the generation.) Then from the definition of the selective values (5) and the considerations expressed back in Table 1,

$$N^2_{AA} = W_{AA,1}(N^1, p^1)(p^1)^2 N^1$$
$$N^2_{AB} = W_{AB,1}(N^1, p^1)2p^1q^1 N^1 \qquad (13)$$
$$N^2_{BB} = W_{BB,1}(N^1, p^1)(q^1)^2 N^1$$

However [see equations (6) and (8)],

$$p^1 = \frac{p^0 W_{AA,0}(N^0, p^0) + q^0 W_{AB,0}(N^0, p^0)}{\overline{W}_0(N^0, p^0)} p^0$$

$$q^1 = \frac{q^0 W_{BB,0}(N^0, p^0) + p^0 W_{AB,0}(N^0, p^0)}{\overline{W}_0(N^0, p^0)} q^0$$

$$(14)$$

$$N^1 = \overline{W}_0(N^0, p^0) N^0$$

Substituting (14) into (13) yields expressions for the genotypic numbers at the end of an environmental period in terms of p and N at the beginning of the period. These expressions are of the form

$$N^2_{AA} = A_{AA,2}(N^0, p^0)(p^0)^2 N^0$$
$$N^2_{AB} = A_{AB,2}(N^0, p^0)2p^0q^0 N^0 \qquad (15)$$
$$N^2_{BB} = A_{BB,2}(N^0, p^0)(q^0)^2 N^0$$

where

$$A_{AA,2}(N^0, p^0) = W_{AA,1}(N^1, p^1) \times$$
$$\frac{[p^0 W_{AA,0}(N^0, p^0) + q^0 W_{AB,0}(N^0, p^0)]^2}{\overline{W}_0(N^0, p^0)}$$

$$A_{AB,2}(N^0, p^0) = W_{AB,1}(N^1, p^1) \times$$
$$\frac{[p^0 W_{AA,0}(N^0, p^0) + q^0 W_{AB,0}(N^0, p^0)] \times}{[q^0 W_{BB,0}(N^0, p^0) + p^0 W_{AB,0}(N^0, p^0)]}$$
$$\overline{W}_0(N^0, p^0)$$

$$A_{BB,2}(N^0, p^0) = W_{BB,1}(N^1, p^1) \times$$
$$\frac{[q^0 W_{BB,0}(N^0, p^0) + p^0 W_{AB,0}(N^0, p^0)]^2}{\overline{W}_0(N^0, p^0)}$$

(16)

The functions defined in (16) are the net selective values for a seasonal environment with a period of two generations. The net selective value (n.s.v.) functions appear in equations for the change in N and p over a seasonal cycle that are exactly parallel to the equations for ΔN and Δp over one generation. By arguments like those used in deriving (6) and (8) one obtains

$$p^2 = \frac{p^0 A_{AA,2}(N^0, p^0) + q^0 A_{AB,2}(N^0, p^0)}{\overline{A}_2(N^0, p^0)} p^0$$

(17)

$$N^2 = \overline{A}_2(N^0, p^0) N^0$$

(18)

Equations (17) and (18) are exactly the same in form as (6) and (8) and, indeed, the similarity is fundamentally due to the parallel form of (13) and (15).

The n.s.v. functions derived here differ systematically from the "coarse grained adaptive function" proposed by Levins (1969). Levins' adaptive function for L generations is a product of L selective values, while equations (16) indicate that with L equaling two, an n.s.v. is the product of three selective values. This discrepancy increases as L becomes larger. It is not clear whether conclusions obtained as the condition for maximizing Levins' adaptive function are compatible with those obtained using the n.s.v.'s in equations (17) and (18).

An interesting property of the n.s.v. functions is that even with full phenotypic dominance the n.s.v. of the heterozygote does not equal that of the dominant homozygote. Full phenotypic dominance entails, assuming A is dominant, that $W_{AA,0}(N, p) = W_{AB,0}(N, p)$ and that $W_{AA,1}(N, p) = W_{AB,1}(N, p)$. Substituting these equalities into (16) indicates that $A_{AA,2}(N, p) \neq A_{AB,2}(N, p)$. These n.s.v.'s are not equal because the expected offspring from heterozygotes differ considerably from those of homozygotes even though their phenotypes otherwise are the same. The n.s.v., in covering more than one generation, takes accounts of this fundamental difference between heterozygotes and homozygotes.

The n.s.v. functions for a seasonal environment with an arbitrary period, say $j+1$, can be generated with the following recursive rules:

$$A_{AA,j+1}(N^0, p^0) = W_{AA,j}(N^j, p^j) \times$$
$$\frac{[p^0 A_{AA,j}(N^0, p^0) + q^0 A_{AB,j}(N^0, p^0)]^2}{\overline{A}_j(N^0, p^0)}$$
$$A_{AB,j+1}(N^0, p^0) = W_{AB,j}(N^j, p^j) \times$$
$$\frac{[p^0 A_{AA,j}(N^0, p^0) + q^0 A_{AB,j}(N^0, p^0)] \times}{[q^0 A_{BB,j}(N^0, p^0) + p^0 A_{AB,j}(N^0, p^0)]}$$
$$\overline{A}_j(N^0, p^0)$$

(19)

$$A_{BB,j+1}(N^0, p^0) = W_{BB,j}(N^j, p^j) \times$$
$$\frac{[q^0 A_{BB,j}(N^0, p^0) + p^0 A_{AB,j}(N^0, p^0)]^2}{\overline{A}_j(N^0, p^0)}$$

The N^j, and p^j appearing in (19) are obtained after j iterations with (14) on N^0 and p^0 using the appropriate selective values in each iteration. An n.s.v. for an environment with a period of 1 generation is simply the usual selective value itself, so, $A_{AA,1}(N, p) = W_{AA,0}(N, p)$, etc.

The n.s.v.'s can be used to solve for a kind of steady state in a seasonal environment, obtained when $p^2 = p^0$ and $N^2 = N^0$ in equations (17) and (18). The defining characteristic of this steady state is that N and p are the same at the start of consecutive seasons, although they change values within a season. Solving for this steady state is especially easy in the sort of seasonal environment we have been considering, where each growing season begins with a population size N^0. In this environment N^0 ceases to be a variable and is instead a parameter, and (18) is then suppressed. The condition for the (nontrivial) steady state from (17) is that $(p A_{AA} + q A_{AB})/\overline{A} = 1$. However $(p A_{AA} + q A_{AB} - \overline{A})$ can be rewritten as $q[p(A_{AA} - A_{AB}) - q(A_{BB} - A_{AB})]$. So an equilibrium is achieved at $\hat{p}^0$ obtained from
$$\hat{p}^0[A_{AA}(N^0, p^0) - A_{AB}(N^0, p^0)]$$
$$- \hat{q}^0[A_{BB}(N^0, p^0) - A_{AB}(N^0, p^0)] = 0 \quad (20)$$

Equation (20) brings out the conclusion that a steady state will occur in a seasonal environment at some frequency only if the heterozygote n.s.v. at that frequency is superior or inferior to both homozygote n.s.v.'s there.

The conditions for stability of the equilibrium in the seasonal environment are not so readily established. Denoting the RHS of (17) by $f(p^0)$, the stability properties at equilibrium are determined from $f'(\hat{p}^0)$, i.e., the derivative of $f(p^0)$ evaluated at equilibrium. If $f'(\hat{p}^0)$ is in $(0, 1)$ perturbations from equilibrium return smoothly, in $(-1, 0)$ the return is with damped oscillations, and if $|f(\hat{p}^0)| > 1$ the equilibrium is unstable. The condition for stability requires, when the selective values do not depend on p, that the heterozygote must be superior, but the situation is more complicated when the selective values do depend on p. We consider later an equilibrium which is stable and where the heterozygote n.s.v. is inferior to both homozygote n.s.v.'s. Two kinds of cases are considered next, one where there is full dominance, the other where the heterozygote is a phenotypic intermediate.

Full dominance

We now show that when the high-r allele, A, is dominant there are certain levels of environmental harshness, measured by N^0, in which the heterozygote n.s.v. is inferior to both homozygote n.s.v.'s. Recall that N_e is the population density at which $W_{AA} = W_{BB}$. If $N^0 < N_e$ and $N^1 > N_e$, then the dominant phenotype, AA or AB, is fittest in the initial generation and the recessive phenotype, BB, is fittest in the next generation of the growing season. When A is dominant

$$A_{AA}(N^0, p^0) - A_{AB}(N^0, p^0) = \{W_{AA}(N^0, p^0) - [q^0 W_{BB}(N^0, p^0) + p^0 W_{AA}(N^0, p^0)]\} \times W_{AA}(N^1, p^1) p^1 / p^0$$

$$A_{BB}(N^0, p^0) - A_{AB}(N^0, p^0) = \{W_{BB}(N^1, p^1) \times [q^0 W_{BB}(N^0, p^0) + p^0 W_{AA}(N^0, p^0)] - W_{AA}(N^1, p^1) W_{AA}(N^0, p^0)\} q^1 / q^0$$

When $N^0 < N_e$ and $N^1 > N_e$, $A_{AA} - A_{AB}$ is positive for all p^0 making the heterozygote n.s.v. always inferior to the AA homozygote n.s.v. If p^0 is close to 1, $A_{BB} - A_{AB}$ is positive if

$$[W_{BB}(N^1, p^1) - W_{AA}(N^1, p^1)] W_{AA}(N^0, p^0)$$

is positive, which is always true when $N^1 > N_e$. If p^0 is close to zero, $A_{BB} - A_{AB}$ is positive if

$$W_{BB}(N^1, p^1) W_{BB}(N^0, p^0) - W_{AA}(N^1, p^1) W_{AA}(N^0, p^0)$$

is positive, and this condition is true when, so to speak, N^0 is not as far below N_e as N^1 is above. So if N^0 is low, the heterozygote n.s.v. is inferior to the BB homozygote n.s.v. only at high p^0, but as N^0 approaches N_e the heterozygote is inferior for an increasingly wider interval of p^0. These results satisfy a necessary condition for the existence of polymorphism caused by the seasonal environment.

By similar reasoning it can be shown that when the high-K allele, B, is dominant, the heterozygote n.s.v. is superior to both homozygote n.s.v.'s when $N^0 < N_e$ and $N^1 > N_e$, and again a necessary condition for the existence of polymorphism is satisfied.

To determine if such polymorphism actually exists (20) was solved numerically for a range of N^0 and the derivative of (17) examined at any equilibrium obtained to see if the equilibrium was stable. The results, presented in Fig. 11, indicate that there is a stable polymorphism over a small range of environmental harshness for both cases of full dominance. The degree of stability, as measured by the rate of return to equilibrium after a small perturbation, is low. The deviation from equilibrium after t growing seasons, $d(t)$, given an initial small displacement d_0, is given by $d(t) = [f'(\hat{p}^0)]^t d_0$ where, as before, $f'(\hat{p}^0)$ is the derivative of (17) evaluated at equilibrium. In the numerical cases presented in Fig. 11, $f'(\hat{p}^0)$ at its stablest was .9989. Fig. 11 shows that where the high-r allele A is recessive its equilibrium frequency, $\hat{p}^0$, is higher for a given N^0 than where A is dominant. Hence for a given harshness, N^0, the observable difference in phenotypic proportions produced by the two genetic mechanisms may be small.

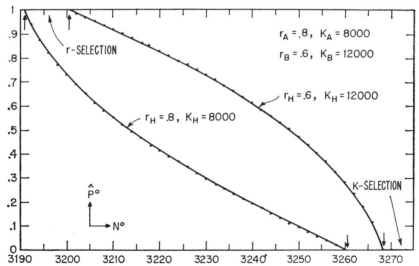

Fig. 11. Polymorphism frequency, $\hat{p}^0$, as a function of environmental harshness, assuming full dominance. Growing season is two generations long. Arrows indicate the end points of the harshness intervals.

The existence of this polymorphism caused by a seasonal environment can be explained by, and could perhaps have been deduced from, an extension of the work by Haldane and Jayakar (1963) on selection in changing environments. They consider a coarse-grained environment where the fitness of the phenotypes varies among different generations, but the fitness for every generation is a constant, i.e., neither frequency nor density dependent. However, in the model considered here, the fitness in the second generation, e.g., $W_{AA}(N^1)$, is a function of the frequency at the beginning of the growing season because $N^1 = \overline{W}(N^0, p^0)N^0$. (Remember, N^0 is a parameter.) So Haldane and Jayakar's results must be extended to frequency-dependent fitnesses, an easy adaptation as it turns out.

Suppose A, the high-r gene is dominant. Let the ratio of p^i to q^i be $u_i = p^i/q^i$. Let F_i be the relative fitness of the recessive in generation i: $F_i(N^i) = W_{BB}(N^i)/W_{AA}(N^i)$. When A is dominant Haldane (1932, p. 181) showed that $u_{i+1} = u_i(u_i + 1)/(u_i + F_i)$. By simple algebraic manipulation Haldane and Jayakar (1963) arrive at two particularly expressive formulas for the difference in u after n generations:

$$u_n - u_o = n - \sum_{i=0}^{n-1} F_i + \sum_{i=0}^{n-1} \frac{F_i(F_i - 1)}{u_i + F_i} \quad (21a)$$

$$\log u_n - \log u_o = -\sum_{i=0}^{n-1} \log F_i$$
$$- \sum_{i=0}^{n-1} \log \left[1 + \frac{(1 - F_i)u_i}{F_i(1 + u_i)} \right] \quad (21b)$$

We consider now what happens to trajectories over n generations starting from various p^0. Suppose p^0 is near zero, so that u_o and all subsequent u_i are near zero. If p^0 is small enough, the second term in (21b) can be ignored and then p^n will be larger than p^0 if $\Sigma \log F_i < 0$ where each F_i is evaluated at the N^i along the trajectory starting at N^0 and $p^0 \approx 0$. $\Sigma \log F_i < 0$ is equivalent to $\Pi F_i < 1$. Thus trajectories starting from very small p^0 will lead to higher p^n provided the geometric mean of the relative fitnesses along the trajectories is less than one. Suppose next that p^0 is near one so that u_o and subsequent u_i are near infinity. If p^0 is close enough to one, then the second term in (21a) can be ignored, and p^n will be less than p_0 if $\Sigma F_i > n$ where the F_i are evaluated at the N^i along the trajectory starting from N^0 and $p^0 \approx 1$. $\Sigma F_i > n$ is equivalent to $(1/n)\Sigma F_i > 1$. Thus trajectories starting from large p^0 lead to lower p^n provided the arithmetic mean of the relative fitnesses along the trajectories is greater than one. The difference between this discussion and Haldane and Jayakar's is simply that here we must be careful to take the means of the F_i along certain

trajectories, while with Haldane and Jayakar's constant fitnesses, the means of the F_i are the same along all trajectories.

The results about the means of the F_i are easily applied to our two-generation seasonal environment. We consider what happens as an environment becomes progressively harsher. First, in a mild environment the recessive high-K allele B is fixed, so the trajectories from any p^0 lead to a lower p^n. But as the environment becomes harsher, at some N_u^0, the trajectories from low p^0 lead to a higher p^n. This N_u^0 is the upper limit of the interval of environmental harshness where the seasonal environment causes polymorphism. N_u^0 is found from the condition on the geometric mean at low p^0:

$$(F_0 F_1)^{\frac{1}{2}} =$$
$$\left[\frac{W_{BB}(N_u^0)}{W_{AA}(N_u^0)} \times \frac{W_{BB}[W_{BB}(N_u^0)N_u^0]}{W_{AA}[W_{BB}(N_u^0)N_u^0]} \right]^{\frac{1}{2}} = 1 \quad (22a)$$

The lower end of this interval occurs when trajectories at high p^0 cease to lead to lower p^n. The lower end, N_l^0, is found from the condition on the arithmetic mean at high p^0:

$$\frac{1}{2}(F_0 + F_1) =$$
$$\frac{1}{2} \left[\frac{W_{BB}(N_l^0)}{W_{AA}(N_l^0)} + \frac{W_{BB}[W_{AA}(N_l^0)N_l^0]}{W_{AA}[W_{AA}(N_l^0)N_l^0]} \right] = 1 \quad (22b)$$

In harsh environments, below N_l^0, the dominant high-r allele, A, is fixed. The equations for N_u^0 and N_l^0 become cubic equations and the arrows in Fig. 11 for the interval end points indicate the solutions of these equations.

Haldane and Jayakar (1963) pointed out, using their constant fitnesses and full dominance, that a changing environment could never lead to an unstable equilibrium. An unstable equilibrium occurs if the geometric mean of the relative fitness is greater than one at low p^0 and the arithmetic mean less than one at high p^0. But since the geometric mean of the fitnesses in a changing environment is necessarily less than the arithmetic mean, the condition for an unstable equilibrium is impossible with constant fitnesses. An unstable equilibrium *is* possible, however, with frequency-dependent fitnesses. We now show that an unstable equilibrium caused by the seasonal environment cannot occur in the model considered here.

Suppose again the high-r allele, A, to be dominant. To obtain any equilibrium we had to assume that $N^0 < N_e$ and $N^1 > N_e$. Then $W_{BB}(N^0)/W_{AA}(N^0) < 1$ and $W_{BB}(N^1)/W_{AA}(N^1) > 1$. By the instability condition the geometric mean at low p^0 is greater than one, and, since the arithmetic mean at high p^0 is less than one, then the geometric mean at high p^0 is also less than one. So a necessary condition of the unstable equilibrium is

$$\frac{W_{BB}(N^0)}{W_{AA}(N^0)} \times \frac{W_{BB}[W_{BB}(N^0)N^0]}{W_{AA}[W_{BB}(N^0)N^0]} > 1$$

and

$$\frac{W_{BB}(N^0)}{W_{AA}(N^0)} \times \frac{W_{BB}[W_{AA}(N^0)N^0]}{W_{AA}[W_{AA}(N^0)N^0]} < 1$$

therefore

$$\frac{W_{BB}[W_{BB}(N^0)N^0]}{W_{AA}[W_{BB}(N^0)N^0]} > \frac{W_{BB}[W_{AA}(N^0)N^0]}{W_{AA}[W_{AA}(N^0)N^0]}$$

which, however, is only true if $W_{BB}(N^0)N^0 > W_{AA}(N^0)N^0$ since B is the high-K allele. But this last inequality entails that B is the high-r allele, which contradicts the premise. So for a two-generation seasonal environment to cause an unstable equilibrium with full dominance requires a more elaborate scheme of frequency-dependent selection.

The case where the high-K allele is dominant, rather than recessive, can most easily be treated by letting p still stand for the dominant allele frequency, now B rather than A. The upper limit, N_u^0, to the harshness interval is found from

$$\frac{1}{2}\left[\frac{W_{AA}(N_u^0)}{W_{BB}(N_u^0)} + \frac{W_{AA}[W_{BB}(N_u^0)N_u^0]}{W_{BB}[W_{BB}(N_u^0)N_u^0]}\right] = 1 \tag{22c}$$

and the lower limit, N_l^0, from

$$\left[\frac{W_{AA}(N_l^0)}{W_{BB}(N_l^0)} \times \frac{W_{AA}(W_{AA}(N_l^0)\ N_l^0)}{W_{BB}(W_{AA}(N_l^0)N_l^0)}\right]^{\frac{1}{2}} = 1 \tag{22d}$$

It can also be shown that the seasonal environment cannot cause an unstable equilibrium with this model.

Phenotypically intermediate heterozygote

The earlier graphical analysis of evolution in the seasonal environment showed that if a phenotypically intermediate heterozygote was superior to both homozygotes at moderate population densities, then a moderately harsh environment would cause polymorphism. Solving (20) for various N^0 revealed the nature of this polymorphism as a function of environmental harshness. The results appear in Fig. 12. Clearly the interval of environmental harshness in which the polymorphism occurs is considerably wider than the interval where there is full dominance as in Fig. 11. Also, the equilibria are somewhat more stable, with $f'(\hat{p}^0)$ at its stablest being .9849.

It should be mentioned that the phenotypically intermediate heterozygote whose trajectories appear in Fig. 6 also had a small interval of harshness causing stable polymorphism. And, as indicated by the earlier graphical analysis, an intermediate heterozygote inferior at moderate densities has an interval of harshness where an unstable equilibrium occurs.

We see that a variety of genetic mechanisms gives rise to a transition zone between K-selection and r-selection within which a stable polymorphism is maintained. However, the mechanism involving a phenotypically intermediate heterozygote superior at moderate population densities shows polymorphism under the widest range of environmental harshness and genetic properties and hence, of those considered, is the mechanism most likely to occur. Perhaps such polymorphism accounts in part for the large amount of heterozygosity observed in many natural populations, e.g., the study of Lewontin and Hubby (1966).

Ecological Energetics and r and K

In this section we explore the phenotypic properties caused to evolve under density-dependent selection. We are not interested in species-specific characters so much as those characteristics applicable to any organism that indicate how an organism participates in the overall energy flow through his population.

We refer to the same population as before but assume that the locus is fixed for either the A or B allele. Therefore the population shows logistic growth as in equation (12). As before, the population is censused at the time of interbreeding.

The strategy of this section is to derive, completely within the realm of ecological energetics, an equation describing population growth. This new equation can then be compared with the logistic equation (12) which was derived from independent premises. The comparison yields expressions for r and K in terms of concepts and measurements belonging to the field of ecological energetics.

Some basic symbols, following Lindeman's (1942) notation are defined:

Λ^i = total standing crop energy of the population at the time of interbreeding. The adults at this time are presumed to have ripe gonads. The caloric content of males and females is identical. As before, the superscript denotes the generation.

λ^i_{in} = total amount of energy received by the population between the interbreeding time of the ith generation and the interbreeding time of the $(i+1)$th generation.

λ^i_{out} = total amount of energy dissipated by the population between the interbreeding time of the ith generation and the interbreeding time of the $(i+1)$th generation.

The fundamental equation in the theory of ecological energetics is the energy balance equation, which follows from the conservation of energy law:

$$\Delta\Lambda^i = \Lambda^{i+1} - \Lambda^i = \lambda^i_{in} - \lambda^i_{out} \tag{23}$$

For the energy balance equation to be of much biological relevance, input and output terms need to be

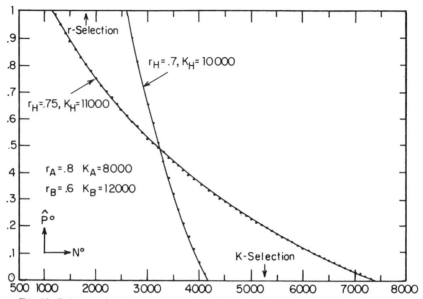

Fig. 12. Polymorphism frequency as a function of environmental harshness, assuming the heterozygote is superior in fitness at medium population sizes and intermediate in fitness at high and low population sizes. Growing season is two generations long.

expanded to indicate in more detail where energy is going and where it is coming from.

The energy dissipated by the population can be subdivided into three categories. The first category is the energy of decomposition of the bodies of the adults who have finished interbreeding. Let E be the average caloric content of an adult with ripe gonads, e be the average caloric content of a zygote, and $2m$ be the average number of zygotes produced in a mating. The average caloric content of an adult after mating can be put as $E - me$. Then the total energy dissipated from the bodies of the adults after mating is $(E - me)N^i$. Second, the organisms which survive from the zygote stage to the time of inter-breeding dissipate energy throughout their lives in respiration. Let H be the average energy respired by an organism surviving from zygote to the time of interbreeding, and f be the fraction of the zygotes which survive to the time of interbreeding. Then the total energy dissipated in this category is $HfmN^i$ assuming there are $N^i/2$ matings. Third, the bodies of the organisms which do not survive from the zygote stage to adulthood dissipate an energy of decomposition. The average energy dissipated per organism in this way can be put as e, the average caloric content of a zygote, on the understanding that any energy received by such organisms before death will not be credited in the energy input computations. The total energy in this category is then $e(1 - f)mN^i$. All the above considerations are summarized as

$$\lambda^i_{\text{out}} = [E - me + Hfm + e(1 - f)m]N^i \quad (24)$$

The energy received by the population is more simply expressed. Let R be the average total energy incident upon or in other ways received by an individual from the zygotic stage until the time of inter-breeding, and t be the fraction of the energy received which is actually transported across the cell membranes and incorporated into the organism. Then

$$\lambda^i_{\text{in}} = RtfmN^i \quad (25)$$

All the previous considerations can be consolidated with the definition of the dissipation factor, D, and the input factor, I:

$$D = E - me + Hfm + e(1 - f)m \quad (26)$$
$$I = Rtfm$$

The actual values of D and I are empirically determined according to formula (26). D and I both have dimensions of energy/organism. The energy balance equation is now

$$\Delta \Lambda^i = IN^i - DN^i \quad (27)$$

Thus I and D are the individual's average contribution, respectively, to the energy input and to the energy dissipation of the population.

At this point we incorporate a specific sort of density dependence into the model. Assume that D and I can be written as

$$D(N) = D_0(1 - dN) \quad (28)$$
$$I(N) = I_0(1 - cN)$$

where D_o and I_o are the energy dissipation and input factors in the absence of crowding, and d and c are crowding coefficients. Substituting (28) into (27) yields a density-dependent energy balance equation,

$$\Delta \Lambda^i = I_o(1 - cN^i)N^i - D_o(1 - dN^i)N^i \quad (29)$$

Equation (29) is the analogue in ecological energetics of the logistic equation of population dynamics:

$$\Delta N^i = rN^i \frac{(K - N^i)}{K} \quad (30)$$

The energy balance equation and the logistic equation are now to be compared. Recall that E refers to the average caloric content of an adult about to interbreed, and therefore

$$\Lambda^i = EN^i \quad (31)$$

If E itself does not change from generation to generation, then

$$\Delta \Lambda^i = E\Delta N^i \quad (32)$$

According to equation (32) the logistic equation, (30), can be multiplied by E and set equal to the energy balance equation, (29). Upon equating coefficients of N and of N^2, one discovers

$$r = (1/E)(I_o - D_o) \quad (33)$$
$$K = \frac{I_o - D_o}{cI_o - dD_o}$$

Equations (33) are expressions for the intrinsic rate of increase and the carrying capacity in terms of quantities customary to ecological energetics.

The expressions for r and K are to some extent interpretable in terms of the concepts of productivity and efficiency. One may assume for most biological systems that the energy input is either subsequently stored in the organization of biomass or variously dissipated. If the productivity of such a system is defined as the amount of energy put into biomass per unit time, then the productivity is clearly equal to the input minus the dissipation. If the efficiency of the system is defined as the amount of energy put into biomass per unit input, then the efficiency equals the difference between the input and dissipation divided by the input.

Inspection of equations (33) reveals the intrinsic rate of increase, r, to be proportional to the individual organism's average contribution to the population's productivity, provided crowding effects are negligible. Furthermore, in the very special circumstance where the population's dissipation is not density dependent, i.e., $d = 0$, the carrying capacity, K, is seen to be proportional to an efficiency of some sort, again provided crowding effects are negligible. However, in general, both c and d are non-zero and the carrying capacity does not correspond to an efficiency.

A perhaps better interpretation of K from an ecological energetics point of view can be developed.

Consider an expression for the individual's average contribution to the population's productivity, including crowding effects:

$$\text{individual productivity} = (I_o - D_o) - (cI_o - dD_o)N \quad (34)$$

Observe the factor $(cI_o - dD_o)$, to be called the crowding increment, because each individual's average contribution to the population productivity is diminished by the amount $(cI_o - dD_o)$ whenever the population size is increased by one member. Indeed, the crowding increment is a measure of the individual's sensitivity to having his average productivity diminished by crowding. Observe next that in equations (33) K is equal to the individual's uncrowded productivity divided by the crowding increment. Thus the carrying capacity equals the individual's average contribution to population productivity under uncrowded conditions divided by a measure of the individual's sensitivity to having that contribution diminished by crowding. Clearly, the smaller the crowding increment, that is, the less sensitive is the individual's average productivity to crowding, then the larger is the carrying capacity.

To conclude, we review some empirical generalizations to which this study lends support. The selection pressure in favor of alleles conferring high-K phenotypes increases, in a seasonal environment, as the dormancy season becomes milder and the growing season longer. Hence proportionately more species and traits responding to K-selection are expected in tropical regions than in temperate regions. In addition, r is proportional to an individual's average productivity under uncrowded conditions, and K is equal to the individual's average uncrowded productivity divided by a measure of sensitivity to crowding. Clearly, a phenotype possessing a high-K might have mediocre productivity in uncrowded conditions offset by a very low sensitivity to having that productivity diminished by crowding. Consequently, one would expect some tropical species to show lower productivity under uncrowded conditions but higher productivity in crowded conditions than comparable temperate species.

In addition, a phenotypically intermediate heterozygote can show a net selective advantage over both homozygotes in a seasonal environment of moderate harshness. If this mechanism for maintaining heterozygosity explains in part the results of electrophoretic analysis of proteins from natural populations, then the loci involved are expected to show fixation for one allele at low latitudes, polymorphism in medium latitudes, and fixation for the other allele at tropical latitudes. However, other causes of heterozygosity might also produce this latitudinal pattern and one would need to confirm that the fitness for the loci involved was density dependent.

ACKNOWLEDGMENTS

I wish to thank W. Bossert, P. Frank, and M. Gadgil for many discussions throughout the course of this study. I also wish to thank H. Lambert and E. O. Wilson for their helpful comments on the manuscript. A suggestion from B. Charlesworth lead to better exposition in part of the section on the mathematical analysis of a seasonal environment.

LITERATURE CITED

Birch, L. C., 1955. Selection in *Drosophila pseudoobscura* in relation to crowding. Evolution **9**: 389–399.

———. 1960. The genetic factor in population ecology. Amer. Natur. **94**: 5–24.

Buzzati-Traverso, A. A. 1955. Evolutionary changes in components of fitness and other polygenic traits in *Drosophila melanogaster* populations. Heredity **9**: 153–186.

Gadgil, M., and W. Bossert. 1970. Life historical consequences of natural selection. Amer. Natur. **104**: 1–24.

Haldane, J. B. S. 1932. The causes of evolution. Cornell University Press, Ithaca, N. Y. 235 p.

Haldane, J. B. S., and S. D. Jayakar. 1963. Polymorphism due to selection of varying direction. J. Genet. **58**: 237–242.

Levins, R. 1968. Evolution in changing environments. Princeton University Press, Princeton, N. J. 120 p.

Lewontin, R. C., and J. L. Hubby. 1966. A molecular approach to the study of genetic heterozygosity in natural populations. II. Amount of variation and degree of heterozygosity in natural populations of *Drosophila pseudoobscura*. Genetics **54**: 595–609.

Lindeman, R. L. 1942. Trophic dynamic aspect of ecology. Ecology **23**: 399–418.

MacArthur, R. H. 1962. Some generalized theorems of natural selection. Proc. Nat. Acad. Sci. **48**: 1893–1897.

MacArthur, R. H., and E. O. Wilson. 1967. The theory of island biogeography. Princeton University Press, Princeton, N. J. 203 p.

Smith, J. M. 1968. Mathematical ideas in biology. Cambridge University Press, London. 152 p.

Wright, S. 1959. Physiological genetics, ecology of populations, and natural selection. Perspect. Biol. and Med. **3**: 107–151.

Social Subordination, Population Density, and Mammalian Evolution

John J. Christian

Natural selection, operating on phenotypic expressions of the genetic material, is generally held to be of fundamental importance in evolution. Natural selection includes environmental selection, by which is meant here the selection of particular correlated phenotypic-genotypic changes or adaptations that confer better survival, including competitive advantages of the genotype, in a new or changing environment. However, the behavioral characteristics of many, if not most, mammals operate to reduce exposure to new or changing habitats. For example, most mammals are reluctant to go beyond the limits of familiar territory—their home range—and generally must be forced to do so. In addition, socially dominant animals usually occupy and hold the most desirable portions of the habitat and are the most successful breeders. In turn, their offspring tend to be dominant. Moreover, mammals generally occupy the habitat for which they are best adapted, and a given species may occupy similar, if not identical, habitats for many millennia. Therefore, if animals occupy a specific habitat and move with it as the habitat shifts with time and climatic change, how can they be subjected to the selective force of environmental changes, and by which mechanisms does mammalian evolution occur? Mammalian evolution might be expected to be conservative and limited to a rate characteristic of the entire ecosystem. However, mammalian evolution has been explosive with respect to rate and diversity.

I suggest that the conservative influence of social dominance is more than offset by other consequences of hierarchical behavior, and that social behavior is a major force in the evolution

of mammals. Mammalian selection and evolution may occur to an important degree through the agency of socially subordinate individuals, and it is these individuals that will provide the genetic material involved in adaptation to new circumstances. An example of the operation of the kinds of force I refer to is provided on the one hand by the social tolerance of bighorn sheep that results in their failure to exploit suitable unoccupied habitats and, on the other hand, by the social intolerance of deer and moose that results in their doing so (*1*). Here I examine some of the methods by which social rank and changes in population density might assume importance in mammalian evolution.

The idea that intra- and interspecific competition are important in evolution is, of course, not new. Interspecific competition generally is assumed to have been, and to be, operative in the evolutionary process, and its possible application to the problem of extinction has received considerable attention (*2*). Intraspecific competition, which must be considered primarily at the local level, often has been considered a factor in evolution in rather general terms. However, the potential importance of social subordination in mammalian evolution has not, to my knowledge, been suggested, although some of Haldane's (*3*) remarks regarding population density are pertinent to some aspects of evolution discussed below. Many authors have commented on the importance of surplus individuals in producing strong pressures for dispersal (*4*). In addition, the proportion of individuals dying is roughly proportional to such pressure. However, the possible role of social rank in evolution has not been mentioned.

Illustrative examples and documentation for this article are taken mainly from the literature on small mammals,

partly because of my personal familiarity with them and partly because they have been studied more than most larger mammals, with notable exceptions such as deer. Usually a single, or only a few, species are mentioned. However, much of what is said may be applicable to many other mammals and inframammalian vertebrates.

First, a brief review of pertinent features of social behavior and related population dynamics of mammals is appropriate. Broadly defined, intraspecific competition refers to direct or indirect competition between members of the same species or population, and also includes reproductive and numerical dominance. However, the term *intraspecific competition* often has been used more restrictively to refer to either overt or less obvious agonistic social behavior of members of the same species or population toward each other. Competition for some environmental factors, such as space, food, or nest site, may or may not be implied when *intraspecific competition* is used in the restrictive sense. Nevertheless, in either usage *intraspecific competition* would reflect the social behavior and organization of populations of a species, although to a greater degree when used in the restrictive sense. However, this difference in usage may be inconsequential because the evolution of intraspecific social competition seems to have resulted in bypassing the inherent dangers of direct competition for environmental necessities (*5*).

Hierarchies of Social Rank

The social organization of most mammals may be based, at least in part, on hierarchies of social rank that consist of a series of dominance-subordination relationships between individuals of a group (usually of the same sex), between families, and between coherent groups having bases of organization other than kinship, or on various combinations of these relationships. It should be noted that, while hierarchical social organization probably is common in mammals and birds and occurs in other vertebrates, it is not universal in animals. Therefore, important as social rank may be in the evolution of mammals and some other vertebrates, it is not a necessary mechanism for speciation or change in adaptation. Such hierarchies may vary considerably in details of arrangement from population to

The author is affiliated with the Research Laboratories, Albert Einstein Medical Center, Philadelphia, Pennsylvania.

population. Social organization of rhesus monkeys provides good examples of such relationships (*6, 7*). Individual aggressiveness toward other members of the same species or toward members of other species (intra- or interspecific competition, respectively) is an important overt manifestation of competitive social behavior. Marked differences in aggressive behavior have been observed between individuals, between strains or species, and between sexes. For example, the common meadow vole (*Microtus pennsylvanicus*) is highly intolerant of, and aggressive toward, other members of its species, whereas the prairie vole (*M. ochrogaster*) is highly tolerant of, and nonaggressive toward, other members of *its* species, but is dominant over *M. pennsylvanicus* (*8*). This difference will be referred to again.

Overt intraspecific aggressiveness is an important factor in establishing and maintaining rank hierarchies in many species—for example, the house mouse and the meadow vole—but it is not mandatory, as the establishment of rank may involve much more subtle behavioral factors. On the other hand there may be a direct relationship between (i) the degree of overt aggressiveness and (ii) the magnitude of dispersal with increased density, and the associated increase in contacts between animals; the general level of aggressive interaction may be greatly increased and, with it, the force to disperse.

Social hierarchies, as such, also constitute a major force for dispersion. Such hierarchies may be of individuals or of groups, but for simplicity I limit discussion here to the former. Low-ranking individuals are generally forced to emigrate from their birthplace and to find space in suitable habitat unoccupied by higher-ranking members of the same species or by members of dominant competing species. It is almost axiomatic that socially subordinate individuals (usually maturing young animals) that are forced to disperse have an extremely high rate of mortality (*9*). Few find suitable vacant niches and survive. However, some of these pioneers will become socially dominant in the new situation.

Gene-flow may be considerably reduced through the function of hierarchies because (i) some aggressive characteristics productive of dominance can be inherited (*10*), (ii) dominant individuals may account for a dis-

proportionately large share of the breeding in a population (*11, 12*), and (iii) there is restricted genetic interchange between subunits, or demes, of a population associated with territoriality and limited home ranges (*11, 13*). For example, if the offspring of dominant animals are dominant, they too will remain in their territory, thus limiting the spread of traits characteristic of dominant animals. Therefore one might expect to find an increase in the prevalence, and a decrease in the variability, of aggressive behavior in succeeding generations. However, progressively greater aggressiveness has not been observed in natural populations. Moreover, there do not appear to be marked differences in aggressiveness between natural populations of a particular subspecies, or of a species, such as might be expected if dominance was determined solely by genetics. There are a number of possible explanations for the apparent absence of unidirectional selection for aggressiveness. (i) Dominance appears to be, at least in part, a function of time and place, particularly at the local level; the first animals born early in the history of a population or early in each breeding season are likely to be dominant over animals born later (*7, 12, 14*). Age and size may largely determine dominance (*7, 10*). (ii) Dominance is relative and not a function of an inborn absolute amount of aggressiveness. If all low-ranking male mice from a number of populations are placed together, a new hierarchy emerges. The same thing occurs if all dominant males are placed together. (iii) There is probably an optimum degree of aggressiveness for a particular species, beyond which increases in aggressiveness may be incompatible with adequate reproduction and survival. Experimental evidence on physiological responses to increased density and aggressiveness suggests that extreme aggressiveness and high reproductive rates may be incompatible (*15*). The essence of the foregoing considerations is that it is not likely that there is continued selection for increasing genetically determined aggressiveness.

These and other components of population dynamics may exhibit wide variations from year to year, so no hard and fast quantitative rules can be derived that will be valid generally. Details of density-dependent inhibition of maturation and the pertinence of the size of overwintering populations

have been discussed in some detail elsewhere (*15, 16*). But, in particular, inhibition of maturation, whether density-dependent or density-independent, of young born late in the breeding season may be a mechanism that has evolved that insures an adequate breeding population an optimum habitat for the succeeding breeding season.

An example of both (i) the relationship between high aggressiveness and inhibition of breeding and (ii) the survival value of inhibition of the maturation of animals born late in the breeding season is given by a population of meadow voles (*Microtus pennsylvanicus*). In the breeding season of a year of high population density, all mature males were scarred from fighting, often severely so, but immature males were completely unscarred (*17*). As soon as the young males began to mature they became scarred, and probably were less likely to breed as a result of such social strife (*15*). Early termination of the breeding season by virtue of inhibition of maturation of animals born late in the season generally accompanies such high densities and the attendant increase in intraspecific strife. Thus, excessive density and aggressiveness may interfere with reproduction. Also it may be surmised from the absence of scars on immature males that they were not attacked and so were not driven to disperse. Thus, males whose maturation is inhibited may remain unattacked in their original territory from the end of one breeding season to the beginning of the next. Such males probably account for 90 percent or more of the breeding population at the beginning of each new season.

Population Densities

Population density acting in concert with intraspecific competition is a major component of the force to disperse (*4, 18*). When densities are very low, a higher proportion of subordinate individuals can find suitable areas in the preferred habitat, the number of dispersing animals is reduced, and the survival of subordinate animals is greatly increased. Broadly speaking, in natural populations of small mammals it is the young each year that comprise the bulk of subordinate animals, and generally only those young that succeed in occupying and establishing themselves in a suitable habitat become dominant. To what degree such dominance is genetically determined is not known.

The dynamics and consequences of hierarchical social organization have been studied for muskrats (9) and woodchucks (*Marmota monax*) (19). In a study of *Peromyscus* it was shown that, when maturing males are released into an incompletely filled habitat, they are able to establish themselves between the areas occupied by resident individuals, but when the available areas are filled, other animals released into the same habitat disperse and usually disappear (20). Comparable results have been obtained from other studies of populations of *Peromyscus* (21). When densities are high, the proportion of dispersing individuals is greater (18). Despite the increase in mortality that accompanies dispersal movements, it also follows that increasingly more marginal and submarginal habitats should become occupied as density increases. The ultimate effects of excessive densities on the involved populations are discussed elsewhere (16, 22) and are not particularly germane to the subject at hand.

The density of many populations of mammals, particularly that of small mammals with high reproductive potential, fluctuates more or less regularly. With each major fluctuation there is a progressive increase in density from year to year, until some maximum is reached; this is followed by a rapid decline. The general magnitudes of the maxima appear to be characteristics of the species and habitat. In addition to fluctuations from year to year (interannual fluctuations) there are increases and decreases in density within a single year (intra-annual fluctuations). Both intra- and interannual increases result in dispersal of subordinate individuals. Overwintering, usually immature, small mammals form the breeding population early in the breeding season each year. However, their mortality is high, and they are replaced by younger recruits, probably drawn from the first wave of litters in the breeding season. The individuals born subsequently comprise the dispersing populations and are, for the most part, probably subordinate. As density increases interannually, the force for dispersal, on the average, increases intra-annually until, in the final year of a "cyclic" increase of some species, the numbers dispersing may reach staggering proportions.

Similar changes may occur in the densities of populations of large mammals, but the time bases are greater. On the other hand, for many species there is little interannual fluctuation in population size, yet dispersal forces still operate intra-annually. The difference between intra-annual and interannual fluctuation in density is primarily quantitative; much wider dispersal accompanies the latter. In either case the vast majority of mammals forced to disperse fail to survive. Nevertheless, once in a great while a dispersing individual may, one would suppose, harbor a mutation or genetic change that increases its ability to adapt to the new surroundings and improves its chances of survival. It is such individuals that should be the basis for evolutionary changes. A suboptimum area could be invaded repeatedly by countless numbers of individuals before a genetic change permitting adaptation occurred. Thus, the dispersal of large numbers of socially *subordinate* individuals into new environments may provide the wherewithal for natural selection, in contrast to the relative conservatism of dominant individuals in an optimum habitat.

In summary, the important points of the foregoing discussion of competitive social behavior are (i) that populations of mammals are generally organized in social hierarchies; (ii) that low-ranking, predominantly young individuals usually are forced to disperse; (iii) that the genetic variability necessary for adaptation and evolution is carried by subordinate individuals; (iv) that the proportion of low-ranking individuals generally increases in association with an increase in the forces for dispersal with increasing density; (v) that the amount of less-than-optimum habitat occupied increases with increased dispersal; (vi) that dispersing individuals may be forced into competition with members of other species as they move into more or less alien habitats; and (vii) that inhibition of maturation late in the breeding season results in insuring the presence of a breeding population in the preferred habitat the following season.

The selective breeding and dispersal that generally accompany social organization should have important genetic and evolutionary consequences. And if genetic change and its correlated phenotypic expression are the basis of evolution (23), genetic change should be an essential component of the mechanism hypothesized here. The role of genetics in evolution and the sources of genetic variation are amply discussed in several recent accounts (23–26).

Optimum Habitat

Most mammals at a given time occupy ecological niches within their general ranges to which they are highly, if not maximally, adapted and for which they show a marked preference. Such a niche may be considered a mammal's optimum habitat, and it can be defined further as that habitat which the mammals in question occupy regularly despite seasonal and other relatively short-term climatic changes and at times of minimum population density. In the last instance, such habitats may be considered refugia. For the most part, they are not the only habitats occupied; mammals may regularly disperse into, and occupy, less advantageous habitats under the pressures of increasing population density and intraspecific social competition. In some instances the optimum habitat may be a temporary seral stage, but one that occurs sufficiently frequently to become available for occupancy regularly as other, similar areas pass into the next seral stage and become uninhabitable for a particular species. A successful, dominant mammal will not leave its preferred habitat, and only animals forced to do so will migrate to other, marginal habitats. This is such a well-known and well-documented fact for many species that it has become axiomatic. Dispersing mammals tend to select habitats as much like the original one as possible, but, as dispersal pressure builds with increasing population density, they tend to disperse into a gradient of habitats differing from the original to increasingly greater degrees. As noted above, life expectancy for the vast majority of such mammals is very poor. Thus it seems unlikely that most mammals are subjected directly to the selective force of a significant change in the environment. Selection among dominant animals in an optimum habitat probably can lead only to further specialization, at a slow rate, for that habitat, since there would be no pressure forcing individuals to go in other directions. Optimum habitats rarely and only gradually change in character and only slowly change in location, and dominant individuals move with geographic shifts. For example, with the advance and retreat of glaciers, it is unlikely that the dominant members of a species were forced into new habitats by climatic or environmental changes, and more likely that they remained in their preferred niche as this niche advanced or retreated with changes in climate, whether in a desert, a tundra, or other type of habitat (2, 27). With advancing glaciation during the Pleistocene there was a general shift southward of whole faunas that have since retreated northward again (27, 28). Arctic shrews (*Sorex arcticus*), northern bog lemmings (*Synaptomys borealis*), and pine martens (*Martes americans*) occurred in Tennessee. Spruce voles (*Phenacomys* cf. *ungava*), yellow-cheeked voles (*Microtus xanthognathous*), and other species of recent boreal mammals were found in Pennsylvania, whereas their present distribution is considerably farther north. During the late Pleistocene period of glacial advance these mammals apparently occupied habitats in Pennsylvania or Tennessee that were quite similar to, but perhaps not completely identical to, those they now occupy either much farther north or as relicts in the boreal zones of higher altitudes. These mammals "moved" with their habitats as the latter changed in latitude (or altitude) with glacial climatic changes. It is reasonable to assume that this continued occupancy of the same type of habitat as it moved, or moves, with major climatic changes was and is the rule for most mammals. Therefore, potential or presumed direct impact of environmental change on a particular species probably is seldom realized, since, as far as the mammals are concerned, significant environmental change probably did not occur. The mammals considered in this article dispersed to areas peripheral to the core of their optimum range instead of changing their habitat.

Social Behavior as a Force
in Mammalian Evolution

However, evolution has taken place, new species have evolved, and the changes do correlate with major environmental changes. What, then, occurs that ultimately results in environmental selection? One must assume that forces for evolution were operative and that these forces promoted exploitation of new environments where

the probability of occurrence of adaptive genetic changes was increased and where environmental selection could operate on preexisting as well as new genetic variability. Subsequently geographic and reproductive isolation would complete the requirements for speciation (see, however, 13). Given the conditions outlined above, it seems reasonable to suggest that social behavior, particularly dominance-subordination hierarchies, leading to dispersal of subordinate individuals constitutes a major force in mammalian evolution and provides a reasonable explanation for a number of features of such evolution, particularly at a local level. The fundamental tenet of the present hypothesis is that dispersal of subordinate individuals into marginal or suboptimum habitats provides an opportunity for environmental selection to operate on the genetic variation in environments which differ from that preferred by the dominant members of a species, and thus to improve chances for species survival and change. If adaptation in a new environment depends on existing genotypic variability in a subordinate animal, the animal might be considered to be "pre-adapted" or prospectively adapted. Since selection operates on phenotypes, any genetic variation or genetic mutation would necessarily have to have an advantageous, or at least neutral, phenotypic expression in order to be selected (23). Once a subordinate individual has moved into a less-than-optimum environment, the processes of natural selection and genetic variation presumably operate, as generally described by modern evolutionary theory (23–26). It seems likely that the dominant core of a population or species is rarely primarily involved in the evolutionary process. Presumably reproductive isolation must eventually isolate the evolving group in order for speciation to occur (but see 13).

Such a behavioral mechanism seems appropriate to explain the evolutionary events that have resulted in the present 14 species of finches in the Galapagos Islands. The original invaders occupied a habitat closest to their original habitat. Subsequently, countless subordinate individuals were driven off into other kinds of habitat, unoccupied by competing passerines. Eventually, a mutation occurred in one of these, which provided improved chances of survival in marginal habitat. The rest of the process logically follows, with succes-

sive occupancy of, and adaptation to, other habitats. What is new in this account of the evolution of these finches is the proposed role of socially subordinate birds and their enforced dispersal into marginal habitats.

The proposed subordination-dispersal force for evolution has a number of other features. First, it would include the flexibility and adaptive characteristics of behavior, defined as the capacity to learn and to adapt to new environmental situations (26). Second, intraspecific competition presumably would become more important as invasion of new environments took place. Interspecific behavioral competition would have the same potentialities, in this situation, as intraspecific competition discussed above. However, interspecific competition has dual potentialities: it may lead to evolution, on the one hand, or to extinction of one of the two competitors, on the other, depending on their relative degree of adaptation and on the intensity of the competition. Intraspecific competition should not lead to extinction unless it became so severe that excessive mortality dominated, rather than reproduction and dispersal. (This conceivably could be the case in some microtines.)

A subordination-dispersal mechanism should affect the rate of evolution, particularly if the rate of mutation is not the limiting factor, which it appears not to be (23). The intensity of dispersive forces in a population undoubtedly varies directly as some function of population density and of the degree of mutual intolerance, short of a degree of intolerance resulting in inadequate reproduction or excessive mortality. Therefore the rate of evolution should be enhanced by greater dispersive forces and increased numbers (29). That is, the opportunity for adaptive genetic change and for environmental selection would be expected to increase progressively as greater numbers of animals dispersed into increasingly diverse habitats. Of course, the rate of evolution would be on a geologic time scale, although the opportunities for genetic change and selection would occur very frequently. An increase in speciation might be expected in association with increased intraspecific intolerance, provided that occupied areas of optimum habitat were within larger areas having sufficient ecological diversity to allow selection to operate on the genotypes of surviving subordinate individuals, and that reproductive isolation from the central (or parental) population was eventually achieved. Continued survival in a new ecological niche or sub-optimum habitat presumably would involve a series of mutations in colonizing animals over a long period. A second, but less rigid, requirement would be that there would be no animals already occupying the ecological habitat into which the migrants moved, or that the migrants with their new adaptive mutations would be dominant over any such occupants. Thus, one would predict a burst of evolutionary change, followed by evolution in situ, upon occupation of a totally new and unoccupied area with suitable and diversified habitats.

Species that experience regular extreme highs in population density might be more likely to exploit, and to evolve in, new environments than those that had evolved sensitive and effective density-dependent mechanisms for limiting population growth at relatively low densities. Also, relatively nonaggressive species would not be expected to evolve as rapidly, or to the same degree, as more aggressive ones. Furthermore, the less aggressive species would be expected to exploit available environments less completely.

Periodic Irruptions

In addition, different species exhibit quantitatively different degrees of population fluctuation. For example, species of the genus *Peromyscus* seldom reach the spectacularly high levels of population that are achieved by some of the microtines. Furthermore, there is great variation among the microtines themselves. Among the most spectacular and most publicized of these upsurges are the great irruptions of lemmings (several species of *Lemmus* and *Dicrostonyx*). Meadow (*Microtus pennsylvanicus*), montane (*M. montanus*), and other voles also may occasionally experience spectacular irruptions and usually exhibit rather marked fluctuations in numbers. On the other hand, prairie voles (*M. ochrogaster*), as well as some other microtines, rarely reach such spectacular numbers, and their reproductive potential appears to be lower than that of many other members of the genus (*30*).

An explanation of the periodic irruptions of a number of microtines may lie in the characteristics of their original habitat prior to deforestation and agricultural use of the land (particularly for hay, fodder, and grains) by man. For example, *Microtus pennsylvanicus*, and possibly some other species of *Microtus* with similar ecological requirements may once have been restricted largely to temporary wet grasslands, such as "beaver meadows" created by abandonment of beaver dams. Such temporary meadows usually progress rapidly through seral stages of reforestation. Therefore a species that depended on such a habitat could not have survived long unless it had evolved mechanisms for discovering and exploiting newly created similar habitats. For example, I recently observed an invasion of a beaver meadow by *M. pennsylvanicus* in less than a week after its creation, in a year when the vole population was very high. Furthermore, the voles had to cross inhospitable forest habitat to invade the meadow. The marked fluctuations in the populations of *M. pennsylvanicus* may have served this end by periodically producing large numbers of migrants. The period between irruptions necessarily would have to be shorter than the temporary grass stage of the beaver meadows. Evolution of a dispersive force would be absolutely essential in such a situation. The development of social hierarchies (with accompanying intraspecific aggressiveness and intolerance) in combination with high population density would create a large number of subordinate individuals that would be driven from the habitat of their birth by dominant individuals, once the available home habitat was filled. The vast majority of such migrants would quickly and inevitably die. A very few would reach unexploited, newly created habitat and survive. Thus, new populations would be established by subordinate migrants, and constant repetition of the process would insure survival of the species as a whole.

Of course, *Microtus pennsylvanicus* has been able to exploit agricultural lands and is extremely common and widely distributed. This vole is much more intolerant of other members of its own species than is *M. ochrogaster*, an inhabitant of the extensive prairie grasslands of North America (*8*), and this may be a pertinent factor. Superficially, this intolerance appears to be a process for selecting social subordination, but, as mentioned above, establishment of social rank in a given population may be more a matter of time and place than of genetics. Similar expansion into man-made habitats has occurred with root voles (*M. oeconomus*) in Finland (*31*). These voles prefer habitats that are very small and discontinuous, but under the pressure of increasing density they may occupy previously uninhabited suboptimum habitat. This dispersion occurs mainly at the time of seasonal change in habitat, and it is young voles that occupy these habitats. In addition, these voles, in years of high population density, drive field voles (*M. agrestis*) from their usual habitats.

In contrast to species such as *Microtus pennsylvanicus*, there are those that originally inhabited continuous and extensive areas of relatively similar habitat, such as the extensive deciduous forest of eastern North America and the great plains of central North America prior to the advent of Europeans. Some members of the genus *Peromyscus* (for example, *P. maniculatus*) may be representative of such species. They seldom irrupt, and their population densities seem to be under much more effective control than is the case with regularly irruptive species. Their reproductive function evidently is much more sensitive to inhibition than is that of those microtines and other small rodents that have been tested (*15, 17*). Differences in food requirements usually appear not to be an important contributing factor in accounting for the different degrees of regulation of population size in *Microtus* and *Peromyscus*. Therefore, we have an interesting situation: one species, *M. pennsylvanicus*, may have survived because it did not evolve an efficient negative-feedback control of population size, whereas *P. maniculatus*, for example, may have survived because it did, and thereby avoided the potential danger of overexploitation of the environment. However, *M. pennsylvanicus* has taken advantage of the vast grasslands created by agriculture, but it has retained its characteristic fluctuations in population, which are inappropriate now. If the foregoing hypothesis is valid, and if present agricultural practices persist long enough, the meadow vole should eventually evolve more effective control of its population densities. In the interim it would face the threat of extinction if it should reach such low numbers after a peak that it was unable to recover from the decline. Social dominance of meadow voles by the grassland inhabitants (*M. ochrogaster*) (*8*) and *Peromyscus maniculatus bairdii* (*32*), by the woodland inhabitants *P. leucopus* (*32*) and *Clethrionomys gapperi* (*33*), and probably by other socially dominant species may limit the expansion of meadow voles into other habitats. *Peromyscus leucopus* also is dominant over *Microtus ochrogaster* and may limit the latter to cultivated fields (*34*).

Somewhere in between the group of mammals represented by *Microtus* and those represented by *Peromyscus maniculatus* are house mice and Norway rats, which appear to be essentially opportunistic in their ability to exploit man-made habitats and to irrupt under favorable conditions. In other words, they are potentially, but not regularly, irruptive.

Arctic and Subarctic Species

In the case of the Microtinae, the problem of those species that inhabit the extensive and relatively uniform arctic and subarctic tundras remains to be considered. Lemmings of the genera *Lemmus, Phenacomys, Synaptomys,* and *Myopus* are generally considered to be among the more primitive of the microtines (*35*). It is conceivable that at one time the ancestral predecessors of today's arctic lemmings —*Lemmus, Dicrostonyx,* and, to a lesser extent, *Phenacomys, Synaptomys (Mictomys),* and *Myopus*—inhabited temporary grasslands and evolved a mechanism for exploiting new and temporary habitats similar to the mechanism proposed for *Microtus.* For example, *Dicrostonyx*, while highly evolved for its arctic habitat, seems to have occupied similar less boreal and discontinuous parklike habitats, and later to have survived only in more purely arctic habitats. There is recent evidence that a relatively boreal parkland-coniferous forest existed in Pennsylvania during the late Pleistocene, and that *Dicrostonyx* did indeed occupy these discontinuous grassy parklands that were interspersed in a region of boreal coniferous forest (*27, 36*). *Dicrostonyx* disappeared with the subsequent gradual transition of boreal parkland to boreal forest and is now restricted to the arctic tundra. If such is the case, then one must assume that the collared lemming (*Dicrostonyx* sp.) adapted to and occupied the arctic much earlier than *Lemmus* and closely related forms did, since the collared lemmings exhibit much greater adaptive specialization for arctic conditions and also appear to have diverged from ancestral microtine-like stock very early in the evolutionary history of the group (*35*). Subsequently they may have survived only in increasingly boreal habitats as a result of competition from more newly evolved, dominant species, possibly microtines, which replaced them in the temporary and more temperate habitats, much as field voles (*Microtus agrestis*) are driven from man-made habitants in Finland by root voles (*M. oeconomus*) in years when population density of the species is high (*31*). Other lemmings, such as species of *Synaptomys*, now occur in several types of habitat other than grassland, but they appear to be subordinate to *Microtus* where the two occur together, and *Synaptomys* is rarely found in the preferred habitats of *Microtus.* The arctic lemmings also may have retained the social characteristics and population responses that are necessary for colonizing new habitats, responses which now are inappropriate for most of their range. However, lemmings (*L. lemmus*) may invade essentially alien habitats and regions in years of high population density, and may even drive regular inhabitants from their customary habitats (*31*). Whether these events actually occurred in the evolution of arctic lemmings may never be determined, but it is a possible explanation for their apparently useless and violent fluctuations in population size. It seems likely that genetic exchange could take place in lemming populations without extreme fluctuations, as it does in many other species.

Density-Dependent

Inhibition of Reproduction

A high reproductive potential and the ability to realize it rapidly are characteristic of microtines and of many other irruptive species of mammals. Nevertheless, reproduction is progressively inhibited as density increases, and may be totally suppressed if densities become sufficiently high (*15*). The inhibition of reproduction is part of a density-dependent endocrine response that also may include increased adrenocortical activity and its sequelae. This density-dependent feedback is believed to operate through social intolerance and aggressiveness in conjunction with increased numbers (*16*). Therefore, if social intolerance and relatively high numbers constitute a force for dispersal and exploitation of new habitats by mammals basically adapted to discontinuous habitats, one might expect more conspicuous changes in adrenocortical and reproductive function with changes in population density in these mammals than in more socially tolerant mammals adapted to continuous habitats. The adrenocortical responses associated with social rank and increased density—and their sequelae, such as decreased resistance to disease—may be considered nonadaptive in terms of individual survival and a disadvantageous by-product of social intolerance and high densities. On the other hand many species adapted to continuous habitats seem to have evolved greater social tolerance, and a more sensitive regulation of reproductive function that serves generally to maintain their populations at relatively lower densities than those of irruptive species. Thus, in mammals adapted to continuous habitat, with their greater social tolerance and more moderate population densities, one might expect much less adrenocortical response to changes in density than one expects in irruptive or potentially irruptive species such as many microtines, house mice, and rats. The lack of correlation between adrenocortical activity and rank or density, together with the marked sensitivity of reproductive function to inhibition by subordinate rank or increased density, in the prairie deer mouse (*Peromyscus maniculatus bairdii*) (*10, 15*) may reflect adaptation to a continuous habitat, with evolution of an efficient feedback mechanism to limit population growth well below irruptive levels. On the other hand, behavioral intolerance and adrenocortical and other density-dependent physiological responses occur in the white-footed mouse (*P. leucopus*) that inhabits serally temporary brushlands and forest edges, as well as in *Microtus pennsylvanicus* and lemmings that, presumably, are basically adapted to discontinuous habitats (*15, 37*). Extensive continuous habitats, which do not impose the necessity for regular dispersal and exploitation of new sites, may have resulted in selection against social intolerance and the adrenocortical and related reactions that accompany it in favor of more benign behavior and more direct and sensitive regulation of reproductive function that usually curtails population growth at levels well below those often achieved by microtines exhibiting extreme fluctuations in population.

A Unifying Hypothesis

Consideration of these problems has led to a unifying hypothesis concerning the direct and important role of social competition—in particular, the role of the subordinate individual—in natural selection and evolution. The essence of this hypothesis is that intraspecific competition is a major force in evolution and that genetic changes on which selection operates are found in these socially subordinate individuals. The probability that an adaptive genetic change will occur will be greater in an environment less favorable than that occupied by the central or parent population. The odds will automatically be increased, in many species of mammals, by the great numerical predominance of subordinate migrants over dominant core residents. Thus, the raw material for speciation by way of natural selection is the rare subordinate migrant that survives in a new habitat. It may be that in these individuals, in suboptimum habitats, genetic change can improve the chances for survival and increase adaptation by selection. It is also suggested that more effective density-dependent control of population growth has evolved in continuous habitats than in discontinuous habitats, where marked social intolerance leads to dispersal and to marked density- and rank-dependent adrenocortical responses. Finally, it is suggested that inhibition of maturation of mammals

born late in the breeding season may have evolved as a means of insuring an adequate breeding population in the following breeding season.

References and Notes

1. V. Geist, Natur. Hist. 76, 24 (1967).
2. J. E. Guilday, in Pleistocene Extinction, The Search for a Cause, P. S. Martin and H. E. Wright, Jr., Eds. (Yale Univ. Press, New Haven, Conn., 1967), p. 121.
3. J. B. S. Haldane, Proc. Roy. Soc. Ser. B Biol. Sci. 145, 306 (1956).
4. P. J. Darlington, Jr., Zoogeography: The Geographical Distribution of Animals (Wiley, New York, 1957).
5. J. J. Christian, Proc. Nat. Acad. Sci. U.S. 47, 428 (1961); V. C. Wynn-Edwards, Animal Dispersion in Relation to Social Behavior (Oliver and Boyd, Edinburgh, 1962).
6. S. A. Altman, Ann. N.Y. Acad. Sci. 102, 338 (1962); C. H. Southwick, ibid., p. 436.
7. C. B. Koford, Science 141, 356 (1963).
8. L. L. Getz, J. Mammalogy 43, 351 (1962).
9. P. L. Errington, Iowa Agr. Exp. Sta. Res. Bull. 320, 797 (1943); Quart. Rev. Biol. 21, 144 (1943); ibid., p. 221; Ecol. Monogr. 24, 377 (1954).
10. A. M. Guhl, J. V. Craig, C. D. Mueller, Poultry Sci. 39, 970 (1960); R. E. Wimer and J. L. Fuller, in Biology of the Laboratory Mouse, E. L. Green, Ed. (McGraw-Hill, New York, ed. 2, 1966), p. 631.
11. P. K. Anderson, in Mutation in Population (Akademia, Prague, 1965), p. 17; C. R. Terman, Ecology 46, 890 (1965).
12. J. A. Lloyd and J. J. Christian, J. Mammalogy 50, 49 (1969).
13. P. R. Ehrlich and P. H. Raven have recently discussed the relationship of gene flow to differentiation of populations and have emphasized the overriding importance of selection [Science 165, 1228 (1969)].
14. J. B. Calhoun, Science 109, 333 (1949); R. Mykytowycz, CSIRO (Commonw. Sci. Ind. Res. Organ.) Wildlife Res. 4, 1 (1959)
15. J. J. Christian, Bull. Sinai Hosp. Detroit 47, 108 (1970).
16. ———, Proc. Nat. Acad. Sci. U.S. 47, 428 (1961); ———, in Physiological Mammalogy, W. V. Mayer and R. G. Van Gelder, Eds. (Academic Press, New York, 1963), vol. 1, p. 189.
17. ———, "Fighting, maturity, and population in Microtus," in preparation.
18. D. B. Van Vleck, J. Mammalogy 49, 92 (1968).
19. F. H. Bronson, thesis, Pennsylvania State University (1961); ———, Ecology 44, 637 (1963); ———, Anim. Behav. 12, 470 (1964); D. E. Davis, J. J. Christian, F. H. Bronson, J. Wildlife Manage. 28, 1 (1964); E. D. Bailey, J. Mammalogy 46, 438 (1965).
20. L. H. Metzger, paper presented before the 47th Meeting of the American Society of Mammalogists (1967).
21. M. C. Healey, Ecology 48, 377 (1967).
22. See, for example, C. Elton, Voles, Mice and Lemmings (Clarendon, Oxford, 1942); D. Chitty, Phil. Trans. Roy. Soc. London Ser. B Biol. Sci. 236, 505 (1952); O. Kalela, Ann. Acad. Sci. Fenn. Ser. A IV Biol. No. 34 (1957); F. Pitelka, "Arctic biology," paper presented at the 18th Biology Colloquium, Corvallis, Oregon, 1957; K. Murray, Ecology 46, 163 (1965).
23. G. G. Simpson, The Major Features of Evolution (Simon & Schuster, New York, 1953).
24. ———, The Geography of Evolution (Capricorn, New York, 1965); E. Mayr, Animal Species and Evolution (Harvard Univ. Press, Cambridge, Mass., 1963).
25. T. Dobzhansky, Genetics and the Origin of Species (Columbia Univ. Press, New York, 1941).
26. J. M. Smith, The Theory of Evolution (Houghton Mifflin, Boston, Mass., 1966).
27. J. E. Guilday, P. S. Martin, A. D. McCrady, Bull. Nat. Speleol. Soc. 26, 121 (1964).
28. C. W. Hibbard, Mich. Acad. Sci. Arts Lett. Rep. 44, 3 (1959); J. E. Guilday and M. S. Bender, Ann. Carnegie Mus. 35, 315 (1960); J. E. Guilday, H. W. Hamilton, A. D. McCrady, Palaeovertebrate (Montpellier) 2, 25 (1969).
29. S. Wright [Genetics 16, 97 (1931); Amer. Natur. 74, 232 (1940); in The New Systemics, J. Huxley, Ed. (Clarendon, Oxford, 1940), pp. 161–183] has pointed out that there is an optimum inter-deme gene flow and population size for evolution. For example, it could be argued that, if Microtus pennsylvanicus, an organism found in discontinuous habitat, did not undergo explosive population growth and dispersion, it would have speciated over its range into a variety of other species, I cannot deal here with the influence of social behavior on these optima; suffice it to say that mechanisms to which I draw attention in this article are assumed to work in parallel with those which classically have been thought to determine evolutionary rates, although I expressly wish to point out the simplifying nature of this assumption.
30. S. A. Asdell, Patterns of Mammalian Reproduction (Comstock, Ithaca, N.Y., 1946).
31. J. Tast, Ann. Zool. Fenn. 3, 127 (1966); ibid. 5, 230 (1968); Ann. Acad. Sci. Fenn. Ser. A IV Biol. No. 136 (1968).
32. J. O. Murie, thesis, Pennsylvania State University (1967).
33. A. W. Cameron, Evolution 18, 630 (1965); G. C. Clough, Can. Field Nat. 78, 80 (1964).
34. J. O. Whitaker, Ecology 48, 867 (1967).
35. E. T. Hooper and B. S. Hart, Misc. Publ. Mus. Zool. Univ. Mich. No. 120 (1962).
36. J. E. Guilday, Amer. Midland Natur. 79, 247 (1968).
37. J. J. Christian and D. E. Davis, J. Mammalogy 47, 1 (1966); ———, in Comparative Endocrinology, A. Gorbman, Ed. (Wiley, New York, 1959), p. 71; R. V. Andrews, Physiol. Zool. 41, 93 (1968).
38. I gratefully acknowledge the helpful comments and criticism in the preparation of this manuscript of George G. Simpson, Carleton S. Coon, John E. Guilday, William H. Burt, J. Kenneth Doutt, David E. Davis, Kenneth Meyers, David Van Vleck, Louise Baenninger, Samuel Horowitz, my students, my associates, and others. However, responsibility for the interpretation and use of these suggestions is mine, and others should not be held accountable for errors of omission or commission. Much of the work on which this article is based was supported by the U.S. Navy, the National Institutes of Health, and C.S.I.R.O. Wildlife Management, Australia.

14

Reprinted from *Proc. Nat. Acad. Sci.*, **66**(3), 780–786 (1970)

Age-Specific Selection*

Wyatt W. Anderson and Charles E. King

DEPARTMENT OF BIOLOGY, YALE UNIVERSITY, NEW HAVEN, CONNECTICUT

Communicated by Theodosius Dobzhansky, April 13, 1970

Abstract. A model is presented for age-specific selection on the genotypes in a population. Each genotype is assigned a life table that specifies the viability and fecundity of its age classes. Breeding and reproduction occur at regular intervals, and generations overlap. Examples were generated on a digital computer. Gene frequencies and the distribution of individuals among the various age classes may oscillate until equilibrium is reached. Moreover, age structure and gene frequencies are intimately related; a change in either factor alone may bring about a change in the other. In an extension of the basic model, the fecundities of the genotypes were regulated by population density. Under the joint action of logistic control and age-specific selection, the growth curve of the population can, for some schedules of selection, show plateaus in an otherwise sigmoid increase. The relevance of the growth patterns obtained in different types of environments to current ideas of "*r*" and "*K*" selection is discussed.

Introduction. Geneticists and ecologists have, until quite recently, tended to view the structure of populations in different ways. Gene frequencies and the forces which bring about changes in them are of primary concern to the geneticist, while the factors determining population size, density, and age structure are of particular importance to the ecologist. It is the purpose of this article to present two simple models for natural selection in which the viability and fecundity of each genotype may differ from one age to another. We shall show that gene frequencies, population size, and age structure are intimately related and together determine the behavior of a population under selection.

Gene frequencies change under natural selection because the various genotypes in a population leave different numbers of offspring. Each genotype is usually assigned a *selective value* that measures this success in transmitting genes to the next generation. Within each genotype, however, are individuals of different age, and the intensity of selection clearly depends on age.[1,2] This complication of selection that varies with age has been neglected in the models of population biology. The only exceptions we have found are the early works of Haldane[3] and of Norton,[4] whose intent was quite different from our own, and a very recent model by Istock.[5]

A Simple Model. Consider a population composed of the genotypes for two alleles at a single autosomal locus. The individuals of each genotype are divided into classes on the basis of age. Each genotype is assigned a schedule of births and deaths at each age. Reproduction occurs at regular intervals, and time is measured in terms of these *breeding intervals*. The viabilities and fecundities are assumed to be alike in males and females. The population size is assumed to be large enough that the effects of sampling error may be ignored. Since organisms may live for a number of time intervals, the model is one of discrete breeding intervals and of overlapping generations. This model fits the breeding regime of many organisms, and by varying the length of the breeding intervals, it can be made to fit virtually all bisexual, diploid organisms. As the time intervals are made shorter, the model becomes in the limit one of continuous change where the viabilities and fecundities are continuous functions of time. As the time intervals are made longer, the model approaches one in which there is a single age class in each genotype and discrete generations.

The action of selection may be visualized as a simple matrix operation by following Leslie's[6,7] approach to population growth. For each genotype there is a "selection matrix" which contains the age-specific fecundities along the top row and the age-specific viabilities along the first subdiagonal. Multiplying this selection matrix by a column vector whose elements are the numbers of individuals in each age class for the genotype under consideration, we obtain a vector of individuals that will begin the next breeding interval. Let the viability of the genotype carrying alleles I and J in age class L be VIJ_L and the corresponding fecundity, FIJ_L. Let the number of individuals of this genotype of age L at the beginning of breeding interval T be NIJ_{LT}. Denote the progeny of genotype IJ as PIJ_T. All individuals of age 1 or older at time $T + 1$ must have advanced from the next lower age within the same genotype at the previous breeding interval, after the selection through mortality. Within each breeding interval we assume that mating is completely random. It is as if each adult sheds gametes according to its genotype and age-specific fecundity, and the new organisms are formed by random combination of pairs of gametes. The progeny generated in breeding interval T will be the individuals of age 0 at time $T + 1$. The progeny of a genotype are not necessarily of the same genotype; the alleles carried by a zygote depend on the genotypes of both parents. The progeny are reassorted to their proper genotypes according to the frequencies expected with Mendelian segregation under random mating. The model is diagrammed in Figure 1 for a genotype with four age classes. This process is repeated for each breeding inter-

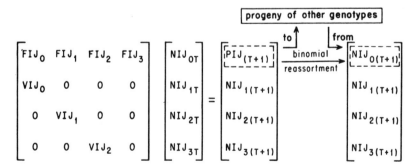

FIG. 1.—The basic model for age-specific selection. See text for explanation of symbols.

val.   Since in the simple model the fitness components are constant and not functions of population number, the population must grow indefinitely large or diminish to extinction according to the fecundities and viabilities.   This first model is thus one of exponential growth.

An Illustration of the Basic Model.   We have investigated the behavior of populations fitting this model on a digital computer.   An example is presented in Figure 2, showing the behavior of a population over 25 breeding intervals.   The viabilities are alike for all genotypes and decrease by half with each age;   hence

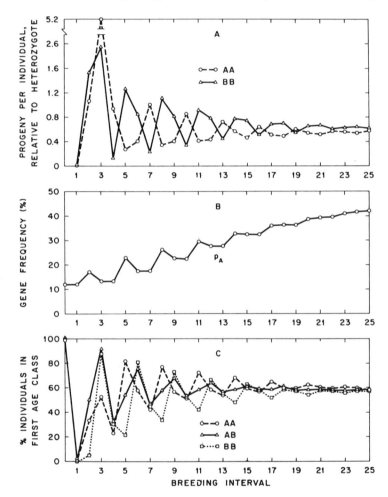

Fig. 2.—An example of age-specific selection under exponential population growth.
   Panel A: Numbers of progeny per individual for the two homozygotes, relative to the heterozygote.
   Panel B: Frequency of allele A.
   Panel C: The age structure as reflected in the percentage of individuals in the first age class.
   Selection parameters for this example: (1) Identical viabilities for the genotypes over the five age classes: VAA = VAB = VBB = 0.5, 0.5, 0.5, 0.5, 0.0; (2) Fecundity schedules: AA = BB = 0,0, 4, 2, 1, and AB = 0, 1, 5, 2, 1; (3) Initial number of individuals, all in the first age class: NAA = NAB = 2, NBB = 21.

the survivorship curve is diagonal.   The heterozygote has a fecundity advantage. The initial frequency of allele A is 0.12, and the population was begun with all individuals in the first age class.   Geneticists usually define selective value as the number of progeny per individual, and so we have plotted this "apparent selective value" in *panel A*.   Since only fecundity differed among the genotypes, the number of progeny is in this case an accurate index of selective value; if deaths had occurred at different rates among the genotypes, births alone would not be sufficient to measure selective value.[8]   The number of progeny is given relative to that of the heterozygotes.   The apparent selective values fluctuated widely for about 15 breeding intervals before stability was reached.   In the example both homozygotes had the same viability and fecundity at each age, so the difference in the number of offspring is clearly related to the changing age structure.   The frequency of allele A (*panel B*) showed an increase toward an equilibrium at 0.5, with an erratic fluctuation which damped slowly.   The over-all course of gene frequency is almost linear and clearly departs from the almost-sigmoid curves obtained with age-free models.   As an index of age structure, we have plotted (*panel C*) the percentage of individuals in the first age class.   Since this age group receives individuals by birth from the other age classes, it is a sensitive measure of changes in the age structure.   The age structure fluctuated widely for about 15 breeding intervals, and then displayed a damped approach to stability.

In examples (not shown) where the age distribution has reached equilibrium and the gene frequency is then changed, large disturbances in the age structure occur as the gene frequency returns to equilibrium.   The two most important components of our models, the age structures and the life history schedules of selection for the genotypes, interact to determine the approach of the population to its equilibrium structure.   The sorts of changes observed under this model of constant selection might well lead an observer to invoke variation in the environment or disturbances within the genome to account for fluctuations in gene frequency or age structure.   Yet these phenomena are natural consequences of a more realistic model, and give us some feeling for the sort of changes that may occur in nature.   A change in the age structure alone is sufficient to produce changes in gene frequency, and conversely, a change in gene frequency alone is sufficient to produce large disturbances in age structure.

A Logistic Growth Model.   The model presented earlier has one major drawback when applied to most populations; it is a model for exponential growth. To remove this unreal assumption, a second model was devised incorporating one kind of logistic control in which reproduction is decreased as the numbers of any genotype approach the carrying capacity of the environment.   Assume that each individual in the population utilizes the same quantity of environmental resources, but that each genotype may have a different carrying capacity, KIJ.   Then, the progeny generated by the M age classes of the NIJ individuals is given by

$$PIJ_{(T+1)} = \sum_{L=1}^{M} NIJ_{LT} \cdot FIJ_L \left[\frac{KIJ\text{-}NTOTAL}{KIJ} \right].$$

179

These progeny are reassorted to the different genotypes with the binomial Mendelian operator used in the exponential model.   The bracketed feedback term in the above expression is limited to an arbitrary minimum value of 0.01 to prevent the generation of negative progeny if KIJ is less than $NTOTAL$, the total number of individuals in the population, and also to permit some reproduction at equilibrium.

The standard logistic model assumes equivalence of all individuals in the population.   However, this assumption is not valid if the population is composed of genotypes with different viabilities and fecundities or of individuals of different ages.   This complication has been considered by Anderson[9] for a population with discrete, nonoverlapping generations, where all members of the population at any time are in the same age class.   We have generalized and extended this type of analysis in the present paper by the age-specific scaling of viability and reproduction for each genotype under a model with overlapping generations.

An Illustration of the Logistic Model.   Most studies of population growth are made under the implicit assumption of no selection acting within the population.   When this assumption is not valid, a large error can result, as indicated in Figure 3.   The values plotted in this figure are for the logistic model when all three genotypes have identical diagonal survivorship curves, but one homozygote, AA, has an advantage in both fecundity and numbers at the start of the population growth.   Counterbalancing these advantages, the AB and BB genotypes have successively higher values of K so that $KAA < KAB < KBB$.   In *panel A*, the growth of the entire population ($NTOTAL$) is initially quite rapid.   If an

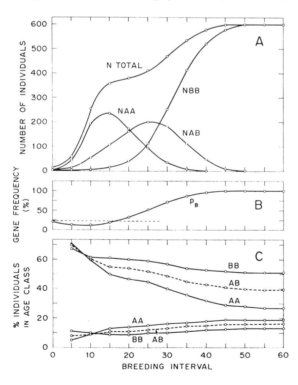

Fig. 3.—An example of age-specific selection under logistic population growth.
Panel A: Changes in total number of individuals and numbers of the genotypes.
Panel B: Change in the frequency of allele B; the dashed line indicates the initial frequency.
Panel C: The age structure as reflected by the percentage of individuals in the first (*top*) and third (*bottom*) age classes.
Selection parameters for these curves are: (1) Identical viabilities for each genotype over the five age classes: $VAA = VAB = VBB = 0.5, 0.5, 0.5, 0.5, 0.0$ (2) Fecundity schedules: $AA = 0, 3, 5, 2, 1$; $AB = BB = 0, 2, 5, 2, 1$. (3) Carrying capacities: $KAA = 500$; $KAB = 750$; $KBB = 1000$. (4) Initial numbers of individuals, all in the first age class: $NAA = 9$; $NAB = 3$; $NBB = 1$.

investigator had no knowledge of the genetic structure, it might be concluded that by the 15th breeding interval the population was approaching an equilibrium population size of about 375 individuals. Subsequently, however, the population increases in size and reaches its true equilibrium level of about 600 individuals in the 50th breeding interval. Again, if observed in nature, it would probably be concluded that the "hump" in the growth curve was caused by an environmental change affecting the carrying capacity for the entire population. However, given a knowledge of the genetic structure of the population, it is apparent that the action of natural selection produces the hump. That is, the AA genotype is replaced by the heterozygote which, in turn, is replaced by the BB homozygote.

There are four major variables that determine the shape of a population growth curve under logistic control. In Figure 3, the rapid initial growth of the AA homozygote is a result of its combined advantages in initial numbers and high fecundity when $NTOTAL$ is small. With the model for exponential growth presented earlier, the AA genotype would rapidly constitute the majority of the individuals and it would remain the most frequent genotype. Under logistic control however, as the population grows the relatively low KAA value chokes the growth of this homozygote while the total population size still permits the AB and BB genotypes, with their higher K values, to increase in number. In the present example, the logistic control acts exclusively on fecundity and, since all genotypes have the same viabilities, the shape of the growth curve is primarily determined by the fecundities, initial numbers, and values of K.

The equilibrium number, about 600 individuals of genotype BB, differs from KBB (1000 individuals) because reproduction and viability are independent in this model. That is, only reproduction is controlled by the $[(KIJ\text{-}NTOTAL)/KIJ]$ term. Death rates are considered to be independent of population size and related only to the viabilities specified at the beginning of a run.

Panel B of Figure 3 presents the change in the frequency of the B allele under this logistic selection. The initial advantage of the AA homozygote is reflected in a decrease in the frequency of the B allele for the first 15 breeding intervals. Subsequently, the B allele increases in frequency until apparent fixation in the 50th breeding interval.

After an initial oscillation similar to that discussed in connection with the exponential model, the age distribution of individuals in each genotype approaches stability (*panel C*). The frequencies of the individuals undergoing selection in each age class and genotype are in the expected orders. That is, a genotype being adversely affected by natural selection is expected to have a lower frequency of young individuals and a higher frequency of old individuals than a genotype with a selective advantage. All three genotypes have approximately the same proportion of individuals in the second age class. The age distributions presented for the AA and AB genotypes become progressively less meaningful as the A allele is eliminated from the population. The point in carrying out this calculation to the 60th breeding interval is to demonstrate that the elimination proceeds without a gross disruption of the age distribution of the eliminated genotypes.

181

Of considerable interest in population biology is the nature of r and K selection. r selection acts in a relatively unrestricted environment by favoring those phenotypes with high fecundity and prereproductive viability. In contrast, K selection acts in limited environments to favor phenotypes that can tolerate high population densities without severe loss of their reproductive potential. The AA homozygote in the example shown in Figure 3 displays a rapid initial increase in numbers. But, as population size increases, the fecundity advantage of this genotype is lost, and it is rapidly eliminated by the BB homozygote, which has an advantage under K selection. However, if in the example shown in Figure 3 the environment cycles with a periodicity of no more than 15 breeding intervals and each genotype is reduced in size in proportion to its numbers at the end of each cycle, the genotypes with an advantage under K selection will become less and less frequent until they are eliminated from the population. Clearly, different population structures may be produced by extreme r or extreme K selection. However, every genotype has both r and K characteristics to varying degrees, and the interaction of the two kinds of selection determines the population structure during the approach to equilibrium.

* This investigation was supported by Biomedical Support Grant FR-07015 from the National Institutes of Health and by grant GB-8191 from the National Science Foundation.

[1] Dobzhansky, Th., R. C. Lewontin, and O. Pavlovsky, *Heredity*, **19**, 597.

[2] Ohba, S., *Heredity*, **22**, 169 (1967).

[3] Haldane, J. B. S., *Proc. Cambridge Phil. Soc.*, **23**, 607 (1927).

[4] Norton, H. T. J., *Proc. London Math. Soc.*, **28**, 1 (1928).

[5] Istock, C. A., *Behav. Sci.*, **15**, 101 (1970). His model differs from ours in that it does not consider distinct genotypes or the approach to equilibrium, but looks at the consequences of age-specific selection on the population as a whole.

In addition, Drs. B. Charlesworth and J. T. Geisel have informed us that they are also studying the problem of age-specific selection, although from a somewhat different viewpoint.

[6] Leslie, P. H., *Biometrika*, **33**, 183 (1945).

[7] Leslie, P. H., *Biometrika*, **35**, 213 (1948).

[8] We have incorporated methods of calculating reproductive values, intrinsic rates of increase, and generation times for the individual genotypes in our computer program. Unfortunately, there are difficulties in defining these quantities for populations whose age structures are not stable. We defer consideration of these matters.

[9] Anderson, W., manuscript in preparation.

15

Reprinted from *Amer. Naturalist*, **106**(949), 388–401 (1972)

SELECTION IN POPULATIONS WITH OVERLAPPING GENERATIONS. II. RELATIONS BETWEEN GENE FREQUENCY AND DEMOGRAPHIC VARIABLES

Brian Charlesworth* and James T. Giesel†

Department of Biology, University of Chicago, Chicago, Illinois 60637

Marked fluctuations in population density and age composition often occur in natural populations. For instance, in insects with many generations in a year, population size is often low over winter and builds up rapidly during summer, declining again in the fall (e.g., Davison and Andrewartha 1948). In some species of insects without such seasonal changes, other factors, such as disease, produce fluctuations in numbers (Ullyett 1945). In vertebrates, irregular changes in density often occur (Errington 1945). Finally, certain mammals exhibit rather regular cycles in density with a periodicity of several years (Elton 1942).

In certain cases, changes in gene frequency in polymorphic systems are correlated with either the season of the year or with nonseasonal fluctuations in density. For example, Timofeef-Ressovsky (1940) described seasonal changes in frequencies of color morphs in a ladybird population. Dobzhansky (1943) found regular seasonal cycles in frequencies of inversion types in *Drosophila pseudoobscura*; similar observations were made by Dubinin and Tiniakov (1945) on *Drosophila funebris*. Gershenson (1945) observed seasonal changes in the frequency of black hamsters in Russia. Changes in gene frequency in protein polymorphisms in voles have recently been described (Semeonoff and Robertson 1968; Tamarin and Krebs 1969; Gaines 1970). Some of these changes are regular and correlated with the cycle of population density of the voles. A substantial number of gene loci in species fluctuating in population density may behave this way.

Seasonal changes in gene frequency may arise because genotypes differ in their sensitivity to temporally changing environmental factors, such as temperature. Another possibility, applicable to both seasonal and nonseasonal changes, is that genotypes differ in their response to population density; one genotype being favored when numbers are low, another when numbers are high. Chitty (1960) proposed that population density cycles are driven by self-perpetuating fluctuations in genotypic composition.

Here we propose an alternative explanation for fluctuations in gene

---

* Present address: Department of Genetics, University of Liverpool, Liverpool L69 3BX, England.

† Present address: Department of Zoology, University of Florida, Gainesville, Florida 32601.

183

frequency in populations with overlapping generations. Our explanation is based on the finding (to be discussed in detail by Charlesworth, in preparation) that genetic equilibrium with overlapping generations is often dependent upon demographic stability. Fluctuations in population age structure due to changes in mortality or fecundity rates, *which may be nonspecific with respect to genotype*, can therefore sometimes result in significant changes in gene frequency in polymorphic systems. If generations overlap, a genotype which reproduces early relative to other genotypes is favored in a population of predominantly young individuals, such as an increasing population. Conversely, such a genotype is at a disadvantage if age structure is weighted toward old individuals, as in a decreasing population. Fluctuations in population age structure, such as accompany population density cycles, therefore alter relative contributions of early and late reproducing genotypes to the gene pool.

In the next section, we present a computer model which illustrates shifts in gene frequency produced by the process just outlined. In the second section, we discuss the theoretical basis of the effect, which is a consequence of the general conditions for genetic equilibrium with overlapping generations derived by Charlesworth (in preparation).

GENETIC CHANGES INDUCED BY CHANGES IN POPULATION GROWTH

In most of this paper we employ Leslie's (1945) model of overlapping generations with discrete age classes. Properties of this model are identical with the continuous time model used by Charlesworth (1970), except that age classes are summed rather than integrated. As discussed by Charlesworth (1970), the models apply to two different reproductive situations: (1) Individuals shed their gametes into a pool from which zygotes are formed by random fusion of gametes. (2) Mating occurs by random "collision" between members of the opposite sex, which are then returned to the pool of potential mates.

Differences in survival and fecundity between sexes can be incorporated into both models, but differences in age-specific fecundities among genotypes cannot be represented with model 2. Whenever fecundity differences between genotypes are assumed, model 1 is implied, although we hope general conclusions apply equally well to both cases.

A single autosomal locus with two alleles A_1 and A_2 was modeled on a computer using a number (n) of discrete but interbreeding age classes for each genotype. Reproduction and passage to the next age class occur at discrete intervals, here called cycles. An individual carrying alleles i and j (where i and j may take values of 1 or 2) and present in age class x at cycle $t-1$ of the population produces $m_{ij}(t,x)$ offspring which enter age class 1 of cycle t. (Genotypes of offspring are determined by segregation, if parents are heterozygous, and by frequencies of A_1 and A_2 among homozygotes.) The individual in question then passes into age class $x+1$ of cycle t, with probability $1-\mu_{ij}(t,x)$, where $\mu_{ij}(t,x)$ is the probability of death between cycles t and $t+1$. If $k_{ij}(t,x)$ is the expected number of offspring produced in cycle t by an ij individual born at cycle $t-x$, we have

$$k_{ij}(t, x) = l_{ij}(t, x) m_{ij}(t, x) \qquad (1)$$

where

$$l_{ij}(t, x) = \prod_{s=1}^{x-1} [1 - \mu_{ij}(t - s, x - s)].$$

Equations for calculating the population's trajectory are described in the Appendix.

If $l(x)$ and $m(x)$ (assumed constant in time) are known for each genotype, together with an initial distribution of genotypes and ages, trajectories of gene frequency and total births (and any functions derived from them) can be calculated by successive application of equations (A1) and (A2) to cycles 1, 2 . . . t. Effects of changing demography on gene frequency can be studied by changing values of $l(x)$ or $m(x)$ functions abruptly at arbitrary times.

In our computer program, the Wrightian fitness w_{ij} of the genotype carrying alleles i and j in an equilibrium population is written as

$$w_{ij} = \sum_{x=1}^{n} e^{-rx} k_{ij}(x) \qquad (2)$$

(after Charlesworth 1970, and in preparation).

Here, r is the rate of population growth of an equilibrium population $(r = \log B(t)/B(t-1))$ in its stable age distribution. This assumption is justified by Charlesworth (in preparation).

Fitness in an equilibrium population is equal to reproductive value at birth (Fisher 1930) for a member of a given genotype (eq. [2]). Since reproductive value measures the proportionate contribution of an individual to later generations (Crow and Kimura 1970, chap. 1), this is reasonable.

To obtain the equilibrium gene frequency (if one exists), we simply substitute w_{ij} into the standard equation (Crow and Kimura 1970):

$$\hat{p} = \frac{w_{12} - w_{22}}{2w_{12} - w_{11} - w_{22}}. \qquad (3)$$

The existence of a polymorphic equilibrium (with $\hat{p}$ between 0 and 1) is guaranteed if $w_{12} > w_{11}, w_{22}$ or vice versa. In our simulations, the heterozygote always had a higher $k(x)$ function for every value of x, guaranteeing heterosis in w_{ij} whatever the value of r. Heterosis in w_{ij} is necessary for preservation of a polymorphism by natural selection when the environment is spatially uniform and no selective differences exist (Norton 1926).

For the equilibrium population to have a stable age distribution, it must satisfy the Lotka equation (Charlesworth 1970). We therefore have

$$\hat{p}^2 w_{11} + 2 \hat{p} \hat{q} w_{12} + \hat{q}^2 w_{22} = 1. \qquad (4)$$

We selected an $l(x)$ and an $m(x)$ function for each genotype such that the expected number of offspring produced by an ij individual at age x was $k_{ij}(x)$. Then $\hat{p}$ for an equilibrium population which grows at rate y is given by equation (3) with y as the value of r. But since $k_{ij}(x)$ functions were chosen without regard to satisfying the Lotka equation, instead of equation (4) we get

$$\hat{p}^2 \sum_{x=1}^{n} e^{-rx} k_{11}(x) + 2\,\hat{p}\,\hat{q} \sum_{x=1}^{n} e^{-rx} k_{12}(x)$$

$$+ \hat{q}^2 \sum_{x=1}^{n} e^{-rx} k_{22}(x) = C_y, \tag{5}$$

where C_y in general differs from 1. If we let $k_{ij}^{(y)}(x) = k_{ij}(x)/C_y$ and substitute functions for the original $k_{ij}(x)$, both equations (3) and (4) are satisfied. We can therefore arrange to have an equilibrium population growing at any predetermined rate y. To examine the effect of changing population growth on gene frequency, simply exchange the initial set of $k_{ij}^{(y)}(x)$ functions for a new set with a different value of y (and also of $\hat{p}$).

The population then starts changing toward the new values of y and $\hat{p}$. Since the relative values of the $k_{ij}^{(y)}(x)$ functions (i.e., $k_{11}^{(y)}(x)/k_{12}^{(y)}(x)$, $k_{11}^{(y)}(x)/k_{22}^{(y)}(x)$ etc.) are not altered by the value of y, there is no change in the direction of selection as measured by the $l(x)$ and $m(x)$ functions with changed y. Any changes in gene frequency are purely responses to changes in the demographic structure of the population.

We simulated a population with a y of zero for 40 cycles (using 13 age classes for each genotype), and then changed y to -10% for 40 cycles, finally raising it to 10% for the last 40 cycles. This pattern of changes was repeated several times to ensure that the pattern of gene frequency change (if any) associated with changes in population growth rate had stabilized. The resultant population cycle resembles vaguely the pattern observed in oscillating vole populations (Tamarin and Krebs 1969) (see fig. 2).

In one set of runs, only differences in $l(x)$ functions among genotypes existed. Changes in gene frequency produced by fluctuating population growth were slight, even when differences in fitness between homozygotes and heterozygotes were as high as 10% of the heterozygote fitness.

In contrast, much greater changes in gene frequency occurred with genotypic differences in $m(x)$. Table 1 summarizes a set of runs in which one homozygote started reproducing one age class earlier than the other. Relative shapes of $k(x)$ functions for homozygotes are shown in figure 1 (numbers in the figure correspond to runs in table 1). In each case, the heterozygote started reproducing at the same stage as the earlier homozygote, and its $k(x)$ function was approximately equal to the higher of

THE AMERICAN NATURALIST

TABLE 1

EXAMPLES OF CHANGES IN GENE FREQUENCY AS FUNCTIONS OF CHANGED RATE OF POPULATION INCREASE

Population	y	w_{11}	w_{22}	$\hat{p}$	p_{obs}	Cycle Number
1	0.0	.8354	.9119	.3486	.3518	40
	−0.10	.8751	.8810	.4879	.3766	60
					.3959	80
	0.10	.7869	.9381	.2249	.3528	100
					.3247	120
	0.0	.8354	.9189	.3486	.3308	140
2	0.0	.6747	.9302	.2169	.2200	40
	−0.10	.8106	.8982	.3496	.2432	60
					.2618	80
	0.10	.6809	.9549	.1239	.2230	100
					.1992	120
	0.0	.6747	.9302	.2169	.2036	140
3	0.0	.8944	.9629	.2601	.2639	40
	−0.10	.9114	.9459	.3790	.2772	60
					.2889	80
	0.10	.8735	.9755	.1625	.2697	100
					.2549	120
	0.0	.8944	.9629	.2601	.2560	140
4	0.0	.9113	.9119	.4983	.5019	40
	−0.10	.9547	.8810	.7241	.5318	60
					.5553	80
	0.10	.8585	.9382	.3040	.5037	100
					.4675	120
	0.0	.9113	.9119	.4983	.4749	140
5	0.0	.8470	.9290	.3169	.3203	40
	−0.10	.9067	.8979	.5225	.3507	60
					.3753	80
	0.10	.7808	.9533	.1757	.3248	100
					.2923	120
	0.0	.8470	.9290	.3169	.2983	140

Population	y	w_{11}	w_{22}'	$\hat{p}$	p_{obs}	Cycle Number
6	0.0	.8752	.9225	.3832	.3867	40
	−0.10	.9282	.8903	.6046	.4188	60
					.4442	80
	0.10	.8145	.9482	.2182	.3904	100
					.3547	120
	0.0	.8752	.9225	.3832	.3616	140
7	0.0	.9448	.9629	.4020	.4060	40
	−0.10	.9627	.9459	.5920	.4234	60
					.4384	80
	0.10	.9227	.9755	.2410	.4139	100
					.3931	120
	0.0	.9448	.9629	.4020	.3950	140
8	0.0	.8161	.9302	.2751	.2783	40
	−0.10	.8842	.8982	.4680	.3080	60
					.3322	80
	0.10	.7428	.9549	.1493	.2825	100
					.2516	120
	0.0	.8161	.9302	.2751	.2570	140
9	0.0	.8083	.9153	.3265	.3297	40
	−0.10	.8753	.8960	.4549	.3514	60
					.3683	80
	0.10	.7680	.9360	.2163	.3254	100
					.2994	120
	0.0	.8083	.9153	.3265	.3068	140

NOTE.—The numbers at the head of each block of data denote the $k(x)$ distributions (fig. 1) characteristic of the population. Fitness values are relative to the heterozygote fitness.

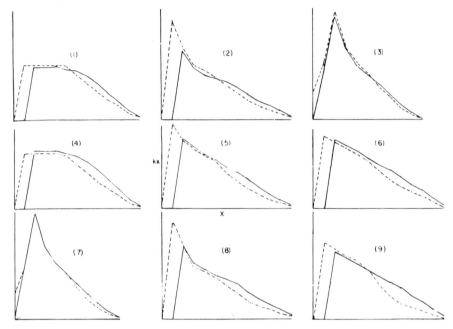

FIG. 1.—Shapes of $k(x)$ functions used in computer calculations. Solid lines represent $k_{11}(x)$, dashed lines $k_{22}(x)$.

the homozygote $k(x)$ functions. Columns 3 and 4 of table 1 give equilibrium fitnesses of A_1A_1 and A_2A_2 (relative to a heterozygote fitness of 1) appropriate for values of y in column 2.

Changes in gene frequency as great as 0.06 occur within 20 cycles of changing from negative to positive y; within 40 cycles, changes of up to 0.09 can occur. In general, on changing to new $k_{ij}^{(y)}(x)$ functions, gene frequency immediately moved toward the new equilibrium frequency, and after some oscillations of small amplitude total birth rate stabilized at an almost constant rate of increase. In figure 2, the logarithm of the number of individuals in age class 1 (on an arbitrary scale) and gene frequency in the first age class are plotted as functions of cycle number. Total num-

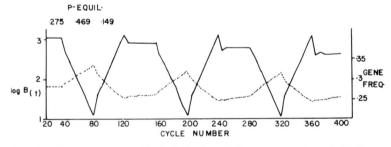

FIG. 2.—Gene frequency (dashed line) and $\log_{10}$ of number of births are shown over 400 cycles for population (8).

ber of individuals of all ages, and gene frequency among all ages, changed similarly.

The form of equation (2) explains these changes in gene frequency If we compare two genotypes, one of whose $k(x)$ function is greater than the other for low values of x, then it will be favored over the second when r is positive, if

$$\sum_{x=1}^{n} k(x)$$

is roughly the same for both genotypes. Conversely, if r is negative, greater weight is given to later age classes and genotypes which reproduce late are favored. In the computer results just described the allele whose homozygote reproduced relatively early was favored in increasing populations, and disfavored in decreasing populations. Changing population growth rate had little effect on gene frequencies when the genotypes differed only in their $l(x)$ functions. Differences in $l(x)$ alone produced little difference in shape of the $k(x)$ functions of the two homozygotes.

In addition, the fact that the changes in gene frequency are almost exactly in phase with changes in population growth rates is explained in these terms. Changing the rate of population growth to a new level also changes the equilibrium gene frequency to a new value (eqq. [2] and [3]). If the new equilibrium is stable, gene frequency starts shifting toward it. When y changes from zero to a negative value, selection immediately starts shifting gene frequency in favor of the allele associated with late reproduction. Changing from a negative to positive rate of growth, favors the "early-reproducing" allele more than in the stationary population. This results in an even bigger difference between the actual gene frequency and the gene frequency toward which the population is moving, and gene frequency change will reverse abruptly. Unless the increase phase is long lasting, gene frequency will not have had time to attain the equilibrium appropriate to an increasing population by the time population growth ceases, but may approach the value for a stationary population. Hence, when the population enters the stationary phase once more relatively little change in gene frequency results.

Changes in gene frequency produced by changes in population growth rates may occur even if the environmental agents which effect population size have no specific effects on genotypes (in terms of age-specific mortality and fecundity factors). The ratio of the $k(x)$ functions of any two genotypes may remain constant throughout the whole cycle of population growth, and still there may be shifts in their relative fitnesses as defined by equation (2). In the computer simulations we changed only the factor C_y in: $k_{ij}{}^{(y)}(x) = k_{ij}(x)/C_y$. Each genotype was affected equally.

Two further conclusions become obvious if equation (2) is rewritten in a slightly different form:

$$w_{ij} = \sum_{x=1}^{n} f_{ij}(x)\, m_{ij}(x),\qquad(6)$$

where

$$f_{ij}(x) = e^{-rx}\, l_{ij}(x).$$

The factor $f_{ij}(x)$ measures the relative frequency of age class x among ij individuals, and reflects the general age distribution of the population. If the shapes of the $m_{ij}(x)$ functions do not change under different conditions, changes in the relative fitnesses of different genotypes depend on changes in the $f_{ij}(x)$ functions, and hence upon age distribution. Thus changes in gene frequency may accompany changes in age distribution even if rate of population growth is constant. If, for example, an additional probability of death per cycle (β) is imposed equally on each genotype and age class, and if its effect on population growth rate is compensated by increased fecundity at each age of each genotype, the new value of w_{ij} in (6) is

$$\overset{*}{w}_{ij} = C \sum_{x=1}^{n} f_{ij}(x)\, [l-\beta]^{x}\, m_{ij}(x),\qquad(7)$$

where C is the factor by which fecundity is increased.

Effects on relative fitness are the same as changing r by $\log(1-\beta)$ (from eq. [4]). Changes in age distribution favoring younger individuals favor early reproducing genotypes; those favoring later ages favor late reproducers.

Conversely, changes in rate of population growth which do not affect the population's age distribution do not alter relative fitnesses. If population growth is diminished by an additional probability of death per cycle (β), r for the equilibrium population must change to $r + \log(1-\beta)$, so that

$$\overset{*}{w}_{ij} = \sum_{x=1}^{n} f_{ij}(x)\, e^{-x\log(1-\beta)}\, [1-\beta]^{x}\, m_{ij}(x)$$
$$= \sum_{x=1}^{n} f_{ij}(x)\, m_{ij}(x).\qquad(8)$$

There is no change in relative fitness. In general, changes in probabilities of death uncompensated by changes in fecundity produce smaller shifts in gene frequency than compensated changes. Changes in fecundity alone produce larger effects than changes in death-rate alone.

Numerical calculations of equilibrium gene frequencies for the same population under different demographic regimes illustrate these points. To facilitate calculations, we used a continuous time model (Charlesworth

190

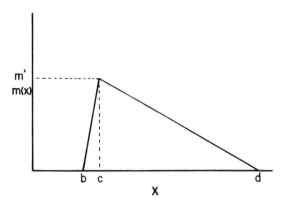

FIG. 3.—The ordinate is $m(x)$; the abscissa is x; m' is the peak fecundity; b, c, and d are ages at first reproduction, peak fecundity, and last reproduction, respectively.

1970) in which summation of age classes is replaced by integration over ages.

Our $m(x)$ function for each genotype is a function of x and the product of the $l(x)$ and $m(x)$ functions (fig. 3) (Lewontin 1965). The $m(x)$ function of the ijth genotype is completely characterized by the peak fecundity m, ages at first and last egg (b and d), and age at peak fecundity (c). Assume for convenience that only b varies between genotypes. We have

$$m_{ij}(x) = \frac{m'(x - b_{ij})}{(c - b_{ij})} \qquad b_{ij} \leqslant x \leqslant c;$$

$$m_{ij}(x) = \frac{m'(d - x)}{(d - c)} \qquad c \leqslant x \leqslant d. \qquad (9)$$

We compared equilibrium gene frequencies in an expanding population with those in stationary and declining populations. Changes in population growth were produced by imposing various combinations of increase in mortality and decrease in fecundity on the $l(x)$ and $m(x)$ functions appropriate to the expanding population. The basic functions for each genotype are of the form

$$l_{ij}(x) = e^{-\mu_{ij}x},$$

where μ_{ij} is the basic death rate of the ijth genotype.

To produce stationary and declining populations, m was reduced from its value in the increasing population, and/or an additional death rate, a, was imposed on each genotype up to age c. Performing the integration equivalent to equation (2), we find

$$w_{ij} = \frac{m'}{(c - b_{ij})} b \int_{ij}^{c} e^{-(r + a + \mu_{ij})x} (x - b_y)\, dx + \frac{m'e^{-ac}}{(d - c)}$$

$$\int_{c}^{d} e^{-(r + \mu_{ij})x} (d - x)\, dx, \qquad (10)$$

so that

$$w_{ij} + m' \left\{ \frac{e^{-(r+a+\mu_{ij})b_{ij}} - e^{-(r+a+\mu_{ij})c}}{(c - b_{ij})(r + a + \mu_{ij})^2} + \frac{e^{-ac}\left[e^{-(r+\mu_{ij})d} - e^{-(r+\mu_{ij})c}\right]}{(d - c)(r + \mu_{ij})^2} \right\}$$
$$+ \frac{m' \, a \, e^{-(r+a+\mu_{ij})c}}{(r + \mu_{ij})(r + a + \mu_{ij})}. \quad (11)$$

Here, r, a, and m have values appropriate to the equilibrium population in question. To determine these values we assumed that the ratio $w_{11}:w_{12}:w_{22}$ has some desired value, giving a certain equilibrium $\hat{p}$ for the basic population, growing at an r of 0.05. The values of the basic death rate of A_1A_2 (μ_{12}) and of b_{11}, b_{12}, b_{22}, c and d were assumed given. Then μ_{11} and μ_{22} were calculated from the ratios w_{11}/w_{12} and w_{22}/w_{12} by Newton-Raphson iteration of equation (11), noting that a is zero. Given the values of the parameters other than m', m' can be found from equation (4). Knowing all the parameters for the basic population, fitnesses and equilibrium frequencies for stationary and declining populations can be calculated, under various combinations of reduction in m' and increase in a. The procedure is to assume an arbitrary value for a and calculate the equilibrium for the appropriate r using equations (3) and (11); m' can then be found as above.

Some results of these calculations are given in tables 2 and 3. In table 2, all genotypes have c and d equal to 20 and 50, respectively, but b varies between genotypes. Results support conclusions developed above: the allele associated with the early-reproducing homozygote is favored in increasing populations and those with high a. Conversely, the biggest change in gene

TABLE 2
CHANGES IN EQUILIBRIUM FREQUENCIES WHEN THERE ARE DIFFERENCES
IN TIMING OF REPRODUCTION

STATIONARY POPULATION ($r = 0$)			DECLINING POPULATION ($r = -0.05$)		
a	m'	$\hat{p}$	a	m'	$\hat{p}$
Model 1: Fitnesses in the expanding population ($r = 0.05$) are as $0.95:1:0.97$; $\mu_{11} = 0.02181$, $\mu_{12} = 0.02000$, $\mu_{22} = 0.02038$; $b_{11} = 10$, $b_{12} = 9.8$, $b_{22} = 10.2$; $\hat{p} = 0.3750$, $m' = 0.2777$.					
0	0.0848	0.3049	0	0.0219	0.2432
0.02	0.1238	0.3170	0.04	0.0476	0.2590
0.04	0.1804	0.3297	0.08	0.1029	0.2788
0.06	0.2623	0.3432	0.12	0.2211	0.3008
Model 2: Fitnesses in the expanding population ($r = 0.05$) are as $0.90:1:0.94$; $\mu_{11} = 0.02227$, $\mu_{12} = 0.02000$, $\mu_{22} = 0.01791$; $b_{11} = 10$, $b_{12} = 9$, $b_{22} = 11$; $\hat{p} = 0.3750$ and $m' = 0.2710$.					
0	0.0822	0.1103	0.12	0.2132	0.0812*
0.02	0.1202	0.1742	...	...	...
0.04	0.1754	0.2309	...	...	...
0.06	0.2553	0.2811	...	...	...

* For $a < 0.12$, the polymorphism is destabilized, and A_2 tends to be fixed.

TABLE 3

CHANGES IN EQUILIBRIUM FREQUENCIES WHEN THERE ARE NO DIFFERENCES
IN TIMING OF REPRODUCTION

STATIONARY POPULATION $(r = 0)$			DECLINING POPULATION $(r = -0.05)$		
α	m'	$\hat{p}$	α	m'	$\hat{p}$
Model 1: Fitnesses in the expanding population $(r = 0.05)$ are as $0.95:1:0.97$; $\mu_{11} = 0.02301$, $\mu_{12} = 0.02000$, $\mu_{22} = 0.02136$; $\hat{p} = 0.3750$ and $m' = 0.2808$.					
0	0.0859	0.3754	0	0.0223	0.3758
0.02	0.1253	0.3753	0.04	0.0483	0.3757
0.04	0.1826	0.3753	0.08	0.1044	0.3757
0.06	0.2655	0.3753	0.12	0.2241	0.3757
Model 2: Fitnesses in the expanding population $(r = 0.05)$ are as $0.90:1:0.94$; $\mu_{11} = 0.02474$, $\mu_{12} = 0.0200$, $\mu_{22} = 0.02278$; $\hat{p} = 0.3750$ and $m' = 0.2862$.					
0	0.0877	0.3757	0	0.0228	0.3766
0.02	0.1280	0.3757	0.04	0.0495	0.3765
0.04	0.1865	0.3756	0.06	0.0728	0.3765
0.06	0.2711	0.3756	0.12	0.2295	0.3764

frequency from the increasing population is present only in the declining population with lowered fecundity. Since the additional death rate, α extends only to the age of peak reproduction, a given α does not compensate fully for a drop in r of the same magnitude. That larger differences in gene frequency associated with different demographic circumstances are present in model 2 than in model 1 reflects the larger differences in b in model 2.

When all genotypes have the same b (c and d and 20 and 50), there is no significant effect of demography on gene frequency (table 3).

Mortality and fecundity factors associated with different demographies are nonspecific with respect to genotype; the ratio of the $k(x)$ functions for any two genotypes is independent of α and m'.

DISCUSSION

We conclude that gene frequencies of polymorphic systems can be altered by natural selection in response to changes in a population's demographic structure without changed selection as measured by genotypic $l(x)$ and $m(x)$ functions. Since many natural populations fluctuate considerably in size and age composition, they are probably rarely demographically stable for more than short periods. The corollary is that gene frequencies in polymorphic systems may seldom be in equilibrium, even without effects of finite population size, immigration, and variations in selection intensity. A small number of loci behaving in this way could stir up a gene pool significantly by dragging along closely linked genes.

Our results are of obvious relevance to fluctuating gene frequencies quoted earlier, especially those related to vole populations, where several protein polymorphisms change in gene frequency with a population cycle similar to figure 2. It is tempting to interpret these changes in terms of genotypic differences in sensitivity to the density-regulating factors producing cycles in numbers, or in terms of Chitty's (1960) hypothesis that oscillations in numbers may be caused by genetic factors. It would, however, be surprising if several different protein polymorphisms, chosen more or less at random, were all related to the mechanism of population fluctuations. Our results suggest that such changes in gene frequency could be purely mechanical side effects of changing demography. Indeed, Gaines (1970) found differences in reproductive parameters between genotypes of *Microtus ochrogaster* and *Microtus pennsylvanicus* which accord well with the way gene frequencies change on our hypothesis.

From an evolutionary viewpoint, the nonspecific dependence of gene frequency on r here described suggests genetic change in response to a shift from saturated environments in which population sizes are stationary to colonizing environments in which they increase rapidly. Such changes are, of course, compounded by founder effects and by changes in selection intensity in a new environment, but might add an additional element to the genetic revolution of Mayr (1963, chap. 17).

Study of natural selection in populations with overlapping generations emphasizes the importance of taking ecological and demographic variables into account in population genetics. Phenomena described here are impossible with discrete generations and depend on explicit incorporation of ecological variables into genetic models.

SUMMARY

Fluctuations in demographic structure of a population with overlapping generations can induce gene frequency changes in polymorphic systems, even if there are no specific relations between genotypic differences and environmental agents which cause demographic changes. The possible relevance of this finding to gene frequency changes in natural populations is discussed.

ACKNOWLEDGMENTS

We thank R. C. Lewontin for discussion of this work and criticism of the manuscript. The work was carried out while the authors were in receipt of postdoctoral fellowships from the Ford Foundation Grant for Population Biology.

APPENDIX

EQUATIONS FOR GENE FREQUENCY CHANGE

If the population is started at cycle zero with Hardy-Weinberg equilibrium in the birth cohort (age class 1), then at an arbitrary cycle t, the state of the population can be described by two equations:

$$p(t) = \left\{ \tfrac{1}{2} f_1(t) + \sum_{x=1}^{t} \right.$$
$$B(t-x)p(t-x)\,[\,p(t-x)k_{11}(t,x) \qquad\qquad (A1)$$
$$\left. + q(t-x)k_{12}(t,x)\,] \right\} \Big/ B(t);$$

$$B(t) = f_2(t) + \sum_{x=1}^{t}$$
$$B(t-x)\,[\,p^2(t-x)k_{11}(t,x)$$
$$+ 2p(t-x)q(t-x)k_{12}(t,x)$$
$$+ q^2(t-x)k_{22}(t,x)\,]. \qquad (A2)$$

Here, $p(t)$ and $q(t)$ are the frequencies of A_1 and A_2 in the birth cohort at cycle t; $B(t)$ is the total number of individuals in the birth cohort; $f_1(t)$ is the contribution from individuals of age greater than 1 who were present at cycle zero to the number of A_1 genes present in the birth cohort at cycle t; $f_2(t)$ is the contribution from such individuals to the total number of individuals in the birth cohort at time t. It should be noted that $k_{ij}(t,x)$ is zero for $x > n$.

If $\phi_{ij}(x)$ is the number of individuals of the ijth genotype aged x $(x > 1)$ at cycle zero, and if $k'_{ij}(t, t+x)$ is the expected number of births each of them produces at cycle t, we have:

$$\tfrac{1}{2} f_1(t) = \sum_{x=1}^{n-(t-1)} [\,\phi_{11}(x)\,k'_{11}(t, t+x) + \tfrac{1}{2}\phi_{12}(x)k'_{12}(t, t+x)\,]; \qquad (A3)$$

$$f_2(t) = \tfrac{1}{2} f_1(t) + \sum_{x=2}^{n-(t-1)}$$
$$[\tfrac{1}{2}\phi_{12}(x)\,k'_{12}(t, t+x) + \phi_{22}(x)\,k'_{22}(t, t+x)\,]. \qquad (A4)$$

Also,

$$k'_{ij}(t, t+x) = m_{ij}(t, t+x-1)\prod_{s=1}^{t-1}[1 - \mu_{ij}(s, x+s-1)]. \qquad (A5)$$

Note that for $t > n$, $f_1(t)$ and $f_2(t)$ are both zero. Also, when death rates and fecundities are independent of time for the period $t < n$ we can simplify (A5) to

$$k'_{ij}(t+x) = m_{ij}(t+x-1)\,l_{ij}(t+x)/l_{ij}(x). \qquad (A6)$$

Equations (A3) and (A4) assume that each genotype has the same number of age classes, which was in fact the case in our calculations. The modifications required for the more general case are trivial.

LITERATURE CITED

Charlesworth, B. 1970. Selection in populations with overlapping generations. I. The use of Malthusian parameters in population genetics. Theoretical Population Biol. 1:352–370.

Chitty, D. 1960. Population processes in the vole and their relevance to general theory. Can. J. Zool. 38:99–113.

Crow, J. F., and M. Kimura. 1970. An introduction to population genetics theory. Harper & Row, New York. 591 p.

Davison, J., and H. G. Andrewartha. 1948. Annual trends in a population of *Thrips imaginis* (Thysanoptera). J. Anim. Ecol. 17:193–199.

Dobzhansky, Th. 1943. Genetics of natural populations. IX. Temporal changes in the composition of populations of *Drosophila pseudoobscura*. Genetics 28:162–186.

Dubinin, N. P., and G. G. Tiniakov. 1945. Seasonal cycles and the concentration of inversions in populations of *Drosophila funebris*. Amer. Natur. 79:570–572.

Elton, C. 1942. Voles, mice and lemmings. Problems in population dynamics. Clarendon, Oxford. 586 p.

Errington, P. L. 1945. Some contributions of a fifteen-year local study of the northern bob-white to our knowledge of population phenomena. Ecol. Monogr. 15:1–34.

Fisher, R. A. 1930. The genetical theory of natural selection. Clarendon, Oxford. 291 p.

Gaines, M. 1970. Genetic changes in fluctuating vole populations. Ph.D. dissertation. Univ. Indiana, Bloomington.

Gershenson, S. 1945. Evolutionary studies on the distribution and dynamics of melanism in the hamster (*Cricetus cricetus*). II. Seasonal and annual changes in the frequency of black hamsters. Genetics 30:233–251.

Leslie, P. H. 1945. On the use of matrices in certain population mathematics. Biometrika 33:183–212.

Lewontin, R. C. 1965. Selection for colonizing ability, p. 79–94. *In*: H. G. Baker and G. L. Stebbins [ed.], The genetics of colonizing species. Academic Press, New York.

Mayr, E. 1963. Animal species and evolution. Belknap, Cambridge, Mass. 797 p.

Norton, H. T. J. 1926. Natural selection and Mendelian variation. London Math. Soc., Proc. 28:1–45.

Semeonoff, R., and F. W. Robertson, 1968. A biochemical and ecological study of plasma esterase polymorphism in natural populations of the field vole, *Microtus agrestis*. Biochem. Genet. 1:205–277.

Tamarin, R. H., and C. J. Krebs. 1969. *Microtus* population biology. II. Genetic changes at the transferin locus in fluctuating populations of two vole species. Evolution 23:183–211.

Timofeef-Ressovsky, N. W. 1940. Zur Analyse des Polymorphismus bei *Adalia bipunctata*. Biol. Zentralbl. 60:130–137.

Ullyett, G. C. 1945. Mortality factors in populations of *Plutella maculipennis*, Curtiss (Tincidae, Lepidoptera) and their relation to the problem of control. Dep. Agr. South Afr. Mem. 2:77–202.

Editors' Comments
on Papers 16 and 17

16   CROW
Some Possibilities for Measuring Selection Intensities in Man

17   BODMER
Demographic Approaches to the Measurement of Differential Selection in Human Populations

SOME EMPIRICAL APPLICATIONS

The empirical use of age-specific birth and death rates, as well as other demographic characteristics of a population, are exemplified by the next two papers. In Paper 16, Crow investigates the potential effects of natural selection on a human population. He asks how much selection could occur in a population with a given set of vital rates if all the variability in those rates were genetically relevant. Crow measures this by his· *index of total selection* (often called the index of the opportunity for selection), which can be related to Fisher's fundamental theorem as expressed by Li (1955). Crow's index, *I*, is a measure of the relative change in fitness of a population that may occur within one generation, suitably defined, and is equal to the variance in fitness over the mean fitness squared. Crow partitions this index into segments that reflect the operation of differential fertility and mortality in the population.

For each genetic type, Crow assigns a mean fitness. If the observed differences in survival and fertility are due only to differences in fitness, selection will operate at the maximal level, *I*. Knowing the survivorship to reproductive age and the distribution of family size of those reaching at least that age, Crow is able to compute the relative contributions of fertility and mortality differences to *I*.

Crow has noted (personal communication) that this index can be expressed in a more instructive way as follows:

$$(1 + I) = (1 + I_m) (1 + I_f)$$

197

since $I_m = [1 - I(x)] / I(x)$. This form shows that, if other stages of mortality are being incorporated into the model, such as, for example, embryonic mortality as well as childhood mortality, the index could be written as

$$(1 + I) = (1 + I_e) (1 + I_m) (1 + I_f)$$

and so on, for combinations of factors. Crow's index can be computed from any data containing the required age-specific vital rates. However, it should be noted that the index does not measure the amount of selection which is actually occurring, but only the maximum that could occur if all variability in vital rates (life history) were due to relevant genetic differences. This index is particularly sensitive to the amount of heritability of fertility and mortality.

The assessment of the actual fitness of a trait in a given age-structured population can be made by use of the schedule of age-specific birth and death rates that trait bearers manifest compared with others in the population. In some cases this information might be available. In Paper 17 Bodmer outlines the simple use of stable population theory for this purpose. He shows that fitness may be measured by either the *net reproductive rate* of various types within a population or, if generation times for the types differ substantially, by the *intrinsic growth rate*, which corresponds to Fisher's Malthusian parameter for the trait. Bodmer considers also the assessment of fitness for those individuals who may be known to be at risk for a trait that has variable age of onset, using schizophrenia as an example. This model assumes that all those with genes for the trait will manifest the trait during their lifetimes; since this may not be true and some gene bearers may live out a normal lifetime with normal vital rates, the net fitness depression due to the trait may be less than computed; even this assumes that all schizophrenia cases are of the same genetic risk group, or at least that the genetic risk group behaves as do all schizophrenics.

This model has been expanded somewhat in Cavalli-Sforza and Bodmer (1971); Nei (personal communication) has pointed out that the model does not consider those who have the trait but do not manifest it and who may never manifest it; the unstated assumption that all trait bearers will eventually die from the trait is necessary for the model to have the genetic relevance claimed for it.

It should also be noted that the use of the Malthusian

parameter for fitness carries with it the assumption of stable population conditions. That is, we assume that the relative fitness difference as measured by differential growth rates will not be affected when the growth of the present population changes, as it must. The fitness measured is only for the specific time and place for which data exist. It would be a significant additional step to assume that the fitness calculated in this way has any generality or that it can be used in the explanation of present gene frequencies by projection from any hypothetical past population. Nonetheless, Bodmer provides an indication of ways to proceed in the analysis of the demographic genetics of present populations, especially as regards the assessment of medical and clinical aspects related to public health.

16

Reprinted from *Human Biol.*, **30**(1), 1–5, 13 (1958)

SOME POSSIBILITIES FOR MEASURING SELECTION INTENSITIES IN MAN [1] *

BY JAMES F. CROW

University of Wisconsin

IT is convenient to consider selection intensity at three levels—total, phenotypic, and genotypic.

There can be selection only if, through differential survival and fertility, individuals of one generation are differentially represented by progeny in succeeding generations. The extent to which this occurs is a measure of *total* selection intensity. It sets an upper limit on the amount of genetically effective selection.

Not all the differential can be associated with differences in phenotype, for there are large environmental and random elements in survival and reproduction. But, to the extent that differences can be associated with phenotype, *phenotypic* selection intensity can be measured.

Selection can be genetically effective only if it is *genotypic*, that is, if different genotypes make differential contributions to future genera-

[1] Paper No. 660 from the Department of Genetics. I should like to acknowledge the help received in numerous discussions with Dr. Newton Morton and Dr. Ove Frydenberg.

* The first four articles in this issue were read before the Wenner-Gren Supper Conference on Natural Selection in Man, held at the University of Michigan on 12 April 1957. These papers are also being published as *Memoir No. 86* of the American Anthropological Association and also as a hard cover book, *Natural Selection in Man*, by the Wayne State University Press.

tions. Thus natural selection is an inefficient process with only a fraction, perhaps a small fraction, of mortality and fertility differences being genetically effective. Furthermore, very little of the genotypic selection effects any permanent biological improvement. As many authors have pointed out (see especially Haldane, '54a), most of the selection is devoted to eliminating recurrent harmful mutants, maintaining systems of balanced polymorphism, and adjusting to momentary fluctuations in the environment—in other words, to maintaining the status quo. Selection is a necessary but not sufficient condition for the directional change of gene frequencies on which evolution depends.

The Total Selection Intensity [2]

The total selection that can occur in man is limited by his low reproductive rate, yet there is still room for fairly intense selection. The most fertile human populations produce 10 or more children per family. This means that if four-fifths of the children were to die prematurely the population number could still be maintained.

I suggest that the total amount of selection is best measured by the ratio of the variance in progeny number to the square of the mean number. (The parents and offspring are assumed to be counted at the same age, perhaps at birth.) This sets an upper limit on the amount of gene frequency change that selection can effect. The reason for the choice of this measure will now be developed.

I shall use *fitness* in a strictly Darwinian sense as the expected number of progeny, counted at the same age as the parents (Fisher, '30). In a precise theoretical analysis of populations, such as man, with overlapping generations it would be better to measure fitness as the natural logarithm of the progeny number, the Malthusian parameter of Fisher ('30: 25). But the cruder formulation is satisfactory for most purposes.

Suppose that the parental generation is counted at birth and consists of a fraction p_1 of individuals with fitness w_1, p_2 with fitness w_2, and so on ($p_1 + p_2 + p_3 + \cdots = 1$). The average fitness of the population is given by

$$\bar{w} = \frac{\sum p_i w_i}{\sum p_i} = \sum p_i w_i.$$

Because of differential productivity, the expected frequency of type

[2] This section has been greatly expanded since the oral presentation as a result of a suggestion of Dr. J. V. Neel, for which I am indeed grateful.

i next generation will be $p_i w_i$, assuming the complete heritability and constancy of fitness. The average fitness next generation will then be

$$\frac{\sum (p_i w_i) w_i}{\sum p_i w_i} = \frac{\sum p_i w_i^2}{\bar{w}}.$$

The relative increment in fitness between the two generations is

$$\frac{\Delta \bar{w}}{\bar{w}} = \frac{\sum p_i w_i^2 - \bar{w}^2}{\bar{w}^2} = \frac{V}{\bar{w}^2} = I \tag{1}$$

where V is the variance in fitness, and I will be called the *Index of Total Selection*.

This means that if fitness is completely heritable, that is, if each offspring has exactly the average of his parents' fitnesses, the fitness of the population will increase at rate I. A trait or a gene that is genetically correlated with fitness will increase in proportion to this correlation. The index therefore provides an upper limit to the rate of change by selection. The actual change in a character will depend also on its heritability and correlation with fitness.

It is of interest to separate I into two components, associated with mortality and differential fertility. This can be done as follows:

Assume p_d parents (counted at birth) die prematurely, and p_s survive until the child bearing period and have varying numbers of births ($p_s + p_d = 1$).

		births (x)	
Premature deaths	p_d	0	$\sum_s p_i = p_s$

Survivals $p_s \begin{cases} p_0 & 0 \\ p_1 & 1 \\ p_2 & 2 \\ p_n & n \end{cases}$

$\bar{x} =$ overall mean number of births

$\bar{x}_s =$ mean number of births per surviving parent

$p_s = \bar{x}/\bar{x}_s$.

The index I, is defined as $V/\bar{x}^2$, where V is the variance in the number of progeny per parent (counting the non-survivors as leaving 0 descendants).

$$I = \frac{V}{\bar{x}^2} = \frac{1}{\bar{x}^2} \left[p_d (0 - \bar{x})^2 + \sum_s p_i (x_i - \bar{x})^2 \right]$$
$$= \frac{1}{\bar{x}^2} \left[p_d (0 - \bar{x})^2 + p_s (\bar{x}_s - \bar{x})^2 + \sum_s p_i (x_i - \bar{x})^2 - p_s (\bar{x}_s - \bar{x})^2 \right]. \tag{2}$$

Now, let V_m be the variance due to differential mortality and V_f be that due to differential fertility.

$$V_m = p_d(0 - \bar{x})^2 + p_s(\bar{x}_s - \bar{x})^2 = \bar{x}^2 p_d/p_s \qquad (3)$$

$$V_f = \frac{1}{p_s} \sum_s p_i(x_i - \bar{x}_s)^2 = \frac{1}{p_s} \left[\sum_s p_i(x_i - \bar{x})^2 - p_s(\bar{x}_s - \bar{x})^2 \right] \qquad (4)$$

thus, from (2), (3) and (4)

$$I = \frac{V_m}{\bar{x}^2} + p_s \frac{V_f}{\bar{x}^2} = \frac{V_m}{\bar{x}^2} + \frac{1}{p_s} \frac{V_f}{\bar{x}_s^2} = I_m + \frac{1}{p_s} I_f \qquad (5)$$

where

$$I_m (= V_m/\bar{x}^2 = p_d/p_s) \text{ and } I_f (= V_f/\bar{x}_s^2)$$

are the indices of total selection due respectively to mortality and fertility. The index due to mortality is especially simple, being the ratio of deaths to survivors.

Some numerical examples based on U. S. Census data are given in table 1. The 1950 and 1910 birth distributions are based on the total children ever born to all women who were of age 45-49 in the census year. The variance is somewhat uncertain because the Census reports lump several classes in the high birth numbers, but this is not likely to introduce a very large error. Another source of error is that no allowance was made for women who died during the childbearing period after having one or more children. It was thought better not to attempt such a correction, since there is almost certainly some lack of independence between fertility and mortality in these ages, and any correction based on an assumed independence might make matters worse. However, the error due to neglecting this correction is not likely to be large, for the death rates at these ages are very low, especially in recent years.

The three mortality values in table 1 are chosen to correspond roughly to the total mortality of women from birth to the end of the child bearing period for three time intervals: at present (10%), during the lifetime of women who were 45-49 in 1950 (30%), and during that of women who were 45-49 in 1910 (50%). Thus the most relevant comparison is between rows (2) and (6). During this 40 year period the index of total selection dropped from 2.6 to 2.1, or by about 20%.

It is interesting to note that the component of selection due to differential fertility actually increased during this period. Although the mean number of children per woman (including unmarried) dropped from 3.9 to 2.3, the index of selection due to fertility differences increased almost 60%. Despite a great lowering of fertility rates the pattern of marriages and births was such as to increase the effective fertility differential.

TABLE 1

The total intensity of selection, I, and its components for several hypothetical populations based on U. S. Census data

					I_m	I_f	I_f/p_s	I
(1)	1950 birth distribution,	10%	mortality		.111	1.143	1.270	1.381
(2)	"	"	30%	"	.429	1.143	1.633	2.062
(3)	"	"	50%	"	1.000	1.143	2.286	3.286
(4)	1910 "	"	10%	"	.111	.784	.871	.982
(5)	"	"	30%	"	.429	.784	1.120	1.549
(6)	"	"	50%	"	1.000	.784	1.568	2.568

The first column, I_m, is the selection intensity due to mortality, the second column, I_f, is that due to differential fertility, and the last gives the total selection intensity from both causes. The third column gives the ratio of I_f to p_s the fraction of survivors.

The column I_m shows what the selective index would be if premature mortality were the only differential factor, *i. e.*, if all surviving women had the same number of children. This shows that with recent low death rates, differential post-natal mortality provides only a small part of the total opportunity for selection.

[*Editors' Note:* Material has been omitted at this point.]

LITERATURE CITED

DOBZHANSKY, TH. 1955 A review of some fundamental concepts and problems of population genetics. Cold Spring Harb. Symp. Quant. Biol., *20*: 1-15.

———— 1957 Genetic loads in natural populations. Science, *126*: 191-194.

FISHER, R. A. 1930 The Genetical Theory of Natural Selection. Oxford.

HALDANE, J. B. S. 1937 The effect of variation on fitness. Am. Naturalist, *71*: 337-349.

———— 1949 Parental and fraternal correlations in fitness. Annals of Eugenics, *14*: 288-292.

———— 1954a The statics of evolution. Evolution as a Process. Ed. by J. Huxley, London.

———— 1954b The measurement of natural selection. Caryologia 6, suppl. 1, 480.

MORTON, N. E., J. F. CROW AND H. J. MULLER 1956 An estimate of the mutational damage in man from data on consanguineous marriages. Proc. Nat. Acad. Sci, *42*: 855-863.

MULLER, H. J. 1950 Our load of mutations. Am. J. Human Genet., *2*: 111-176.

WRIGHT, S. 1931 Evolution in Mendelian populations. Genetics, *16*: 97-159.

———— 1949 Adaptation and selection. Genetics, Paleontology and Evolution. Ed. by G. L. Jepson, G. G. Simpson and E. Mayr, Princeton.

17

Reprinted from *Proc. Nat. Acad. Sci.*, **59** (3), 690–699 (1968)

DEMOGRAPHIC APPROACHES TO THE MEASUREMENT OF DIFFERENTIAL SELECTION IN HUMAN POPULATIONS*

By Walter F. Bodmer

DEPARTMENT OF GENETICS, STANFORD MEDICAL CENTER, PALO ALTO, CALIFORNIA

The existence of extraordinarily high levels of genetic heterogeneity in natural populations is one of the most striking discoveries of experimental population genetics. The evolutionary mechanisms underlying this heterogeneity are not clearly understood and remain a challenge to population genetic theory. Recent studies by Lewontin and Hubby[1] on *Drosophila pseudoobscura* and by Harris[2] in man have re-emphasized the large number of loci at which, on the average, individuals are heterozygous with respect to polymorphisms and suggest that up to 30 per cent of all loci may, in fact, be polymorphic. If all such loci were maintained polymorphic independently by heterozygote advantage, there would be an inconceivably large difference in fitness between the multiple heterozygote and the population mean, and an enormous implied selective burden on the population.[1, 3] These difficulties can be essentially eliminated, as discussed by Sved, Reed, and Bodmer,[4] King,[5] and Milkman,[6] if the condition of independence between loci is removed. Specifically these authors have pointed out that if there is an upper limit, or threshold, to the fitness, a much larger number of polymorphic loci can be maintained for a given level of selective elimination in the population. However, even this important, though simple, modification of classical models for the maintenance for polymorphisms is probably not adequate by itself to explain observed levels of heterogeneity.

It is important to bear in mind that those selective forces which may have been necessary for the initial establishment of a polymorphism may no longer exist later. Once established in a large population, such as the human population, a polymorphism can be maintained by very much smaller selective differences than may have been needed to establish it within a reasonable time span. Even if selection were completely relaxed, the rate of loss of a polymorphism by random genetic drift would be very slow in a reasonably large population. In some cases, reversed selection may, of course, ultimately eliminate one of the alleles. These considerations suggest that the number of polymorphisms now existing is the result of a dynamic equilibrium between the rate of formation of new polymorphisms and the rate at which they are lost by reverse selection or drift. This balance is very hard to quantitate, since the rate of formation or loss of polymorphisms by selection depends on a complex interaction between the occurrence of particular alleles in the population and the current selection environments. It does not seem unreasonable to suppose that the rate of formation exceeds the rate of loss, in which case the numbers of polymorphisms would increase essentially without limit. Thus, the majority of observed polymorphisms may be "relics"[7] of previous "selective crises." The detection of small selective effects associated with such polymorphisms may be almost impossible and in any case will have little relevance to the reasons for their establishment. At any given time in evolutionary history rather few polymorphisms may be in the process of becoming established by relatively powerful selective forces, and it is these differences that will be most easily discernible. Transient polymorphisms will, of course, be included in this category and may be essentially indistinguishable from those destined to become "stable."

A major key to unraveling the basis for widespread polymorphism clearly lies in an understanding of the nature of selective forces. The biological fitness of an individual is measured by his over-all contribution to future generations, which depends on all the many interacting factors that contribute to his reproductivity.[8] These include, as major components, not only viability, but also fertility as measured both by the probability of marriage and the distribution of the numbers of children, all as a function of the age of the individual. Many studies on selection, even in human populations, have concentrated on measuring only one component of reproductivity, such as, for example, the viability as determined by departures from expected proportions of offspring, assuming Mendelian segregation. Very large numbers of offspring are needed to detect small differences from Mendelian expectations. For example, about 300,000 offspring need to be observed to be reasonably sure of detecting a 1 per cent departure from a 1-to-1 segregation due to selection.[9] Measurement of mortality alone or fertility alone may be quite misleading and cannot, in general, lead to proper estimates of selective differences.

The demographer measures the rate of increase or decrease of a population by a quantity r, the intrinsic rate of increase, which is calculated from a knowledge of the age-specific birth and death rates. This relatively simple information, which indirectly takes into account all the factors affecting reproductivity, is generally adequate for the quantitative measurement of the magnitude of selective differences, though the determination of their causes may require much more complete specification of the relevant factors. Though population genetic models generally assume that fitness can be measured simply by the relative numbers of offspring of different genotypes which survive to maturity, studies which also take into account other factors, such as fertility,[10] indicate that provided the selective parameter used is representative of the over-all reproductive rate, the qualitative conclusions of the simpler models remain valid.

Patterns of reproductivity may change radically with social and cultural changes in the structure of the human population. In the remainder of this paper I shall concentrate on illustrating the use of the intrinsic rate of increase for the measurement of selective differences and for assessing the effects of changes in social customs, especially changes in the age of marriage and childbearing, on these forces.

Measurement of Fitness by the Intrinsic Rate of Population Increase.—The intrinsic rate of population increase, r, is generally obtained as the solution of Lotka's fundamental equation

$$\sum_x e^{-rx} l_x b_x = 1, \tag{1}$$

where x measures time, in units of one or more years; l_x is the probability of survival from birth to age x; and b_x, the age-specific birth rate, is one half the number of births to individuals in the age group x to $x + 1$. The evaluation of r is usually done in terms of the females, and assumes that the population has achieved a stable age distribution as determined by l_x and b_x, and is increasing at a rate e^{rx}. The values of l_x and b_x for the general population are usually obtained from a combination of census and vital statistics data.

The number of children born per person born, or net reproductive rate, is given by

$$R_0 = \sum_x l_x b_x \tag{2a}$$

$$R_0 = l \sum_x b_x \tag{2b}$$

if mortality during the reproductive years is negligible and l is the probability of surviving to the reproductive age. R_0 is a measure of the rate of population increase per generation, and so is related to r and T, the length of a generation, by the equation

$$R_0 = e^{rT}. \tag{3}$$

If r is sufficiently small, $e^{-rx} \sim 1 - rx$ so that equation (1) becomes, approximately,

$$\sum_x (1 - rx)\, l_x b_x = 1, \tag{4}$$

giving the following approximate solution for r:

$$r = (R_0 - 1)/R_1, \tag{5}$$

where

$$R_1 = \sum_x x\, l_x b_x.$$

The quantity R_1/R_0 is the mean age at childbearing, since $l_x b_x$ is the number of births to individuals of age x, so that, approximately,

$$T = R_1/R_0. \tag{6}$$

The above equations constitute a simple statement of Lotka's[11] classic work on population growth.

The net reproductive rate, R_0, is only a satisfactory measure of selection if there are no differences in the generation time T. Two genotypes with the same R_0 but different values of T clearly have different fitnesses, since from equations (5) and (6), the genotype with the smaller generation time has a higher intrinsic rate of increase. To relate the intrinsic rate of increase to the usual measures of selection per generation used in simple population genetic models, fitness must be measured by $e^{r_i \bar{T}}$, where $\bar{T}$ is an average generation time for the whole population and r_i is the intrinsic rate of increase of some particular subsection of this population. The relative fitness of two genotypes will then be measured by

$$w = \frac{e^{r_1 \bar{T}}}{e^{r_2 \bar{T}}} = e^{(r_1 - r_2)\bar{T}}, \tag{7}$$

where r_1 and r_2 are the two respective intrinsic rates of increase and $\bar{T}$ is either the mean generation time of the population from which the genotypes were drawn or the mean generation time of the two genotypes.   Small differences in $\overline{T}$ will not have much effect on the relative fitness.   When $(r_1 - r_2)\overline{T}$ is small, equation (7) gives, approximately,

$$w = 1 - (r_1 - r_2)\overline{T}. \tag{8}$$

If the two components of the population corresponding to r_1 and r_2 have approximately the same generation length T, then their relative selection coefficient, $(r_1 - r_2)T$, takes the approximate form

$$(r_1 - r_2)T = \left(\frac{R_0^{(1)} - 1}{R_1^{(1)}} - \frac{R_0^{(2)} - 1}{R_1^{(2)}}\right) T = \frac{R_0^{(1)} - R_0^{(2)}}{R_0^{(1)}R_0^{(2)}} \tag{9}$$

from equations (8), (5), and (6), where $R_0^{(1)}$ and $R_0^{(2)}$ are the net reproductive rates corresponding to r_1 and r_2.

Fitness of a trait with a variable age of onset: A variable age of onset of a genetic trait poses special problems for the estimation of fitness, since individuals with the "abnormal" genotype who have not yet developed the trait are necessarily included among the normal part of the population.   Changes in the distribution of the age of childbearing may have marked effects on the fitness of such genotypes.   If we assume that the individuals at risk with respect to the trait have a normal mortality and fertility pattern before they show it, then a plausible model for estimating their fitness in terms of observable parameters can be constructed. Normal fertility (b_x) and mortality (l_x) rates are taken from the relevant general population.   We require in addition the probability χ_x that individuals first develop the trait at age x or the probability L_x that they have not yet developed it by age x, both of which can be obtained from the distribution of the age of onset of the trait.   Also required are the age-specific fertility (b_x') and mortality (l_x') rates of people who already have the trait.   Ideally these rates should be obtained as a function of the age of onset, but in practice there will usually not be enough data for such a detailed breakdown, so that it will have to be assumed that the age-specific rates are independent of the age of onset.   It can then be shown[12] that the number of children born to surviving individuals, "at risk," of age x is

$$p(x) = l_x b_x \prod_{i=1}^{x-1}(1 - \chi_i) + l_x'b_x'\sum_{r=1}^{x}\frac{l_r}{l_r'}\chi_r\prod_{i=1}^{r-1}(1 - \chi_i) = l_x b_x L_x + l_x'b_x'\sum_{r=1}^{x}\frac{l_r}{l_r'}\chi_r L_r. \tag{10}$$

The first term is the contribution from those who do not yet show the trait, and the second term is the sum of the contributions from those individuals who developed the trait at all ages up to age x.   The net reproductive rate is then $\sum_x p(x)$ and, from equation (1), the intrinsic rate of increase is the solution of the equation

$$\sum_x e^{-rx}p(x) = 1. \tag{11}$$

A simple upper limit for the fitness of a debilitating disease with a late age of onset can be obtained if it is assumed that $b_x' = 0$, i.e., that individuals are no longer fertile once they have contracted the disease.   A lower limit is obtained if χ_x is the relative probability of death at age x, rather than the probability of onset of the disease at age x, in which case $l_x' = 0$.   In either case, only the first term of equation (10) is nonzero, and $l_x L_x$, the probability that individuals destined to get the disease survived to age x without getting it, simply replaces l_x in the standard formulae for the calculation of r.

Fitness of schizophrenics: Relatively few adequate attempts have been made to measure differential selection in human populations, taking account of demographic principles.   Notable exceptions are the studies of Reed, Neel, and Chandler[13] on Huntingdon's chorea, of Bajema[14] on the relationship between IQ and fitness, and of Erlenmeyer-Kimling and her colleagues[15, 16] on schizophrenia. In this section we shall consider the application of some of the above formulae to predicting the effects of demographic changes on the fitness of schizophrenics.

About half of the patients in mental hospitals are classified as schizophrenics, the incidence of the disease being around 1 per cent in the United States population. There has been much discussion over many years as to the extent of a possible genetic basis for schizophrenia (for a critique see ref. 17).   Recent evidence favors at least a major polygenic component for the disease,[18, 19] interpreted by Gottesman and Shields[20] in terms of a threshold model.   It should be emphasized, of course, that differential selection will only affect the frequency of a trait insofar as it has a heritable component, which may be either genetic or "sociocultural." The strong familial concentration of schizophrenia (about 16.5% of offspring with one schizophrenic parent,[19] and 35% with both parents schizophrenic[21]) leaves no doubt that changes in the fitness of schizophrenics may, to some extent, affect their frequency in the population, whether or not the basis for the familial concentration is genetic.

Studies by Goldfarb and Erlenmeyer-Kimling,[15] and others earlier, indicate that the major effect of schizophrenia on fitness is through a reduction of the probability of marriage after onset of the disease, mainly during the period between onset and first admission to a mental institution.   The model developed above would seem, therefore, to be reasonably applicable on the assumption that mortality and fertility are normal before admission to an institution, but fertility essentially zero afterwards (i.e., $b_x' = 0$).

Data, distinguished according to sex, on the distribution of the age at first admission of schizophrenics to an institution, were taken from Rosenthal.[22]   The cumulative probabilities derived from these distributions give L_x, the probability of not having developed the disease by age x.   Age-specific birth rates for the general population were taken from *Vital Statistics of the United States, 1963.*[23] Mortality during the reproductive period was ignored, and a value of $l = l_x = 0.94$ for all x was assumed, based on recent mortality statistics.   An illustrative calculation of the intrinsic rates of increase and generation lengths for the general population and for female schizophrenics, assuming 1960 birth rates, and hence of the selective disadvantage of schizophrenia, is shown in Table 1.   There is a relatively small difference between the results obtained by the use of equation (8) and its approximation equation (9), which assumes equal generation lengths.

TABLE 1. *Illustrative calculation of approximate selective disadvantages of schizophrenic females, assuming 1960 birth rates.*

Age group	Proportion first admitted at age x, a_x[*]	$\sum_x a_y$	$L_x = 1 - \sum_x a_y$	b_x[†]	Median age x
<15	0.003	0.003	0.997	0.008	14
15–19	0.052	0.055	0.945	0.0891	17.5
20–24	0.109	0.164	0.836	0.2575	22.5
25–29	0.131	0.295	0.705	0.1986	27.5
30–34	0.145	0.440	0.560	0.1144	32.5
35–39	0.146	0.586	0.414	0.0573	37.5
40–44	0.119	0.705	0.295	0.0153	42.5
45–49	0.099	0.804	0.196	0.009	47.5
50–54	0.076	0.880	0.120	0	52.5
55–59	0.055	0.935	0.065	0	57.5
60–46	0.031	0.966	0.034	0	62.5
>65	0.034	1.000	0	0	—

General populations: R_0[‡] $= 5 \times 0.94 \times 0.5 \times \sum_x b_x = 2.35\sum_x b_x = 1.744$; $R_1 = 2.35\sum_x x b_x = 45.559$

$$T = R_1/R_0 = 26.12; \quad r = R_0 - 1/R_1 = 0.016$$

Schizophrenics: $R_0' = 2.35\sum_x L_x b_x = 1.252$; $R_1' = 2.35\sum_x x L_x b_x = 31.354$

$$T' = R_1'/R_0' = 25.04; \quad r' = R_0' - 1/R_0' = 0.008$$

Selective disadvantage $= (r - r')\bar{T} = 0.205$ (from equation (8)) (approximate formula gives $(R_0 - R_0')/(R_0 R_0') = 0.225$) from equation (9).

[*] These values were taken from the data for females given by Rosenthal[22] (Fig. 2).
[†] These values are taken from ref. 23, Tables 1–6, for 1960. Since they are based only on the female population and relate to 5-year intervals, they must be corrected by a factor $5 \times 1/2$.
[‡] It is assumed that $l_x = l = 0.94$ for all values of x.

TABLE 2. *Net reproductive rates* (a) *and approximate selective disadvantages* (b) *of schizophrenics for different birth rate distributions.*

	(a) Year			(b) Year		
	1940	1950	1960	1940	1950	1960
Males	0.56	0.77	0.92	0.86	0.61	0.51
Females	0.76	1.04	1.25	0.40	0.27	0.22

The figures in (a) are the estimated net reproductive rates R_0', using equation (10), and those in (b) are the selective disadvantages, calculated from equation (9), using birth-rate distributions for the years indicated in column heads (*Vital Statistics of the United States, 1963*, vol. 1) and age of admission distributions (from Rosenthal[22]) as indicated for the rows. The male age-specific birth rates, b_x^*, were obtained from the female rates b_x according to the formula $b_x^* = 1/2(b_x + b_{x+1})$ (see text for further details).

The net reproductive rates, calculated using equation (10), and the approximate selective disadvantages of male and female schizophrenics assuming 1940, 1950, and 1960 birth-rate distributions are shown in Table 2. Age-specific birth rates for males are not readily available from standard sources of vital statistics. It was therefore assumed that the b_x values for females applied to males, age $x +$ 2.5, since the average difference in age at marriage between males and females is about 2.5 years. Thus, for 5-year intervals, the age-specific birth rate for males, b_x^*, was given in terms of that for females, b_x by

$$b_x^* = 1/2 (b_x + b_{x+1}).$$

The over-all lower fitness of males than females is due mainly to the higher proportion of males admitted to institutions at an earlier age,[22] which may reflect an earlier age of onset of the disease in males or a tendency for males to be institutionalized at an earlier age because of social pressures.   The female selective disadvantages agree reasonably with those suggested by Erlenmeyer-Kimling and Paradowski's[16] more direct study, while the male values are a little higher.   For females, some differences due to the assumption of no fertility after admission to an institution are to be expected, while for males the marriage rate before admission is probably somewhat lower than that for the general population.[15] The increasing use of drugs to mitigate the effects of schizophrenia, allowing more home remissions after first admission to an institution, may of course accentuate the decreased selective disadvantage of schizophrenics by permitting higher levels of fertility after onset of the disease.

The estimated decrease in the selective disadvantages of both males and females from 1940 to 1960 is almost twofold.   This is the effect expected from a trend toward an earlier average age of childbearing in the general population applied to schizophrenics before onset of the disease and shows how significant the effect of such a demographic change can be, even over a relatively short period of time.

Discussion.—Though the methods we have discussed for measuring selective differences are still approximate and subject to a number of simplifying assumptions, they do at least make some attempt to take into account the combined effects of fertility and mortality.   The demographic approach is essential for the proper measurement of selective differences in human populations and clearly indicates the type of data needed for their measurement.   A lack of appreciation of these needs may severely limit the value of otherwise comprehensive studies on particular genetic traits.

The use of demographic data can, as we have indicated, be very valuable for predicting future changes due to demographic trends and may be more sensitive for the determination of small selective differences.   Thus, Bodmer and Cavalli-Sforza[9] have calculated the approximate selective difference between blood groups O and A resulting from their association with duodenal ulcer, using data on the incidence of this disease and on the distribution of the age at death due to duodenal ulcer, together with vital statistics estimates for l_x and b_x.   Their estimated effective selection coefficients of 5×10^{-6} for females and 9.5×10^{-5} for males show, as has been previously emphasized by others,[24] that the effect of this association is much too small to be of any significance in counteracting incompatibility selection.   Two main factors contribute to these low selective values, firstly the low incidence of duodenal ulcer, and secondly the late onset of the disease relative to the reproductive period.[24]

The interpretation of the genetic effects of a change in the selection coefficients for schizophrenia is undoubtedly very complex.   At most, 5 to 10 per cent of schizophrenics in the population come from the families of schizophrenics, so that their increasing fitness will, at least in the short term, have only a marginal effect on the over-all incidence of the disease.   If Gottesman and Shields[20] are correct in their interpretation in terms of a polygenic threshold model, the genes which jointly increase the predisposition to the disease may be maintained polymorphic by their separate effects in other "unaffected" individuals. In this case, the frequency of schizophrenia, insofar as it is controlled by genetic factors, will be largely determined by the relative reproductive performance of those "unaffected" individuals who carry one or more of the relevant genes. Anecdotal information[19] suggests that nonschizophrenic close relatives of schizophrenics may have intellectually desirable attributes, whereas Erlenmeyer-Kimling and Paradowski[16] indicate in addition that these same individuals may have relatively high reproductivity. Thus an increase in the fertility of schizophrenics, while perhaps increasing slightly the incidence of the disease, may also conceivably increase the frequency of some desirable genetic attributes in other individuals. These suggestions, which at the present time are, of course, quite hypothetical, are simply made to emphasize the problem of interpreting the effects of selection, conscious or unconscious, in the face of tremendous, underlying and unknown genetic heterogeneity.   Even the most rapid genetic changes, however, take place slowly relative to the time span of man's history and the pace of social changes. Hopefully, the basis for a successful treatment for schizophrenia will be found within one or two generations at most, well before there has been time for any significant genetic change by natural selection.

If one accepts the fact that behavioral attributes, like all other characteristics, have a genetic component, then it is clear that there may exist many genetic polymorphisms for genes affecting human behavior. Though hard to identify, these may, in fact, be the polymorphisms subject to the strongest selective pressures in our modern society.

In this paper, the effect of demographic changes on one particular aspect of differential selection has been emphasized as an illustration of the general problem of the interaction of genetics and demography.   There are, of course, many other important interactions, such as changes in mating patterns (mainly, perhaps, an increase in assortative mating with respect to some socioeconomic and cultural attributes), the effect of increased geographical and social mobility on marriage stability, and the effects of family planning.   Population projections, which are an essential part of both population genetics and demography, require for their complete specification a knowledge of demographic parameters as a function of genetic relationship.   There is a tremendous dearth of such information.[25] Undoubtedly the collection, analysis, and interpretation of such data would be of great value to our present-day society.

Summary.—The problem of explaining the extraordinarily high observed numbers of polymorphisms is discussed.   Understanding the nature of selective forces must be a major key to the explanation.   In human populations a demographic approach is essential for the proper estimation of selective differences. The calculation and use of the intrinsic rate of increase for the measurement of natural selection is discussed with special reference to a trait with a variable age of onset.   Application to data on schizophrenia shows that the selective disadvantage due to this disease has probably almost halved in the period from 1940 to 1960 because of a decrease in the average age of childbearing.   The problems of interpretating such effects in the face of large, unknown levels of genetic heterogeneity and the need to collect and analyze more demographic data collected with the family as the unit are emphasized.

Many of the ideas discussed in this paper were developed in collaboration with Professor Luigi Cavalli-Sforza and will be discussed in a forthcoming book.[9] I am grateful to Dr. S. Kessler for his help with data on schizophrenia.

* The investigation was supported, in part by a Public Health Service Rearch Career Program Award (GM 35002–01) and a research grant (GM 10452) from the Public Health Service.

[1] Lewontin, R. C., and J. L. Hubby, *Genetics*, **54**, 595 (1966).

[2] Harris, H., *Proc. Roy. Soc. (London)*, Ser. B, **164**, 298 (1966).

[3] Kimura, M., and J. F. Crow, *Genetics*, **49**, 725 (1964).

[4] Sved, J. A., T. E. Reed, and W. F. Bodmer, *Genetics*, **55**, 469 (1967).

[5] King, J. L., *Genetics*, **55**, 483 (1967).

[6] Milkman, R. D., *Genetics*, **55**, 493 (1967).

[7] Epstein, C. J., and A. G. Motulsky, in *Progress in Medical Genetics*, ed. A. G. Steinberg and A. G. Bearn (New York: Grune & Stratton, 1965), vol. 4, p. 85.

[8] Bodmer, W. F., and L. L. Cavalli-Sforza, *World Population Conference, 1965* (United Nations, 1967), p. 455.

[9] Bodmer, W. F., and L. L. Cavalli-Sforza, in preparation.

[10] Bodmer, W. F., *Genetics*, **51**, 411 (1965).

[11] Lotka, A. J., *Elements of Mathematical Biology* (New York: Dover Publications, Inc., 1956).

[12] The probability of not developing the trait by age x is

$$L_x = (1 - \chi_1)(1 - \chi_2)\ldots(1 - \chi_{x-1}) = \prod_{i=1}^{x-1}(1 - \chi_i).$$

The probability of surviving to age x, before onset, is l_x. Thus the number of births to individuals of age x who have not yet developed the trait, but are destined to, is

$$L_x l_x b_x.$$

The probability of first contracting the disease at age r is

$$\chi_r \prod_{i=1}^{r-1}(1 - \chi_i) = \chi_r L_r.$$

The probability of surviving to age x having first shown the trait at age $r(<x)$ is

$$l_r \frac{l_x'}{l_r'},$$

The number of births to individuals of age x who first showed the trait at age $r(<x)$ is therefore

$$(b_x')\left(l_r \frac{l_x'}{l_r'}\right)(\chi_r L_r).$$

The total number of births to individuals of age x who have already developed the trait is thus

$$l_x' b_x' \sum_{r=1}^{x} \frac{l_r}{l_r'} \chi_r L_r.$$

The total number of births to individuals of age x destined to develop the trait is therefore

$$p(x) = b_x l_x L_x + l_x' b_x' \sum_{r=1}^{x} \frac{l_r}{l_r'} \chi_r L_r.$$

[13] Reed, T. E., J. H. Chandler, E. M. Hughes, and R. T. Davidson, *Am. J. Human Genetics*, **10**, 201 (1958).

[14] Bajema, C. J., *Eugenics Quarterly*, **10**, 175 (1963).

[15] Goldfarb, C., and L. Erlenmeyer-Kimling, in *Expanding Goals of Genetics in Psychiatry*, ed. F. J. Kallmann (New York: Grune and Stratton, 1962).

213

[16] Erlenmeyer-Kimling, L., and W. Paradowski, *Am. Naturalist*, **100**, 651 (1966).

[17] Jackson, D. D., *The Etiology of Schizophrenia* (New York: Basic Books, Inc., 1960).

[18] Gottesman, I. I., and J. Shields, *Brit. J. Psychiat.*, **112**, 809 (1966).

[19] Heston, L. L., *Brit. J. Psychiat.*, **112**, 819 (1966).

[20] Gottesman, I. I., and J. Shields, these PROCEEDINGS, **58**, 199 (1967).

[21] Rosenthal, D., *J. Psychiat. Res.*, **4**, 169 (1966).

[22] *Ibid.*, **1**, 26 (1961).

[23] U.S. Public Health Service, *Vital Statistics of the United States, 1963* (Washington, D.C.: Government Printing Office, 1964), vol. 1, *Natality*.

[24] Reed, T. E., in *Proceedings of the Conference on Genetic Polymorphisms and Geographic Variations in Disease*, ed. B. S. Blumberg (New York: Grune & Stratton, 1960), p. 80.

[25] Bodmer, W. F., and J. Lederberg, *Proceedings of the Third International Conference of Human Genetics*, ed. J. F. Crow and J. V. Neel (Baltimore: Johns Hopkins Press, 1964), pp. 459–471.

214

Part III

DISPERSAL AND POPULATION DISTRIBUTION

Editors' Comments
on Papers 18, 19, and 20

18 WAHLUND
*Composition of Populations and of Phenotypic Correlations
from the Viewpoint of Population Genetics*

19 WRIGHT
Isolation by Distance

20 BATEMAN
*Contamination in Seed Crops: III. Relation with Isolation
Distance*

Since population genetics is the study of the dynamics of a gene pool, which are basically reproductive processes, the patterns of mating must be assessed if we are to understand what is going on in a particular population. Two aspects of mating structure are quite important in determining the distribution of genes through a population and over time. These are the size of the mating group and the association of mates with each other as this relates to their genes. The standard terms for these two aspects of population structure are the *effective population size* and the *degree of inbreeding*.

We have seen earlier that aspects of overlapping generations are important from the point of view of assessing the means of action of natural selection, density, and size regulation. If the age-specific vital rates are known, evolutionary (long-term) aspects of the population can be predicted from discrete generation models with parameters that are condensations of the information contained in the age-specific vital rates. This is true also for the mating structure of a population. Once we know the breeding pattern and the effective population size, we can use discrete generation, or demography-free, models for long-term behavioral analysis of the genetics. However, the way in which these condensed values are produced and the shorter-term consequences of the behavior of specific populations must be studied by looking in detail at the mating patterns. It is to this that we now turn.

In discussing these aspects of what we have broadly called "mating behavior," we are perforce almost completely restricted to human populations. Little is known of these aspects from wild populations of most other species. This area of population genetics has been developed theoretically to a far greater extent than have the aspects of demographic genetics that we have covered so far. Therefore, before beginning our review, we would like to refer to many of the other available reviews. Since we shall only highlight a small part of the work that has been done in this area, the reader is strongly urged to consult other sources for further details.

Human population structure has been surveyed in historical context by Schull and MacCluer (1968) and by Schull (1972); both stress mating patterns. Harrison and Boyce (1972a) survey migration studies. Morton (1969) reviews concepts of genetic kinship, relating them to work on migration and subdivision of a population. Cannings and Cavalli-Sforza (1973) cover these topics and examine their effects on measures of genetic "distance." Several symposium volumes provide excellent collections of papers related to these topics, notably Morton (1973), Harrison and Boyce (1972b), Sutter (1962), and Crawford and Workman (1973).

Major textbooks and treatises on these topics should also be cited here. The best overall summaries are in Crow and Kimura (1970), Cavalli-Sforza and Bodmer (1971), and Jacquard (1974). Reviews of theoretical topics more restricted to the personal investigations of the author are Malecot (1969), Wright (1969), Kimura and Ohta (1971), Ewens (1969), Karlin (1968), Moran (1962), and Nei (1975).

For most population genetic computations, the actual census size of a population is of less importance than the quantity known as its effective population size, generally denoted N_e. This quantity is the size that the given population would have if it consisted only of randomly mating adults with discrete generations. Its determination from a real population under various demographic assumptions is extensively considered in Crow and Kimura (1970), which is based on their original work published earlier. There are two commonly used "types" of effective size: one treats N_e as that number of randomly mating adults which would produce the observed variance in gene frequency owing to genetic drift in a finite population; the other treats N_e as that size which would produce the same inbreeding coefficient or genetic homozygosity. Both require the mean and variance of family sizes, one reason why that distribution is important.

For an age-structured population in which one can assume fixed age-specific birth and death rates, there are several means of computing N_e. These are discussed fully in Cavalli-Sforza and Bodmer (1971) and Crow and Kimura (1972). Other approaches have been suggested (Nei and Imaizumi, 1966; Felsenstein, 1971; Hill, 1972). These measures are all various ways of condensing what is known about mortality and fertility into a single number that can be used in the well-understood discrete generation models.

Finite population size, as well as the way in which mates are chosen, can affect the genetics of a population. Both factors can be measured by the degree of association of genes in individuals or by the genetic variability in the population; both measures have common expressions in many cases, and it is necessary to review these topics briefly.

It has long been known that some simply inherited traits appear with increased frequency in matings between kin. Studies early in this century dealt with this (e.g., Jacob, 1911), but only until after about 1910 did authors begin to look at quantitative theoretical aspects of this question as related to gene frequencies in panmictic populations. Most notable of these early studies were the papers by Pearl (1913, 1914), Jennings (1914, 1916), Robbins (1917, 1918a, b), and Fish (1914). The genetics of *inbreeding* were really developed by Wright in 1921 (and in many later papers). If we have pedigrees for a population, we can compute the probability that both alleles at any given locus in an individual are identical by descent from a common ancestor; this is the inbreeding coefficient *f*.

Wright's initial development of the theory of inbreeding was largely based on his use of path coefficients and concepts of correlation. Later, the same problems were studied from the standpoint of probability by Haldane and Moshinsky (1939), Cotterman (1940), and Malecot (1948), although the conclusions are of course the same. Some treatments of this problem have been expressed in terms of formal algebraic theory (Cotterman, 1940; Etherington 1939, 1941); these papers and a discussion of them can be found in Ballonoff (1974a).

Inbreeding affects the genotype frequencies within a population by increasing the frequency of homozygotes by a factor determined by the degree of inbreeding and the gene frequencies. In a finite population, even if it is mating at random, something similar occurs. If there are $2N$ genes (N individuals), there is a probability of $1/2N$ that a gene drawn once and passed on to an individual in the next generation will be drawn twice; some genes will not be drawn at all. In the next generation, therefore, some genes will be identical by descent. This process is known as genetic drift and is discussed in many works (see Crow and Kimura, 1970). Over time, and in the absence of mutation, a population becomes more and more homozygous as alleles are randomly eliminated. This is true even though the population is mating at random and, within the limits of finite size, the genotype frequencies at any generation will approximate those of the Hardy–Weinberg equilibrium.

Thus, the same effect that inbreeding has on genotype frequencies within a population is produced by random mating in a series of small populations. In any such series, the probability of identity by descent at any time *t* is expressed as an inbreeding coefficient, although in any particular population in the series genotype frequencies will be in Hardy–Weinberg proportions. See Jacquard (1974) or Crow and Kimura (1970) for a discussion of this concept in detail, and especially Jacquard (1975).

In the absence of migration and mutation, the processes just outlined will lead to an increased number of homozygous individuals at an increased number of loci from what would be found in a randomly mating large population. Among other things, this will cause recessive traits to be manifest in higher frequency, and many studies have demonstrated that biometric measurements, fitness, and so on, are lowered by inbreeding. For humans, the most comprehensive search for inbreeding effects was by Schull and Neel (1965) in Japan; very little of significance was found. Discussions of this topic are also in Sutter and Tabah (1951, 1952), Jacquard (1974), and Cavalli-Sforza and Bodmer (1971) where data are given, to mention but a few.

Of course if any species exists for hundreds of thousands of generations, theory would predict complete homozygosity and identity by descent even in a large population . This is not observed, although homozygosity in wild populations is usually above 90 percent. Obviously, mutations, selection, and other factors prevent this total inbreeding from occurring so that we are not all identical to the same primordial bacterium! What is important about inbreeding is that (1) it presents a way of measuring the mating behavior of a population to the depth of available data, and (2) that it relates to some possible clinical problems with regard to recessive traits in humans.

POPULATION SUBDIVISION

Looking at genotype frequencies, the Swedish geneticist Sten Wahlund (1928) first demonstrated that subdividing a population into random-mating subpopulations *by itself* produced a deviation from Hardy-Weinberg genotype proportions in the whole population. This *Wahlund effect* is equivalent to de facto inbreeding within the subdivisions. Paper 18 was a fundamental contribution to demographic genetics, but has never been reprinted nor has it previously appeared in English.

Wahlund shows that the increase in homozygotes occasioned by population subdivision is by an amount σ^2, which is the variance in gene frequencies among subdivisions. That is, the frequency of A_1A_1 homozygotes is

$$Pr(A_1A_1) = p^2 + \sigma^2$$

Unless each subpopulation is the genetic image of the total population, this variance will be strictly positive, and it can be shown that the genotype frequencies will be those which would occur in an undivided population with inbreeding coefficient equal to

$$f = \frac{\sigma^2}{pq}$$

This value is called the *standardized* or *Wahlund's variance*.

Wahlund also deals with the various ramifications of genetic linkage on the correlations between traits that result from the structuring of the population, at all times using a random-mating model. First he discusses the return of a population from linkage disequilibrium to equilibrium as linked genes are increasingly recombined over time. He looks at the immediate, and subsequent, effects of subdivision with and without migration on genotype frequencies of individual loci and of linked loci. Recently, Nei (1965) has expanded Wahlund's results for an arbitrary number of alleles and for nonrandom mating within isolates.

MIGRATION AND ISOLATION

Real biological populations are structured in any number of ways; one of the most important of these is the isolation between subdivisions of a geographically dispersed population or species. This is discussed very briefly at the end of Paper 18. Given some knowledge of the way in which populations are subdivided and the patterns of migration among them, degrees of genetic identity between any two subpopulations can be computed. Populations separated by distance r but connected by migration have some probability that any two alleles drawn at random from those populations will be identical by descent. The problem of assaying this genetic identity depends on the population model being used. The development and testing of alternative models has been one of the more active areas in population genetics history.

The first comprehensive attempt at a model for measuring the genetic effects of subdivision with migration was due to Wright (1943), which laid the groundwork for extensive studies to come later by other authors. Paper 19 is a portion of this paper. Wright begins by outlining the simplest possible model of subdivision, which he calls the "island model" (first mentioned in Wright, 1931). The subdivisions of a population are assumed to be random-mating isolates that exchange mates, at a certain rate (m), which are drawn at random from the other isolates. Thus, the probability that a migrant comes from any other subpopulation is independent of any aspect of the location of the other subpopulation. For such a model, Wright derives the equilibrium genetic variance produced, and hence the local inbreeding coefficient, as

$$f = \frac{\sigma^2}{pq} \approx \frac{1}{1 + 4Nm}$$

Wright acknowledges that this model is unlikely to apply to many real populations, since in reality the likelihood of mate exchange depends on the distance between subpopulations. Several alternative models have been proposed by various authors, and vigorous work has been done in deriving and testing them against data. In Paper 19, Wright proposes a continuous model in which migration is restricted to small distances in a population of continuously distributed individuals of assumed density. Wright proposed to analyze this by looking at the inbreeding coefficient of an individual at any place in the distribution of the population, viewing this coefficient as the correlation between uniting gametes. Wright's basic result is that the inbreeding coefficient in a local population of size kN is given by the sum of $k - 1$ terms of his equation (21). Wright's model is sensitive to the size of the neighborhood from which mates are drawn, that is, the effective population size; he finds that two-dimensional populations produce less local differentiation than do one-dimensional ones, such as a string of populations along a shoreline, at least for a model of density and distance such as he uses.

Several other models have since been proposed to increase the number of situations for which we can estimate the genetic effects of isolation by distance. A discontinuous model has been developed by Malecot (1950) and by Kimura and Weiss (1964); it is generally called the *stepping-stone model,* since in it populations are conceived of as being located at discrete points on a lattice of one, two, or three dimensions, with migration between them at given rates (and generally assumed to be the same in all directions). It is not appropriate to go into details of this and other models here as they become very complicated; however, we can note that the stepping-stone model and recently developed continuous models obtain compatible results: that is, given various parameters of migration and distance, identity decreases roughly exponentially with distance.

For further reading, one could see the references cited previously, plus Weiss and Kimura (1965), Cavalli-Sforza and Bodmer (1971), Morton et al. (1971), Malecot (1969), Morton (1969, 1973), and Nei and Imaizumi (1966), which demonstrate the usefulness of some of these models in analyzing empirical situations. For other comments on these methods, see Cannings and Cavalli-Sforza (1973).

The work on migration and subdivision is related directly to the concept of genetic identity, either by way of the inbreeding coefficient or the coefficient of kinship, and measures either the probability that randomly drawn genes are identical or the correlation between those genes.

Since the continuous models of isolation by distance require the parameters of the migration distribution, it is important to look at studies that have investigated migration patterns. Among the first was Brownlee (1911), who examined the distribution and dispersion by carriers of epidemic diseases. Brownlee was primarily interested in the rate of spread of an initially localized epidemic by the movement of exposed individuals to and from the area in a uniformly populated space. He had theorized that the migration pattern observed would approximate that derived from the assumption of random movement of individuals to various distances, but the distributions he observed were leptokurtic, or more like a negative exponential distribution. This applied to the temporal spread of epidemics as well as to their spread in space, and also to the dispersion pattern over time of insects released from a central point.

In 1947, Bateman examined empirical aspects of isolation distance and the contamination of seed-crop stands by wind- and insect-borne pollen (Paper 20). He found migration relationships similar to Brownlee's, in which the contamination at distance D, $F(D)$, can be fitted to a function of the form

$$F(D) = \frac{Y}{D}\, e^{-kD}$$

where the parameters Y and k are related to the specific contamination problem. The rapid falloff in contamination with distance is similar to the rapid decline in the amount of migration with distance. Bateman notes that a genetic correlation between plants with their distances apart will arise due to this contamination process. Paper 20 combines several aspects of the effect of migration and isolation and foreshadows much work on the migration problem among human populations. We have omitted the sections dealing with data fitting. Later papers by Bateman (1950, 1962) deal directly with the universality of gene dispersion patterns and are recommended.

18

COMPOSITION OF POPULATIONS AND OF GENOTYPIC CORRELATIONS FROM THE VIEWPOINT OF POPULATION GENETICS

Sten Wahlund

National Institute for Racial Biology, Uppsala

A translation of the article
ZUSAMMENSETZUNG VON POPULATIONEN UND KORRELATIONSERSCHEINUNGEN
VOM STANDPUNKT DER VERERBUNGSLEHRE AUS BETRACHTET
Hereditas XI:65-106, 1928

NOTE ON TRANSLATION

This article was translated by Marianne Langenbucher Rowe, Department of Germanics, Rice University, with editorial and technical assistance from Paul Ballonoff. In most places, we have kept the translation literally close to the original. However, in a few places, we interpreted more freely, to make otherwise difficult academic German phrases more intelligible. We generally followed Wahlund's practice of leaving alternative expressions of an idea in parentheses.

Heredity has been greatly neglected by researchers in demography, in spite of the fact that pronounced biological features, such as mortality and fertility, have been dealt with at length. The omission is surprising on first sight; one would expect more attention to heredity. This neglect seems to result from statisticians' preference for assuming homogeneity in factors not being considered directly, an assumption generally applied to hereditary factors. Although this attitude is convenient, we should not infer from it that heredity has no effect either for individuals or for populations.

The chief concern of research in genetics has been to determine how individual traits are controlled by hereditary factors. Not until basic principles regarding individual heredity were generally adopted has it been possible to study population effects. The main subject of concern in this area has been population composition under monohybrid heredity. Various researchers have had differing viewpoints, but a comprehensive statement has been put forth by Hultkrantz and Dahlberg (1927). These workers assumed that random mating occurs within the population, i.e., that all gametes have an equal probability of mating with each other to form zygotes. However, such an assumption is never completely realized in practice, a point that must be recognized even if the technical detail of its use in obtaining a result is of little importance.

That the idea of random mating is invalid rests on a number of assumptions. First, cases of deviation are caused by selection or differences in mortality and birth relationships [see Dahlberg (1926) and Hultkrantz and Dahlberg (1927)]. Another reason is the appearance of incest to different degrees, a problem that has been studied by Dahlberg with results soon to be published. A third factor is the appearance of more-or-less well-defined isolation boundaries within the population. Such boundaries may be geographical or social. I shall attempt to describe these mathematically later in this work.

In dihybrid, trihybrid and higher polyhybrid heredity, the meeting of genes "at random" is prevented by the linkage of genes on the chromosome. Previous research in this area has been carried out by Weinberg (1909) and Philiptschenko (1924). Their discussions are incomplete in some important aspects. Therefore, I would like to present a better-rounded presentation of the problem.

In practice it is possible to follow one population from one point in time to another, but the various generations cannot be distinguished from each other. In theoretical discussions on heredity it is necessary to go from the children to the grandchildren, to the great-grandchildren, etc. The fact that in practice the generations overlap obviously does not disturb the applicability of the rules proposed. The process only becomes more balanced and proceeds with more continuity than it should according to the theoretical formulas.

DEVELOPMENT OF POLYHYBRID TRAITS IN A POPULATION IN PANMIXIA

To begin, we shall study population evolution as related to heredity in random crossing and in the absence of all selective forces, i.e., under the conditions which Weismann collectively calls panmixia.

S. Wahlund

First, we shall observe relationships in monohybrid heredity. As mentioned previously, various researchers have dealt with this subject and proved that in monohybrid heredity and panmixia a stable zygotic relationship results which is independent of the initial zygotic distribution of a population. This is due to the fact that in monohybrid heredity every parent supplies only one gene to the respective trait of the offspring. Since we assume that crossing occurs at random, the gene combinations must therefore also occur at random.

If, in monohybrid heredity, we assume that the recessive tendency R occurs with gametic frequency r, and if the dominant gamete D occurs with d frequency, then the incidence of recessive homozygotes is $RR = r^2$, the frequency of heterozygotes is $RD = 2rd$, and the incidence of dominant homozygotes is $DD = d^2$. Three zygote proportions must remain constant under panmixia in all generations.

In monohybrid heredity, random crossing results in random combinations of genes (zygotes) in the offspring and a consequent constancy of the population. However, in polyhybrid heredity, panmixia results in random combinations of gene groups (gametes). Thus random crossing is ultimately capable of breaking gene combinations so that they become restricted by chance. Only after infinitely many generations does the population attain a state of equilibrium in which the frequency of gene combinations (gametes) is in agreement with that value which is expected by chance from the gene incidences, i.e., the product from the frequencies of the genes which entered the combination in question.

Weinberg stated almost 20 years ago that after an infinite number of generations polymerism and panmixia will structure a population in the manner expected from gene frequencies by chance. He further postulated that after a relatively small number of generations, this limiting value will be approximated in practice, if the hereditary factors are not too numerous. Weinberg needed this conclusion to be able to assume genetic stability in the population, in his theoretical explorations of the genetic correlation between related individuals. In this study he did not have occasion to examine more closely the laws of the genetic changes in a population brought about as a result of redistribution of genes.

After Weinberg, the person most concerned with the question of the redistribution of genes in polymerism and panmixia was Philiptschenko (1924). He dealt with population changes in panmixia only, however, in di- and trihybrid heredity, for which he established the condition for genetic equilibrium to be a gametic "concordance". This occurs in a dihybrid population if the recessive genes (traits) R_1 and R_2, corresponding to the dominant genes D_1 and D_2, respectively, are found in combinations (gametes) R_1R_2, R_1D_2, D_1R_2, D_1D_2, with incidence a:b:c:d, such that the product of the outer members equals the product of the inner members; i.e., ad = bc. If ad < bc, the proportion of the exterior gametes, i.e., R_1R_2 and D_1D_2, becomes successively greater in every generation. If ad > bc, the proportion of the interior gametes, i.e., R_1D_2 and D_1R_2, becomes successively greater in every generation. With the aid of numerical examples Philiptschenko further determined that gametic discordance ultimately changes toward the closest concordant relationship according to "its own sense of possibility of transmutation." In trihybrid heredity, Philiptschenko again found a concordant gamete relationship, determined by the use of four equations, which determines the condition for the stability of the population. With the exception of this special case, however, he failed to determine the conditions for increase or decrease in the incidence of various gametes. Neither does he specify any general formulas.

The conditions established by Philiptschenko for the equilibrium of a population in di- and trihybrid heredity can be traced back to Weinberg's conditions— especially that polymeric equilibrium occurs in a population if it is so composed that it may occur randomly on the basis of gene frequencies alone. This condition is to be preferred, since it is uniform for all types of polymerism, and in part makes possible the direct retracing of the proportions of gametes and zygotes from gene frequencies in a population assumed to be stable, i.e., a population presumed to have developed for a sufficient period of time under polymerism. This means more than being able to count on the equilibrium being produced in a population after several generations, as Weinberg has emphasized. It is equally important that during an observation period changes may occur not only through selection, intermingling of foreign hereditary elements, or change in environment, but in polymerism also as a result of the recombination of the genes (change in the zygote proportions) insofar as the condition for stability is not satisfied. This aspect of the problem will be discussed in the following section.

The recessive genes in dihybrid heredity will be denoted R_1 and R_2, corresponding to the dominant genes D_1 and D_2. The frequency of the former in the population will be designated r_1, r_2, and the frequency of the latter, d_1, d_2; thus, $r_1 + d_1 = 1$ and $r_2 + d_2 = 1$. We assume that no linkage of these hereditary factors is present. Both genes, which are transmitted separately to the offspring, can combine together in four ways: R_1R_2, R_1D_2, D_1R_2 and D_1D_2. We assume that these combinations (types of gametes) occur with a frequency (r_1r_2), (r_1d_2), (d_1r_2), (d_1d_2). If a gene were to be distributed randomly in the population, then (r_1r_2) would be the same as r_1r_2, $(r_1d_2) = r_1d_2$, etc. This is, however, not the case. The gene combinations (gametes), not the genes contained in them, randomly combine to form zygotes. In this way a certain gene combination (gamete) R_1R_2 can deviate in its frequency (r_1r_2) from the value r_1r_2 randomly expected from the gene frequencies. It can deviate, in fact, around a value that we shall designate as $\Delta_{R_1R_2}$ (positive or negative).

Thus

$$(r_1r_2) = r_1r_2 + \Delta_{R_1R_2} \qquad (1a)$$

analogously,

$$(r_1d_2) = r_1d_2 + \Delta_{R_1D_2} \qquad (1b)$$

$$(d_1r_2) = d_1r_2 + \Delta_{D_1R_2} \qquad (1c)$$

$$(d_1d_2) = d_1d_2 + \Delta_{D_1D_2} \qquad (1d)$$

Because $(r_1r_2) + (r_1d_2)$ and $r_1r_2 + r_1d_2$ are the same as r_1 by addition of equations (1a) and (1b), one obtains $\Delta_{R_1R_2} + \Delta_{R_1D_2} = 0$; therefore, $\Delta_{R_1D_2} = -\Delta_{R_1R_2}$. From equations (1a) and (1c), one obtains $\Delta_{D_1R_2} = -\Delta_{R_1R_2}$. Analogously, equations (1c) and (1d) yield $\Delta_{D_1D_2} = -\Delta_{D_1R_2} = +\Delta_{R_1R_2}$. If we insert these expressions in equations (1b) to (1d), we have

$$(r_1r_2) = r_1r_2 + \Delta_{R_1R_2} \qquad (2a)$$

$$(r_1d_2) = r_1d_2 - \Delta_{R_1R_2} \qquad (2b)$$

$$(d_1r_2) = d_1r_2 - \Delta_{R_1R_2} \qquad (2c)$$

$$(d_1 d_2) = d_1 d_2 + {}^\Delta R_1 R_2 \qquad\qquad (2d)$$

In dihybrid heredity and panmixia all gametic types with equal value (they need not have the same sign, however) must be differentiated from those values which they would have if there were no linkage between their inherent genes: $R_1 R_2$ and $D_1 D_2$ in the same direction; $R_1 R_2$ and $D_1 R_2$ in the same direction.

If crossing now occurs in the population, it is possible that in one half of the children one gene combination (gamete) will receive both genes from the same parent. In the other half, the gene combination will receive a gene from each parent. In the first case in which a gene combination, e.g., $R_1 R_2$, is maintained, the frequency remains unchanged; $(r_1 r_2) = r_1 r_2 + {}^\Delta R_1 R_2$. In the latter case, R_1 and R_2 combine with a frequency $r_1 r_2$, which is directly conditioned by the frequency of the genes. Thus we can say that in half the cases the "unbroken" gene combination (the gamete) $R_1 R_2 = (r_1 r_2)$ is retained; in the other half of the cases, it is broken to $r_1 r_2$.

After a crossing with frequency $(r_1 r_2)$, the gamete frequency $R_1 R_2$ is obtained from the following formula:

$$(r_1 r_2)_1 = \frac{1}{2}(r_1 r_2) + \frac{1}{2} r_1 r_2$$

$$= \frac{1}{2}(r_1 r_2 + {}^\Delta R_1 R_2) + \frac{1}{2} r_1 r_2 = r_1 r_2 + \frac{1}{2} {}^\Delta R_1 R_2$$

The frequencies for the gametes $R_1 D_2$, $D_1 R_2$ and $D_1 D_2$ after crossing are likewise found from the following formulas:

$$(r_1 d_2)_1 = \frac{1}{2}(r_1 d_2) + \frac{1}{2} r_1 d_2 = r_1 d_2 - \frac{1}{2} {}^\Delta R_1 R_2$$

$$(d_1 r_2)_1 = \frac{1}{2}(d_1 r_2) + \frac{1}{2} d_1 r_2 = d_1 r_2 - \frac{1}{2} {}^\Delta R_1 R_2$$

$$(d_1 d_2)_1 = \frac{1}{2}(d_1 d_2) + \frac{1}{2} d_1 r_2 = d_1 d_2 + \frac{1}{2} {}^\Delta R_1 R_2$$

If the calculations from generation to generation are continued, the frequencies of the combinations $R_1 R_2$, $R_1 D_2$, $D_1 R_2$ and $D_1 D_2$ are obtained after an arbitrary number of crossings, n, from the following formulas:

S. Wahlund

$$(r_1 r_2)_n = r_1 r_2 + \frac{1}{2^n} \Delta_{R_1 R_2} = \frac{1}{2^n}(r_1 r_2) + \frac{2^n - 1}{2^n} r_1 r_2 \qquad (3a)$$

$$(r_1 d_2)_n = r_1 d_2 - \frac{1}{2^n} \Delta_{R_1 R_2} = \frac{1}{2^n}(r_1 d_2) + \frac{2^n - 1}{2^n} r_1 d_2 \qquad (3b)$$

$$(d_1 r_2)_n = d_1 r_2 - \frac{1}{2^n} \Delta_{R_1 R_2} = \frac{1}{2^n}(d_1 r_2) + \frac{2^n - 1}{2^n} d_1 r_2 \qquad (3c)$$

$$(d_1 d_2)_n = d_1 d_2 + \frac{1}{2^n} \Delta_{R_1 R_2} = \frac{1}{2^n}(d_1 d_2) + \frac{2^n - 1}{2^n} d_1 d_2 \qquad (3d)$$

Equations (3a) to (3d) indicate the frequency of the gametes $R_1 R_2$, $R_1 D_2$, $D_1 R_2$ and $D_1 D_2$ after an arbitrary number of generations if they have been identified in a specified generation. Thus the frequency of the gametes will ultimately be transformed into the frequencies that they would have had if the genes had not been linked but had randomly combined (if the gametes were ruptured). This shift proceeds in such a way that the difference between the initial frequency and limiting frequency in each generation disappears. The changes are equal for the frequency of the four gametes. $R_1 R_2$ and $D_1 D_2$ change in the same direction as do $R_1 D_2$ and $D_1 R_2$, but the development of the first two takes the reverse direction from that of the last two. The condition for stability in the population must be one in which the frequency of the gametes is in agreement with the value which they would have if no linkage existed $[(r_1 r_2) = r_1 r_2, (r_1 d_2) = r_1 d_2,$ etc., and therefore $\Delta_{R_1 R_2} = 0]$.

As mentioned previously, the population changes in panmixia are connected with the eventual breaking of the gene combination that comprises a gamete. If we now want to deal with the relationships in polyhybrid heredity in general, we must differentiate among unbroken, once-broken, twice-broken, etc., gametes. By unbroken gametes we mean those which have received all their genes from one and the same gamete in the initial population. These genes have remained together in the hereditary process. Once-broken gametes are those which received their genes from two gametes of the initial population and which therefore, through heredity, occurred only once. Twice-broken gametes are those whose genes came from three gametes of the initial population and which were constructed from two recombinations of genes from different gametes.

We shall now ascertain which portions of unbroken, once-broken, twice-broken, etc., gametes a population exhibits after an indefinite number of n crossings.

Every gamete of the nth generation can be traced back to two gametes in the (n - 1)th generation, to four gametes in the (n - 2)th, etc., that is, from 2^n gametes in the initial generation. The probability that a gene of the gametes in the nth generation is derived from a certain gamete in the initial generation is thus $(1/2)^n$. If a characteristic is conditioned by m hereditary traits and every gamete constructed from m genes, the probability that all these genes stem from a certain gamete in the initial generation becomes $[(1/2)^n]^m = (1/2)^{mn}$. The probability that all genes of an nth-generation gamete stem from one and the same arbitrary gamete of the initial generation becomes $2^n/2^{mn}$, in that one now has to deal with a possible 2^n gametes in this generation. This expression presents the probability that a certain gamete will be transmitted intact for n generations in m-hybrid heredity.

When a gamete is broken, the following can happen in various ways. Each gene from the broken gamete— an optional two, three, etc.— can originate from either of one parental gamete and the rest from the other. Because the genes of a broken gamete can be assigned in various ways to the genes of the initial generation from which they came, we can distinguish different types of broken gametes. To maintain simplicity, we shall assume that we are dealing with only one of these types.

A certain gene of an nth-generation gamete, as we have shown, originates with a $(1/2)^n$ probability from a certain gene of an initial-generation gamete, and the genes from initial-generation gametes, in the case of m-hybrid heredity, with a probability of $(1/2)^{mn}$.

Once-broken gametes are descended from two initial-generation gametes, as was stated. Their 2^n gametes can combine in pairs in $2^n(2^n - 1)$ different ways. It is clear then that a certain type of once-broken gamete appears in the nth generation with a probability of $2^n(2^n - 1)(1/2)^{mn}$. Since three gametes in the initial generation can combine in $2^n(2^n - 1)(2^n - 2)$ ways, it is understood that twice-broken gametes of a certain type appear in the nth generation with a probability of $2^n(2^n - 1)(2^n - 2)(1/2)^{mn}$. In general, a certain type of i-times-broken gamete in the nth generation appears with a probability of $2^n(2^n - 1)(2^n - 2) \ldots (2^n - i)(1/2)^{mn}$.

A gamete that has been broken m - 1 times is completely broken. All its genes are descended from different gametes of the initial population. (We are ignoring the possibility of incest here.) The expression for the part of the completely broken gametes of a population after n crossings becomes $2^n(2^n - 1)(2^n - 2) \ldots (2^n - m + 1)(1/2)^{mn}$. If a characteristic in a population is conditioned in panmixia through an arbitrary number, m, of hereditary factors, and we can proceed from the genetic composition of a certain generation, we shall be able to determine the total composition of the population after an arbitrary number, n, of generations by use of the following:

For every gamete of every type of unbroken gamete, $\dfrac{2^n}{2^{mn}}$

For once-broken gametes, $\dfrac{2^n(2^n - 1)}{2^{mn}}$

For twice-broken gametes, $\dfrac{2^n(2^n - 1)(2^n - 2)}{2^{mn}}$

For i-times-broken gametes, $\dfrac{2^n(2^n - 1)(2^n - 2)(2^n - i)}{2^{mn}}$

For (m - 2)-times-broken gametes, $\dfrac{2^n(2^n - 1)(2^n - 2) \ldots (2^n - m + 2)}{2^{mn}}$

For (m - 1)-times-broken gametes (completely broken gametes),

$$\frac{2^n(2^n - 1)(2^n - 2) \ldots (2^n - m + 1)}{2^{mn}}$$

Thus we have arrived at a general rule for genetic changes in polymerism and panmixia. (Weinberg dealt mathematically with the frequency of completely broken genes.)

If a certain gamete, e.g., the totally recessive $R_1 R_2 R_3 \ldots R_m$, is further transmitted unbroken in the population, its frequency $(r_1 r_2 r_3 \ldots r_m)$ remains unchanged. This value can deviate from that expected at random from the product of the gene frequencies r_1, r_2, r_3, ..., r_m. If, however, the gametes are broken once— we assume in the groups $(R_1 R_2 \ldots R_i)(R_{i+1} R_{i+2} \ldots R_m)$— their frequency will be determined by the frequency of groups in the initial population multiplied by each other, using our notation $(r_1 r_2 \ldots r_i)(r_{i+1} r_{i+2} \ldots r_m)$. These groups have been constructed randomly. As in the unbroken-gamete frequencies, these two group frequencies need not equal those frequencies which are expected at random from the frequency of the genes that comprise the gametes, i.e., with $r_1 r_2 \ldots r_i$, $r_{i+1} r_{i+2} \ldots r_m$. If a gamete is broken numerous times, its frequency is generally determined by the frequencies of the parts multiplied with one another. Only a completely-broken gametic frequency agrees with the value expected at random from the gene frequencies, i.e., with $r_1 r_2 r_3 \ldots r_m$. Because the frequency of unbroken, once-, or numerous-times-broken genes can be derived from the gametic construction of the initial population in this way, we reach the following conclusion: that in panmixia it is theoretically possible to determine the gametic constitution of a population after an arbitrary number of generations in every stage of polyhybrid heredity, if the hereditary composition of a population is known at a certain point in time. (It is naturally also possible to derive the earlier composition of the population.)

.After one crossing, it is obvious that only unbroken or once-broken gametes are possible. The proportion of all the others is zero. In general, an i-times-broken gamete can arise only if, through panmixia, so many generations have passed that the value 2^n surpasses i. Thus, the greater n is, i.e., the more panmictic crossings there are, the more repression of the unbroken or relatively seldom broken gametes takes place.

Completely broken gametes can be produced only after $2^n + 1$ surpasses m. We find that when this has happened the proportion of the completely broken gamete type is related to the (m − 2)-times broken gametes as $(2^n - m + 1):1$. The proportion of the (m − 2)-times broken types to the (m − 3)-times broken types is $(2^n - m + 2):1$, and so on. It is understood, therefore, that the completely broken gametes will determine the value of the gamete frequency relatively soon, and that the changes of the (m − 2)-times-broken combination must dominate in such a way that they will determine the evolution. In the latter case, however, only combinations tied in pairs are retained. This is why the gamete frequency changes as in dihybrid heredity, i.e., continually in the same direction and in such a way that half the deviation from the limiting value disappears in every generation. It is obvious that the limiting value is that value which is expected at random from the product of the gene frequencies r_1, r_2, r_3, $\ldots$, r_m.

The speed of the gametic changes is clearly shown in Table 1, which for various types of polyhybrid heredity, shows the proportions of gametes broken in various generation times.

Where the table shows products, the first factor is the number of the types, which we have not discussed previously; the second factor is the proportion of types derived from the formulas found on pages 231 and 232.

It must be stressed that it is not certain how much closer in value a more-frequently broken gametic frequency is to the completely broken frequency than to a less-often-broken gametic frequency. Because of this we are certain only in dihybrid heredity that the gamete frequencies always approximate their respective limiting values successively. Only in dihybrid heredity do the gamete changes always take the same direction.

Even if it is impossible to give a general theory of the evolution of the gamete frequencies, the following can be deduced from the results of pages 232 and 233 and Table 1. If the gamete proportions of a population deviate from the values expected at random from the gene frequencies, these values will accrue rapidly— at least when the hereditary factors are not too numerous.

If a population, whose gamete frequencies are in random genetic equilibrium for polyhybrid heredity, is mixed with hereditary elements of another population, the state of equilibrium is disturbed and the population develops quickly in the direction of a state of equilibrium determined by the new gene frequencies. In monohybrid heredity the mixture occurs immediately; in polyhybrid heredity it stretches over several generations. In other words, in polyhybrid heredity we have to count not only on a direct change in the population from racial mixing, but also on its continual change in following generations. It is, therefore, of importance that in a statistical study of phenomena where hereditary factors are involved one consider the occurrence of mingling of foreign hereditary elements during the last generation and before the observation period. It is best if a demographic examination of the population and especially of its migratory relationships is made. These problems will be looked at more closely in the following section.

Table 1. Proportions in various grades of broken gametes in various generations after an initial population.

Number of generations after an initial generation	Dihybrid heredity		Trihybrid heredity			Tetrahybrid heredity			
	Unbroken gametes	Completely broken gametes	Unbroken gametes	Once-broken gametes	Completely broken gametes	Unbroken gametes	Once-broken gametes	Twice-broken gametes	Completely broken gametes
1	0.5	0.5	0.25	$3 \cdot 0.25$	—	0.125	$7 \cdot 0.125$	—	—
2	0.25	0.75	0.0625	$3 \cdot 0.1875$	0.375	0.0156	$7 \cdot 0.0469$	$6 \cdot 0.0938$	0.0938
3	0.125	0.875	0.0156	$3 \cdot 0.1094$	0.6563	0.0020	$7 \cdot 0.0137$	$6 \cdot 0.0820$	0.4106
4	0.0625	0.9375	0.0039	$3 \cdot 0.0586$	0.8203	0.0002	$7 \cdot 0.0037$	$6 \cdot 0.0513$	0.6665
5	0.0313	0.9687	0.0010	$3 \cdot 0.0303$	0.9082	0.0000	$7 \cdot 0.0009$	$6 \cdot 0.0284$	0.8231
10	0.0010	0.9990	0.0000	$3 \cdot 0.0010$	0.9971	0.0000	$7 \cdot 0.0000$	$6 \cdot 0.0001$	0.9941

continued...

Table 1 continued

Number of generations after an initial generation	Pentahybrid heredity				
	Unbroken gametes	Once-broken gametes	Twice-broken gametes	Thrice-broken gametes	Completely broken gametes
1	0.0625	15·0.0625	—	—	—
2	0.0039	15·0.0117	25·0.0234	10·0.0234	—
3	0.0003	15·0.0017	25·0.0103	10·0.0513	0.2051
4	0.0000	15·0.0002	25·0.0032	10·0.0417	0.4999
5	0.0000	15·0.0000	25·0.0009	10·0.0257	0.7202
10	0.0000	15·0.0000	25·0.0000	10·0.0010	0.9904

We have dealt thus far only with the changes of frequencies of gametes in polyhybrid heredity and panmixia. However, the zygote and phenotype proportions of which they are composed determine the population changes which will occur.

In monohybrid heredity with complete dominance two phenotypes occur, the recessive and dominant whose frequencies we designate (R_1) and (D_1), respectively. The first phenotype is composed of recessive homozygotes R_1R_2, and the latter of the heterozygote R_1D_2 and the dominant homozygotes D_1D_2. Therefore,

$$(R_1) = R_1R_2$$
$$(D_1) = R_1D_2 + D_1D_2$$

From this it follows that these monohybrid phenotypes must be constant under panmixia.

In dihybrid heredity four phenotypes are found:

236

$$(R_1R_2) = R_1R_1R_2R_2$$

$$(R_1D_2) = R_1R_1R_2D_2 + R_1R_1D_2D_2$$

$$(D_1R_2) = R_1D_1R_2R_2 + D_1D_1R_2R_2$$

$$(D_1D_2) = R_1D_1R_2D_2 + R_1D_1D_2D_2 + D_1D_1R_2D_2 + D_1D_1D_2D_2$$

The phenotypes of the left sides of the equation and the zygotes of the right sides have been given notation analogous with the monohybrid notation.

We see from these formulas that the phenotypes add up in the same way as the gamete proportions (this also applies to higher degrees of polymorphism):

$$(R_1R_2) + (R_1D_2) = (R_1)$$

$$(R_1R_2) + (D_1R_2) = (R_2)$$

$$(D_1R_2) + (D_1D_2) = (D_1)$$

$$(R_1D_2) + (D_1D_2) = (D_2)$$

The phenotype (R_1R_2), i.e., a zygote $R_1R_1R_2R_2$, can only be produced through the union of two R_1R_2 gametes. The frequency of the (R_1R_2) phenotype is given by the following formula:

$$(R_1R_2) = (r_1r_2)(r_1r_2) = r_1^2r_2^2 + 2r_1r_2\Delta_{R_1R_2} + (\Delta_{R_1R_2})^2$$

For the nth generation,

$$r_1^2r_2^2 + \frac{2r_1r_2\Delta_{R_1R_2}}{2^n} + \frac{(\Delta_{R_1R_2})^2}{2^n}$$

Furthermore, from this formula we obtain

$$(R_1D_2) = (R_1) - (R_1R_2)$$

$$(D_1R_2) = (R_2) - (R_1R_2)$$

$$(D_1D_2) = 1 - (R_1R_2) - (R_1D_2) - (D_1R_2) = 1 - (R_1) - (R_2) + (R_1R_2)$$

In this way we find that the phenotype proportions in dihybrid heredity can be traced back to the gamete proportions in a simple manner. From these formulas it follows that the frequencies of the four phenotypes differ in the same amounts from those values which they should have had if no linkage existed—

237

(R_1R_2) and (D_1D_2) in the same direction, and (R_1D_2) and (D_1R_2) in the opposite direction. When crossing occurs in the population, the frequencies of the phenotypes of the same value and in the same direction, (R_1R_2) and (D_1D_2), (R_1D_2) and (D_1R_2), are changed. The last of these change in an opposite way to the first. The evolutionary course of the phenotype frequencies takes the same direction in dihybrid heredity and approaches those values which they would have had if no linkage existed.

We do not have the space to deal with the changes of phenotype proportions in higher grades of polyhybrid heredity. In general, what has been said concerning the changes of gametic proportions applies to changes in higher grades of polyhybrid heredity. It is obvious that the phenotypes, as well as the gametes which comprise them, change relatively fast toward the direction of the limiting values which would be randomly produced by the gene frequencies inherent in them. Naturally, less can be said about the evolutionary change in the phenotype frequencies that are constructed of various gametes (with the exception of the totally recessive phenotypes), because this cannot be done for the gametes. Consequently, one knows for certain only in dihybrid heredity that the frequencies of the phenotypes constantly approach their respective limiting values. Thus only in dihybrid heredity do the changes of phenotypes always take the same direction.

GENETIC CORRELATIONS IN PANMIXIA

A statistically confirmed change in a population may relate not only to environmental changes, selection or similar phenomena, but may also be due to recombinations of the genes in polymorphism, as discussed earlier. How is one to establish when the latter case has occurred? There are difficulties associated with this in practice, especially in attempting to confirm a population change in a certain material. These difficulties exist regardless of whether recombinations of the genes we are dealing with or other factors were decisive. More importantly, we have ascertained that recombinations of genes can be found only if a relatively new mixture with foreign hereditary elements has taken place. It may be established in practice that interbreeding has taken place, but only with great difficulty can it be ascertained in what way the mixture is different from that of the population, and to what extent the interbreeding destroyed equilibrium. The results of this section, which deals with correlation changes in panmixia, provide a possible means for solving this problem. A fundamental clarification of the rules pertaining to the correlation of hereditary traits is understandably of great and independent significance. This is because, as is well known, almost every causal analysis is conducted statistically by examining correlations. The concept of correlation is used here in its broadest sense.

The preceding deductions form a suitable basis for the clarification of correlations of hereditary traits in panmixia.

We shall assume that two monohybrid traits are determined by the unlinked genes R_1 and D_1, and R_2 and D_2. The gene frequencies of a certain examined population will be designated r_1 and d_1, and r_2 and d_2. If correlation exists between the genes of both traits, the probability that an allele of one trait will be linked with an allele of the other trait must deviate from the probability expected at random. The probability that alleles R_1 and R_2 will meet in an individual deviates, for instance, from the values expected at random, $r_1 r_2$, on the basis of the gene proportions by a value which will be designated $\Delta_{R_1 R_2}$ (positive or negative).

Using notation analogous to the three-gene combinations, we have the following relationships, which correspond to the frequencies of the gene combinations [see formulas (1a) to (1d)]:

$$(r_1 r_2) = r_1 r_2 + \Delta_{R_1 R_2}$$

$$(r_1 d_2) = r_1 d_2 + \Delta_{R_1 D_2}$$

$$(d_1 r_2) = d_1 r_2 + \Delta_{D_1 R_2}$$

$$(d_1 d_2) = d_1 d_2 + \Delta_{D_1 D_2}$$

The correlations between both traits are changed by crossing. After an arbitrary number, n, of crossings, the following relationships between the gametes of both traits, which agree with formulas (3a) to (3d), are obtained.

$$(r_1 r_2)_n = r_1 r_2 + \frac{1}{2^n} \Delta_{R_1 R_2}$$

$$(r_1 d_2)_n = r_1 d_2 - \frac{1}{2^n} \Delta_{R_1 R_2} = r_1 d_2 + \frac{\Delta_{R_1 R_2}}{2^n}$$

$$(d_1 r_2)_n = r_1 d_2 - \frac{1}{2^n} \Delta_{R_1 R_2} = d_1 r_2 + \frac{\Delta_{R_1 D_2}}{2^n}$$

$$(d_1 d_2)_n = d_1 d_2 + \frac{1}{2^n} \Delta_{R_1 R_2} = d_1 d_2 + \frac{\Delta_{D_1 D_2}}{2^n}$$

It is evident that in monohybrid heredity and panmixia the correlation between two traits which are restricted by different and unlinked hereditary tendencies is reduced to the extent that half the difference between the actual frequency of the gamete combination and the frequency with which it would occur in case of a correlation eventually disappears in every generation.

We saw previously that in polyhybrid heredity the frequencies of the gene combinations (gamete frequency) change in the direction of increasingly-more-broken combinations. After an infinite number of generations, the frequency attains that value which it whould have had if its genes had randomly crossed. Thus the correlation between the genes of a combination becomes increasingly more broken. The genes thus independently produce one and the same trait as has been assumed. If we assume now that a part of the genes contained in the combination determine a trait, and the other part another trait, it is then understood that the correlation of such a gene combination (gametes) of these traits will also eventually be broken, becoming zero after many generations. These correlation changes (mutations) need not always be in the same direction. The correlation can arise in one evolutionary phase and sink in another. Even if we cannot count on a uniform developmental slope in the correlation, the fact remains that if hereditary elements are bred into a population and they differ from it in two of its traits, correlation arises between them. It disappears eventually, however, under panmixia.

We have assumed here that both traits are determined by different genes. If they are dependent in part on the same genes, it is immediately clear that the correlation has developed toward a randomly-determined limiting value.

EFFECT OF ISOLATES ON ZYGOTE PROPORTIONS AND
THE GENE FREQUENCIES OF THE POPULATION

In a previous section (p. 226), formulas were presented for the composition of a population in monohybrid heredity under the assumption of random crossing. This hypothesis at times produces results that deviate from reality. A cause of such deviations in a population, which has up to now not been of fundamental concern, is the occurrence of limits, which will be designated as isolation boundaries. In this and the following sections, we shall attempt to determine how important such boundaries are and to elucidate under which circumstances one is allowed to work with the formulas and results that are built upon the presupposition of complete panmixia. Furthermore, we shall allude to the effect of isolation upon the population's evolution and correlation relationships.

The isolation can be geographical. Regions whose inhabitants are viewed in total as a population can be more or less completely separated by natural barriers, such as mountains, forests, etc., into sections called isolates. The isolation can also be social. A person belonging to a particular social class marries within his own class more frequently than into a higher or lower class.

To calculate the effect of isolation, we shall at first proceed from the premise that there are two isolates within a population. In each of them we assume that random crossing takes place. Between both isolates, however, no crossing occurs. We presuppose no differences in birth and mortality conditions. The zygotic composition of each isolate, designated G and H, will be determined in monohybrid heredity according to the formulas presented on page

We assume the proportionate sizes of the component population to be $g:h$, such that $g + h = 1$. The frequency of the gametes of the recessive trait R is r_g and r_h. The frequency of the corresponding dominant trait D is d_g and d_h, such that $r_g + d_g = 1$ and $r_h + d_h = 1$.

If we assume that random crossing takes place in G and H, we obtain the zygote proportions shown in Table 2.

The zygotic proportions in the total population (G + H) are obtained by weighting the zygotic proportions of G and H with their fraction of the total population, designated g,h [see Table 3, column (1)].

Column (2) of Table 3 specifies the zygotic proportions calculated from the average gametic proportions of the total population. These zygotic proportions would have obtained if random crossing had taken place in the entire population, rather than if resulting from two isolates.

Table 2

Zygote types	Proportions	
	in G	in H
Recessive homozygotes, RR	r_g^2	r_h^2
Heterozygotes, RD	$2r_g d_g$	$2r_h d_h$
Dominant homozygotes, DD	d_g^2	d_h^2
Recessive genes, RR	r_g^2	r_h^2
Dominant genes, RD + DD	$1 - r_g^2$	$1 - r_h^2$

Table 3

Zygote types	Proportions of gamete population (G + H)		
	Actual (1)	Calculated from gamete proportion (2)	Remainder (1) − (2) = (3)
Recessive homozygotes, RR	$gr_g^2 + hr_h^2$	$(gr_g + hr_h)^2$	$+ gh(r_g - r_h)^2$
Heterozygotes, RD	$2gr_g d_g + 2hr_h d_h$	$2(gr_g + hr_h)$ $\cdot (gd_g + hd_h)$	$- 2gh(r_g - r_h)^2$
Dominant homozygotes, DD	$gd_g^2 + hd_h^2$	$(gd_g + hd_h)^2$	$+ gh(r_g - r_h)^2$
Recessive genes, RR	$gr_g^2 + hr_h^2$	$(gr_g + hr_h)^2$	$+ gh(r_g - r_h)^2$
Dominant genes, RD + DD	$1 - gr_g^2 - hr_h^2$	$1 - (gr_g + hr_h)^2$	$- gh(r_g - r_h)^2$

242

We see that the homozygotes RR and DD of the entire population exhibit an excess of $gh(r_g - r_h)^2$ [since gh and $(r_g - r_h)^2$ are positive] on the basis of the two isolates present, compared with those values which they would have had in complete panmixia. The heterozygotes show a deficit that is twice as great as this value. The recessive genes thus exhibit a surplus of $gh(r_g - r_h)^2$ and the dominant genes an equally great deficit.

Figure 1 represents the size of the homozygote surplus which results after the formation of both isolates whose presence in the population we have assumed. The abscissa gives the positive difference between the gametic proportions $(r_g - r_h)$ of the partial population; the ordinate gives the homozygote increase $gh(r_g - r_h)^2$. The reading on the y axis is given for various scales, depending on how the portions of the partial populations g and h vary from the total population. The variable $gh(r_g - r_h)^2$ can, moreover, be obtained for every value from g and h in that the result for g,h = 1/2 is multiplied with the value 4gh.

The larger the difference between the gamete proportions of both isolates, the greater the homozygote excess is.* If the gametic composition as well as the zygotic composition of both isolates are identical, whereby $r_g = r_h$, then the increase in homozygosis equals 0. The occurrence of isolates has no influence. Homozygote increase achieves its maximum for both types of homozygotes equal to gh if the difference between the gamete proportions of the isolate equals 1. Thus, if one of the proportions r_g and $r_h = 1$, the other is 0. Heterozygotes are then not present in the population.

The same conditions which determine the increase of homozygosis apply to the decrease of heterozygote frequencies, with the difference that this decrease is twice as great as the increase of both types of homozygotes.

We shall now deal with this problem in a more general manner. We shall assume that the population which we are studying with regard to a simple Mendelian trait is composed of an arbitrary number, m, of more or less isolated partial populations (isolates). We further assume that random

---

* The derivative $\dfrac{d[gh(r_g - r_h)^2]}{d(r_g - r_h)} = 2gh(r_g - r_h)$ and increases linearly with $(r_g - r_h)$. The greater the difference between the gametic proportions of the isolates, the stronger, therefore, the homozygote surplus is.

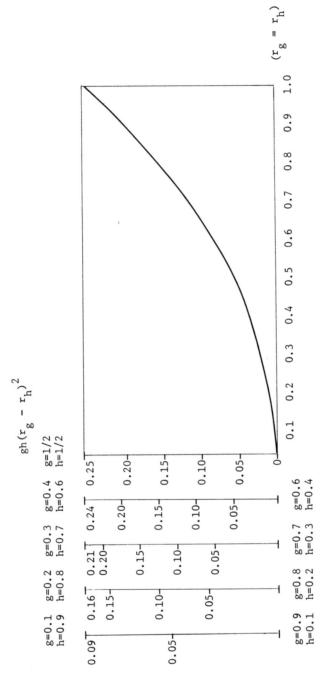

Figure 1. Difference $gh(r_g - r_h)^2$ between the actual and those homozygotes RR or DD which are calculated from gamete proportions whereby the difference $(r_g - r_h)$ varies between the gamete frequencies of two isolates contained in a population and the proportions g and h of both isolates in the total population.

crossing has taken place within each population. Moreover, we now assume that there is no selection for the m isolates, called I_1, I_2, I_3, ..., I_m. It will be assumed that the frequency of gametes for the recessive trait R is r_1, r_2, r_3, ..., r_m, the frequency of gametes for the dominant trait D is d_1, d_2, d_3, ..., d_m, and the portions of the total population are i_1, i_2, i_3, ..., i_m.

The zygotic proportions of the partial population are

Recessive homozygotes: $\quad RR = r_1^2,\ r_2^2,\ r_3^2,\ ...,\ r_m^2$

Heterozygotes: $\quad RD = 2r_1d_1,\ 2r_2d_2,\ 2r_3d_3,\ ...,\ 2r_md_m$

Dominant homozygotes: $\quad DD = d_1^2,\ d_2^2,\ d_3^2,\ ...,\ d_m^2$

Recessive genes: $\quad RR = r_1^2,\ r_2^2,\ r_3^2,\ ...,\ r_m^2$

Dominant genes: $\quad RD + DD = 1 - r_1^2,\ 1 - r_2^2,\ 1 - r_3^2,\ ...,\ 1 - r_m^2$

The zygotic proportions of the total population are found since (as on pp. 241–242) the zygotic proportions of the partial population (isolate) are weighted by the parts i_1, i_2, i_3, ..., i_m that constitute these in the total population [see Table 4, column (1)].

Column (2) of Table 4 presents the zygotic proportions that are calculated from the average gametic proportions of the total population (the proportions we call r and d), and which we would have found if this population had not consisted of m isolates but had been subject to random crossing. The increase of the proportion of recessive homozygotes RR on the basis of the isolates found is obtained by subtracting the value of column (2) from column (1):

$$(i_1r_2^2 + i_2r_2^2 + i_3r_3^3 + ... + i_mr_m^2) - (i_1r_1 + i_2r_2 + i_3r_3 + ... + i_mr_m)^2$$
$$= i_1r_1^2 + i_2r_2^2 + i_3r_3^2 + ... + i_mr_m^2 - r^2$$
$$= r^2 + i_1r_1^2 + i_2r_2^2 + i_3r_3^2 + ... + i_mr_m^2 - 2r^2$$
$$= r^2 + i_1r_1^2 + i_2r_2^2 + i_3r_3^2 + ... + i_mr_m^2$$
$$= 2r(i_1r_1 + i_2r_2 + i_3r_3 + ... + i_mr_m)$$

thus becomes $i_1r_1 + i_2r_2 + i_3r_3 + ... + i_mr_m$

Because $i_1 + i_2 + ... + i_m = 1$, the equation can be written as

$$i_1r^2 + i_2r^2 + i_3r^2 + ... + i_mr^2 + i_1r_1^2 + i_2r_2^2 + i_3r_3^2$$
$$+ ... + i_mr_m^2 - 2i_1rr_1 = 2i_2rr_2 - 2i_3rr_3 - ... - 2i_mrr_m$$

Table 4

Zygote types	Proportions of gamete population $(I_1 + I_2 + I_3, \ldots, I_m)$		
	Actual (1)	Calculated from gamete proportions (2)	Remainder (1) − (2) = (3)
Recessive homozygotes, RR	$i_1 r_1^2 + i_2 r_2^2 + i_3 r_3^2 + \ldots + i_m r_m^2$	$(i_1 r_1 + i_2 r_2 + i_3 \cdot r_3 + \ldots + i_m r_m)^2$	$i_1(r - r_1)^2 + i_2 \cdot (r - r_2)^2 + i_3 \cdot (r - r_3)^2 + \ldots + i_m(r - r_m)^2$
Heterozygotes, RD	$2 i_1 r_1 d_1 + 2 i_2 r_2 d_2 + 2 i_3 r_3 d_3 + \ldots + 2 i_m r_m d_m$	$2(i_1 r_1 + i_2 r_2 + i_3 r_3 + \ldots + i_m \cdot r_m)(i_1 d_1 + i_2 d_2 + i_3 d_3 + \ldots + i_m \cdot d_m)$	$-2[i_1(r - r_1)^2 + i_2 \cdot (r - r_2)^2 + i_3 \cdot (r - r_3)^2 + \ldots + i_m(r - r_m)^2]$
Dominant homozygotes, DD	$i_1 d_1^2 + i_2 d_2^2 + i_3 d_3^2 + \ldots + i_m d_m^2$	$(i_1 d_1 + i_2 d_2 + i_3 \cdot d_3 + \ldots + i_m d_m)^2$	$i_1(r - r_1)^2 + i_2 \cdot (r - r_2)^2 + i_3 \cdot (r - r_3)^2 + \ldots + i_m(r - r_m)^2$
Recessive genes, RR	$i_1 r_1^2 + i_2 r_2^2 + i_3 r_3^2 + \ldots + i_m r_m^2$	$i_1 r_1 + i_2 r_2 + i_3 r_3 + \ldots + i_m r_m$	$i_1(r - r_1)^2 + i_2 \cdot (r - r_2)^2 + i_3 \cdot (r - r_3)^2 + \ldots + i_m(r - r_m)^2$
Dominant genes, RD + DD	$i_1(1 - r_1^2) + i_2 \cdot (1 - r_2^2) + i_3 \cdot (1 - r_3^2) + \ldots + i_m(1 - r_m^2)$	$1 - (i_1 r_1 + i_2 r_2 + i_3 r_3 + \ldots + i_m \cdot r_m)^2$	$-[i_1(r - r_1)^2 + i_2(r - r_2)^2 + i_3(r - r_3)^2 + \ldots + i_m \cdot (r - r_m)^2]$

and we immediately obtain

$$i_1(r - r_1)^2 + i_2(r - r_2)^2 + i_3(r - r_3)^2 + \ldots + i_m(r - r_m)^2 = \sigma^2 \quad (1)$$

This expression, which we call σ^2, is positive because all terms are positive, an indication that because of the isolates the proportion of recessive homozygotes shows a surplus over the proportion that would have occurred in random crossing. [It can easily be shown that the equation $gh(r_g - r_h)^2$ on p. 243 is a special case of σ^2.]

It can likewise be proved that the proportion of dominant homozygotes is increased to the value

$$i_1(d - d_1)^2 + i_2(d - d_2)^2 + i_3(d - d_3)^2 + \ldots + i_m(d - d_m)^2$$

Since $d = 1 - r$ and, furthermore, $d_1 = 1 - r_1$, $d_2 = 1 - r_2$, $d_3 = 1 - r_3$, $\ldots$, $d_m = 1 - r_m$, it is seen that the preceding equation is the same as

$$i_1(r - r_1)^2 + i_2(r - r_2)^2 + i_3(r - r_3)^2 + \ldots + i_m(r - r_m)^2 = \sigma^2$$

Because both homozygote proportions increase by the value σ^2, the heterozygote proportions must decrease by twice this value. The recessive genotypes increase by σ^2; the dominants decrease by σ^2.

The expression σ, as was just shown, is identical to the standard deviation of a series made up of the gametic proportions of various isolates and their proportional representation of the total population.

Thus, if an investigated population in which no selection takes place is composed of a number of more or less isolated partial populations, and in each individual isolate random crossing has taken place, then in the total population the homozygote proportions exceed by a certain amount the homozygote proportions calculated from the random-mating frequency of zygotes. This value is equal to the square of the standard deviation of a series that is constructed from the gametic proportions of the isolates with a representation proportional to the respective strata of the population. The heterozygote proportion is less than that calculated, by twice this amount. The proportion of recessive genes indicates a surplus; the proportion of dominant genes a deficit of the same amount.

The proportions of the recessive homozygotes (RR), the heterozygotes (RD) and the dominant homozygotes (DD) in our hypothetical population are expressed through the values $r^2 + \sigma^2$, $2(rd - \sigma^2)$ and $d^2 + \sigma^2$. If simple Mendelian

S. Wahlund

heredity is assumed and one attempts to calculate the remaining zygotic proportions from the proportion of recessive genotypes (RR), i.e., from the
recessive homozygotes, and furthermore counts on their frequency as being that
of random crossing, then these calculations will lead to false results. The
proportion of the recessive gametes R becomes $\sqrt{r^2 + \sigma^2}$ instead of r. The
proportion of the dominant gametes D becomes $1 - \sqrt{r^2 + \sigma^2}$ instead of $1 - r$.
The proportion of dominant homozygotes (DD) becomes $(1 - \sqrt{r^2 + \sigma^2})^2$ or reduces
to $1 + r^2 + \sigma^2 - 2\sqrt{r^2 + \sigma^2}$ instead of $d^2 + \sigma^2 = 1 + r^2 + \sigma^2 - 2r$. The difference between the computed and the actual proportion of dominant homozygotes
(DD) thus becomes $-2(\sqrt{r^2 + \sigma^2} - r)$. A proportion results which is less than
the actual one by a value twice as great as the difference between the calculated and the actual gamete proportion R. The calculated proportion of the
heterozygote (RD) surpasses the actual proportion by the same value.

The homozygote surplus (the heterozygote deficit) in the present isolate,
as well as the difference between the computed and the actual zygotic proportions, becomes larger in proportion to the increase of the above-mentioned
standard deviation (σ) of the series constructed of gametic proportions of
the isolate. Only if this standard deviation equals 0, which only occurs when
the gametic proportions of all isolates are identical, and thus when the total
population is homogeneous, do the conditions correspond completely to those of
random crossing.

It is, of course, very important to be able to calculate in practice how
great the overrepresentation of the homozygotes is on the basis of isolation
boundaries in a particular case. How are these isolation boundaries to be
evaluated? If the material can be divided into isolates with internal random
crossing, one need only apply our formulas, with a correction for variations
of gamete proportions of the isolates selected at random.

Things are, however, never so simple. In the preceding calculations we
considered the isolates to be tightly defined. We imagined that a discontinuous
transition existed from one area with random crossing to the next area. (The
terminology of geographic isolation is used here.) In practice, one should
be able to depend on the fact that such well-defined isolates occur only
rarely. Real isolates grade continuously into one another. In geographical
isolation one could conceive of the unlikely possibility that a population is
equally divided over an area and that the probability of marriage (crossing)

decreases with distance according to a definite law equally applicable to all. In this condition, which may be called isolation, the above-mentioned effects take place. As a result one is not capable of identifying a single isolate. Isolation exists, but no isolates.

In practice, the effect of the isolates can never be determined exactly on the basis of the described conditions. It can easily be shown that if one attempts to determine σ by dividing a material into isolates, values too small will be achieved under the usual conditions, instead of exact results.

It is assumed that t can be differentiated from more or less isolated groups in a population with an average proportion of recessive homozygotes of r_1^2, r_2^2, ..., r_s^2, ..., r_t^2, and a part of the total population of i_1, i_2, ..., i_s, ..., i_t.

If in the total population the proportion of recessive gametes is r, then

$$\sigma^2 = i_1(r - r_1)^2 + i_2(r - r_2)^2 + \ldots + i_s(r - r_s)^2 + \ldots + i_t(r - r_t)^2 \quad (1a)$$

σ_I does not agree with the homozygote increase as long as it is not known that random crossing in the sth isolate takes place. Thus σ can only be viewed as a first approximation.

We now assume that a certain group, e.g., the sth, in which random crossing does not take place is divided into k parts within which crossing occurs more frequently. In other words, the group is divided into k isolates.

The average proportion of recessive homozygotes of these k isolates is taken as r_{s1}^2, r_{s2}^2, r_{s3}^2, ..., r_{sk}^2 and their respective parts in the sth group as i_{s1}, i_{s2}, i_{s3}, ..., i_{sk}. When we replace $i_s(r - r_s)^2$ in the expression $_1^2$ [i.e., in formula (1a)] by $i_s(r - r_s)^2 + i_s i_{s1}(r - r_{s1})^2 + i_s i_{s2}(r - r_{s2})^2 + i_s i_{s3}(r - r_{s3})^2 + \ldots + i_s i_{sk}(r - r_{sk})^2$, we know that in our calculations we are dealing with the isolates comprising the group rather than the sth group itself. The expression $i_s i_{s1}(r - r_{s1})^2 + i_s i_{s2}(r - r_{s2})^2 + i_s i_{s3}(r - r_{s3})^2 + \ldots + i_s i_{sk}(r - r_{sk})^2$ is equal to the expression

$$i_s[i_{s1}(r - r_{s1})^2 + i_{s2}(r - r_{s2})^2 + i_{s3}(r - r_{s3})^2 + \ldots + i_{sk}(r - r_{sk})^2]$$

The expression within brackets is equal to the square of the standard deviation of a provisional mean value r of the series constructed from the gamete proportions of the k isolates comprising the sth group, and where the representation of the gamete proportions of the series is proportional to the respective part of these isolates in the sth group. If we designate the

S. Wahlund

standard deviation of the actual mean value r_s as σ_s, the expression becomes

$$i_s[\sigma_s^2 + (r - r_s)^2] = i_s(r - r_s)^2 + i_s\sigma_s^2$$

In regard to the k isolates of the sth group, $i_s(r - r_2)^2$ is replaced by $i_s(r - r_s)^2 + i_s\sigma_s^2$.

If all groups are divided up in this way, one obtains a second approximation of the homozygote increase, which will be called σ_{II}^2, whose value is derived from the following formula (analogous notation):

$$\sigma_{II}^2 = i_1(r - r_1)^2 + i_2(r - r_2)^2 + \ldots + i_s(r - r_s)^2 + \ldots$$
$$+ i_t(r - r_t)^2 + i_1\sigma_1^2 + i_2\sigma_2^2 + \ldots + i_s\sigma_s^2 + \ldots + i_t\sigma_t^2$$

where

$$\sigma_{II}^2 = \sigma_I^2 + i_1\sigma_1^2 + i_2\sigma_2^2 + \ldots + i_s\sigma_s^2 + \ldots + i_t\sigma_t^2$$

If a new division of the small isolates took place based on the same principle, we would have a third approximation, σ_{III}^2. The series constructed from σ_I^2, σ_{II}^2, σ_{III}^2, etc., is continually increasing. For every approximation are added proportional, squared values resulting from the standard deviations of gametic proportions of the new isolates. Such a series must have a limiting value. This limit is approached when the conditions of the partial isolates approach those of isolates with random crossing; when their standard deviations approach zero and the partial isolates are homogeneous, the limit value will be σ^2.

Theoretically, one can more or less expect separate partial populations (population strata) in which random crossing takes place. In practice, one obtains a mean value of σ^2 from the separation into partial isolates.

Only under the assumption that the variance σ^2 of the gamete proportion is negligibly small can the exact zygotic proportion of a population be calculated. In doing so one can use theoretical values based on random crossing. When does this happen? This question is related to another, perhaps more important, one: When may the standard deviation of the series of gamete proportions be ignored; or when can the population be considered homogeneous from the viewpoint of genetics? In other words, when can one say that there are no groups which differ from each other in relationship to the hereditary composition of the population? We shall find in the following section that the condition for which the standard deviation 0 is found also constitutes the criterion for equilibrium in the population.

EFFECT OF ISOLATES ON CHANGES IN THE POPULATION

It is obvious that the standard deviation of the series of gametic proportions has a tendency to disappear due to migration within the population. The
populations evolve toward the direction of greater homogeneity. The migration
has a leveling effect on differences in the gametic proportions. With the
disappearance of this standard deviation, the population evolves toward an
equilibrium determined solely by its gametic frequencies.

We therefore want to study under certain assumptions the change in genetic
character from generation to generation in a population that has no random
crossing. A more definite knowledge of the changes of the genetic character
of a certain material requires exhaustive demographical information, above all
in reference to migration in coneection with marriages. We suppose that there
is no selection in the population.

A random migration from an isolate has no effect upon its gametic composition. Immigration changes it, however, if the hereditary composition deviates
from that of the isolates. We assume that in a definite, arbitrarily chosen
isolate, e.g., the sth, a certain number of individuals comprising a part a_{s1}
of the isolate immigrate. We assume further that the i_sth isolate is stable
in the total population so that an equal number of individuals migrate. (As
indicated before, the gametic proportions are not changed through random migration. Thus the last supposition has no major significance.) One assumes,
therefore, that a part a_{s1} of the sth isolates is exchanged for the same number
of immigrants. The proportion of recessive gametes (R) of the immigrants is
called r_{x1}; the corresponding proportion of the isolate before the immigration
is r_s, and after the immigration r_{s1}. The proportion r_{s1} is obtained from the
formula

$$r_{s1} = (1 - a_{s1})r_s + a_{s1}r_{x1} \qquad (2a)$$

Call the proportion of dominant gametes (D) of the immigrants d_{x1}, where
$r_{x1} + d_{x1} = 1$. Let d_s be the corresponding proportion of the isolate before
immigration, where $r_s + d_s = 1$, and let d_{s1} be the proportion after immigration,
where $r_{s1} + d_{s1} = 1$. Then the following formula is obtained analogously

$$d_{s1} = (1 - a_{s1})d_s + a_{s1}d_{x1} \qquad (2b)$$

Since random crossing takes place within the isolate, the zygotes of the

251

S. Wahlund

first offspring generation resulting from migration occur in the following proportions: $(r_{s1})^2$ from RR, $2(r_{s1})(d_{s1})$ from RD, $(d_{s1})^2$ from DD.

In further migration and crossing we assume that in the 1st, 2nd, 3rd, ..., (n − 1)th generation a part a_{s2}, a_{s3}, a_{s4}, ..., a_{sn} will be exchanged for the same number of immigrants. The proportions of their recessive gametes (R) are designated by r_{x2}, r_{x3}, r_{x4}, ..., r_{xn}, and the proportions of their dominant gametes by d_{x2}, d_{x3}, d_{x4}, ..., d_{xn}. The portion of the nth migration of the isolate is expressed by a_{sn}, and the respective gamete proportions of the immigrants by r_{xn} and d_{xn}. The recessive and dominant gametes (R and D) of the first generation of the population have already been designated r_{s1}, and d_{s1}. The corresponding proportions of the 2nd, 3rd, 4th, ..., nth generation will be designated by r_{s2} and d_{s2}, r_{s3} and d_{s3}, r_{s4} and d_{s4}, ..., r_{sn} and d_{sn}. The gametic proportions of the sth isolates after n migrations and crossings are expressed by r_{sn} and d_{sn}. The proportions of zygotes RR, RD and DD after n migrations and crossings are expressed by $(r_{sn})^2$, $2(r_{sn})(d_{sn})$ and $(d_{sn})^2$.

The gametic proportions of the population after the 2nd, 3rd, 4th, ..., nth migrations and crossings are given by the following general formulas, which are analogous to formulas (2a) and (2b):

$$r_{s2} = (1 - a_{s2})r_{s1} + a_{s2}r_{x2}$$

$$d_{s2} = (1 - a_{s2})d_{s1} + a_{s2}d_{x2}$$

$$r_{s3} = (1 - a_{s3})r_{s2} + a_{s3}r_{x3}$$

$$d_{s3} = (1 - a_{s3})r_{s2} + a_{s3}d_{x3}$$

$$r_{s4} = (1 - a_{s4})r_{s3} + a_{s4}r_{x4}$$

$$d_{s4} = (1 - a_{s4})d_{s3} + a_{s4}d_{x4}$$

$$\text{etc.}$$

$$r_{sn} = (1 - a_{sn})r_{s(n-1)} + a_{xn}r_{xn} \tag{2c}$$

$$d_{sn} = (1 - a_{sn})d_{s(n-1)} + a_{sn}d_{xn} \tag{2d}$$

When the immigrating contribution from generation to generation is taken as a stable gamete composition r_x from R, and d_x from D, the following equations derived from formula (2) result:

$$r_{s1} - r_x = (1 - a_{s1})(r_s - r_x)$$

$$d_{s1} - d_x = (1 - a_{s1})(d_s - d_x)$$

$$r_{s2} - r_x = (1 - a_{s2})(r_{s1} - r_x)$$

$$d_{s2} - d_x = (1 - a_{s2})(r_{s1} - d_x)$$

$$r_{s3} - r_x = (1 - a_{s3})(r_{s2} - r_x)$$

$$d_{s3} - d_x = (1 - a_{s3})(d_{s2} - d_x)$$

etc.

$$r_{sn} - r_x = (1 - a_{sn})(r_{s(n-1)} - r_x)$$

$$d_{sn} - d_x = (1 - a_{sn})(d_{s(n-1)} - d_x)$$

When the values of the first set of equations, $r_{s1} - r_x$ and $d_{s1} - d_x$ are inserted into the second, and the resultant values, $r_{s2} - r_x$ and $d_{s2} - d_x$, are inserted into the third, etc., one ultimately obtains

$$r_{sn} - r_x = (1 - a_{s1})(1 - a_{s2})(1 - a_{s3}) \ldots (1 - a_{sn})(r_s - r_x) \qquad (3a)$$

$$d_{sn} - d_x = (1 - a_{s1})(1 - a_{s2})(1 - a_{s3}) \ldots (1 - a_{sn})(d_s - d_x) \qquad (3b)$$

When the value $(1 - a_{s1})(1 - a_{s2})(1 - a_{s3}) \ldots (1 - a_{sn})$ is replaced by $(1 - a_s)^n$ so

$$1 - a_s = \sqrt[n]{(1 - a_{s1})(1 - a_{s2})(1 - a_{s3}) \ldots (1 - a_{sn})}$$

(a_s is a geometric mean constructed from the various migrations), then one obtains

$$r_{sn} - r_x = (1 - a_s)^n(r_s - r_x) \qquad (4a)$$

$$d_{sn} - d_x = (1 - a_s)^n(d_s - d_x) \qquad (4b)$$

Since a is an actual fraction, $(1 - a)^n$ approaches 0 with increasing n; i.e., r_{sn}, d_{sn} are increasingly closer to r_x, d_x. The gametic proportions of the isolates eventually approach those of the immigrants— more quickly in the beginning through constant migration, then increasingly more slowly.

In reality, however, the gametic proportions of the immigrants are seldom constant. The immigration consists of influx from regions that themselves show changes in gametic proportions as a result of migration. The changes in gametic proportions of a region caused by migration influence the development in those regions where migratory flow originated. If, in addition,

S. Wahlund

changes during migration are often of unequal magnitude, it is impossible to solve this confusion of effects and retroactive effects theoretically.* One must be satisfied with handling more or less schematic cases.

Wherever migration takes place, a leveling process of hereditary differences occurs. We assume that this process is uniform within a population of t isolates, so that in every isolate the immigration presents a representative selection of the remaining isolates, with compositions corresponding to their gametic proportions. We further hypothesize that the population is externally isolated during the observation period. (In practice, one can differentiate the immigrants and their offspring in a population or can establish the effect of an eventual immigration in other ways.)

In a specific isolate, e.g., in the sth, whose part in the total population is taken to be i_s, and whose original recessive gamete (R) proportion is designated r_s, the proportion of recessive gametes of the first immigration consistent with the above assumptions on leveling will appear as follows:**

$$\frac{1}{1 - i_s} r - \frac{i_s}{1 - i_s} r_s$$

The proportion of the second immigration becomes

$$\frac{1}{1 - i_s} r - \frac{i_s}{1 - i_s} r_{s1}$$

the proportion of the third immigration becomes

$$\frac{1}{1 - i_s} r - \frac{i_s}{1 - i_s} r_{s2}$$

etc. The proportion of the nth immigration becomes

$$\frac{1}{1 - i_s} r - \frac{i_s}{1 - i_s} r_{s(n-1)}$$

---

* Here we are dealing with an important, previously unworked, and difficult area for demographic research. The intent is to study inner migration and, in doing so (avoiding as much as possible administrative divisions) to determine the isolates of the population in detail and to elucidate their origins and effects.

** The immigrated population should show the same average gamete proportions as the remaining isolates. If its proportion of recessive gametes is r_y, the equation $i_s r_s + (1 - l_s) r_y = r$ is obtained. The proportion of a population's recessive gametes is equal to the respective gamete proportions that are weighted with the fractions of both groups. The above expression for r_y is derived from this equation.

From these formulas it follows that the effect of the immigration is the same if a greater immigration is substituted as a part $1/(1 - i_s)$ of the actual immigration and with a recessive gamete proportion r equivalent to that of the total population. A further migration of actual size, $1 - [1/(1 - i_s)] = [1/(1 - 1_s)]$, or of the same size as the theoretical immigration, may also occur. This migration is of the same composition as the isolates, so we need not consider it.

For immigration into the isolate in the proportions a_{s1}, a_{s2}, a_{s3}, ..., a_{sn}, we are thus able to substitute an equivalent constant immigration showing the proportions

$$\frac{a_{s1}}{1 - i_s}, \frac{a_{s2}}{1 - i_s}, \frac{a_{s3}}{1 - i_s}, \ldots, \frac{a_{sn}}{1 - i_s}$$

whose gametic composition is equal to the average of the total population (r from R, d from D). In the following, the former expressions will be designated b_{s1}, b_{s2}, b_{s3}, ..., b_{sn}. The following formulas for the gametic composition of the sth isolate are the result of having substituted b for a and r for r_x in equations (3) and (4):

$$(r_{sn} - r) = (1 - b_{s1})(1 - b_{s2})(1 - b_{s3}) \ldots (1 - b_{sn})(r_s - r) \quad (5a)$$

$$(d_{sn} - d) = (1 - b_{s1})(1 - b_{s2})(1 - b_{s3}) \ldots (1 - b_{sn})(d_s - d) \quad (5b)$$

If the average coefficient

$$b_s = 1 - \sqrt[n]{(1 - b_{s1})(1 - b_{s2})(1 - b_{s3}) \ldots (1 - b_{sn})}$$

is inserted, then

$$(r_{sn} - r) = (1 - b_s)^n (r_s - r) \quad (6a)$$

$$(d_{sn} - d) = (1 - b_s)^n (d_s - d) \quad (6b)$$

It must be noted that formulas (6a) and (6b) are also applicable to a constant coefficient b_s.

We shall now examine more closely the significance of coefficient b_s, which we assume is equal to $a_s/(1 - i_s)$. When that part which constitutes the immigration and emigration of isolates in the total population is p_s, then $a_s = p_s/i_s$ and therefore

$$b_s = \frac{p_s}{i_s(1 - i_s)} \quad (7)$$

The individuals that marry in a specific generation belong to a part i_s of the sth isolate. The probability of crossing between the isolate and the

255

S. Wahlund

remaining populations must always be $2_{is}(1 - i_s)$ in random crossing. Furthermore, if we retain the assumption that the size of the population remains unchanged and suppose that emigration and immigration are equally strong, then
one half the marriages in question must come to the sth isolate, the other
half to the remaining population.* For a moment we shall associate the concept
of "random crossing" with the idea that no migration takes place in the population except in marriage. The proportion of the total population which
constitutes the migrated (out of) as well as immigrated (into) individuals
becomes $i_s(1 - i_s)$ in random crossing. Since p designates the actual proportion, formula (7) shows that the coefficient b_s stands for the part which each
proportion of immigrants (or emigrants) constitutes compared to the proportion
that would be present in random crossing. Clearly, the coefficient comprises
that part of the isolate which actually married outsiders, compared to the
part which married outsiders in random crossing if all migrations occurred in
connection with marriages. The coefficient b_s is therefore a direct measurement for the number of those who actually marry outside of the isolate in
comparison to those who do this in random crossing. When this type of crossing
takes place, the coefficient is 0. If we designate the difference between the
coefficient and 1, by β_s (thus $1 - b_s = \beta_s$), it becomes evident that this value,
which we call the isolation factor, presents a direct measure of the isolation's
strength. If no isolation is present (random crossing), the isolation factor
is 0. The more strongly the relationships deviate from those of random crossing,
i.e., the stronger the isolation becomes, the greater the factor becomes.
When isolation is total, it attains the value 0.**

---

* We need only assume that the division of the marriages in question over the
isolate and remaining population occurs according to a specific principle,
which in the following case is the same as in the assumed random crossing,
whereby $b_s = 2_{ps}/[2_{is}(1 - i_s)]$.

** In the case of heavy migration one can imagine that the isolation factor is
smaller than the value 0. For example, one can assume that i_s is the smallest
of the values i_s and $1 - i_s$ (one could also choose $1 - i_s$). Then $1 - i_s \geq 1/2$
and $1/(1 - i_s) \leq 2$. However, p_s is $< i_s$; therefore, $p_s/[i_s(1 - i_s)] < 2$.
That the expression is ≥ 0 is self-evident. Consequently,

$$-1 < \beta_s = 1 - p_s/[i_s(1 - i_s)] \leq +1$$

Theoretically, then, the isolation factor can lie between -1 and +1. However,
we may in practice neglect the possibility that $-1 < \beta_s < 0$, which would mean
such a heavy migration that the balance would be upset.

If the isolation factor is varied from generation to generation, one obtains the effective average isolation factor from the geometric mean of the
component factors in the sth isolate (see p. 255).

Formulas (6a) and (6b) can be written as follows:

$$r_{sn} - r = (\beta_s)^n (r_s - r) \tag{8a}$$

$$d_{sn} - d = (\beta_s)^n (d_s - d) \tag{8b}$$

The greater the isolation factor the longer the leveling proceeds. If
the isolation factor is 0 (random crossing), average gametic proportions set
in.

Since analogous formulas can be devised for the remaining isolates, it is
possible, by substitution of the values obtained for $r_{1n} - r$, $r_{2n} - r$, $\ldots$,
$r_{sn} - r$, $\ldots$, $r_{tn} - r$ in formula (4a), to construct the following equation
for the standard deviation of the series of gametic proportions within the
entire population after n migrations and crossings (σ_n):

$$(\sigma_n)^2 = i_1 (\beta_1)^{2n} (r_1 - r)^2 + i_2 (\beta_2)^{2n} (r_2 - r)^2 + \ldots + i_s (\beta_s)^{2n}$$
$$\cdot (r_s - r)^2 + \ldots + i_t (\beta_t)^{2n} (r_t - r)^2 \tag{9}$$

We supposed that a randomly chosen number of individuals from the remaining
isolates would immigrate into every isolate. This can only be achieved if the
strength of the isolation (the isolation factor, e.g., $\beta_s = 1 - b_s$) has no
relationship to the deviation of gametic proportions from the average population.
Isolates whose gametic composition deviates more severely from the average
composition of the population may be no more (and no less) isolated than those
closer to the average of the population. We can, therefore, exchange the
isolation factors β_1, β_2, $\ldots$, β_s, $\ldots$, β_t for a factor called β that is
average (or constant) for all isolates.

Formula (9) then appears as follows:

$$(\sigma_n)^2 = (\beta)^{2n} [i_1 (r_1 - r)^2 + i_2 (r_2 - r)^2 + \ldots$$
$$+ i_s (r_s - r)^2 + \ldots + i_t (r_t - r)^2]$$
$$(\sigma_n)^2 = (\beta)^{2n} (\sigma)^2 \tag{10}$$

After the root has been found, one obtains

$$\sigma_n = (\beta)^n \tag{11}$$

S. Wahlund

The standard deviation of the series of gametic proportions (hereditary diversity) between the various isolates decreases more and more when more migrations and crossings have taken place. This occurs quickly in the beginning, then more slowly and drops steadily to a proportion like that of the isolation factor (β) of the remaining standard deviations. The zygotic changes $(\sigma_n)^2$ resulting from the isolates are reduced in the same way, proportional to the same value squared, but more rapidly.

Having given an expression for the decrease in the standard deviation in gametic proportions from generation to generation, we return to the following problem: When can $(\sigma_n)^2$ be disregarded and when can we count on the presence of the zygotic proportions expected from random crossing; when can σ_n be disregarded and the hereditary composition of the population viewed as homogeneous?

We shall let σ_n represent the standard deviation in the sample and σ represent the standard deviation of n previous generations. How is σ_n related to results based on σ? If the sampled standard deviation is not known, it follows that the standard deviation of previous generations is even less known.

Formula (11) shows that σ_n constitutes a part $(\beta)^n$ of σ. Since the values of gametic proportions vary between 0 and 1, under the most favorable conditions we can arrive at the value 0.5. Usually it remains much lower, usually well below 0.25. In reference to a certain material it is always possible to determine the maximum possible value of the standard deviation of the series of gametic proportions after n generations. If the type (β) of isolation from this point in time is known, it is possible to calculate the maximum value for the standard deviation of the series of gametic proportions, according to the formula. The extent to which the value is sufficiently small can now be determined, i.e., in what respect the standard deviation makes up a sufficiently small fraction of its primary value.

Thus we obtain means to determine how a certain material can be seen as hereditarily homogeneous or not, and how well the zygotic proportions agree with those values theoretically derived from random crossing.

We assume that an extensive knowledge of the demographic renewal process of the population, and especially the migration and marriage conditions, must be achieved in order to have a somewhat reliable idea of the order of magnitude of the isolation factor (β). Furthermore, this investigation must be carried out most thoroughly, since a superficial one would determine β values which are too small.

It is obviously of great interest to be able to establish how many generations must be investigated demographically in order to determine to what extent a material satisfies the demands of sufficient hereditary homogeneity. Also, if the material is too heterogeneous, how far back must one go to be able to divide the material into sufficiently homogeneous groups? From formula (11), one obtains

$$n = \frac{\log\left(\frac{\sigma_n}{\sigma}\right)}{\log \beta} \tag{12}$$

In this formula σ_n/σ is included so that the material can be viewed as homogeneous, and is that part of the standard deviation of the series of gamétic proportions to which one considers the resulting standard deviation to have decreased. β forms the average isolation factor. A value of n is obtained that corresponds to the number of generations which must be investigated.

To show how quickly the differences in gametic proportions are balanced, we present Table 5, which indicates, for varying values of the isolation factor, how soon the composition of a population can be considered homogeneous, assuming a decrease of the standard deviation by 1/5, 1/10 and 1/20.

Table 5. Balancing speed of gamete proportions of a population in various values of the isolation factor.

If a population is regarded as homogeneous during a decrease of the standard deviation by:	Homogeneity occurs after the following number of generations if the isolation factor =								
	0.1	0.2	0.3	0.4	0.5	0.6	0.7	0.8	0.9
$\frac{1}{5}$	1	1	1	2	2	3	5	7	15
$\frac{1}{10}$	1	1	2	3	3	5	6	10	22
$\frac{1}{20}$	1	2	3	3	4	6	8	13	29

From formula (11),

$$n = \frac{1}{2}\frac{\log\left(\frac{\sigma_n}{\sigma}\right)}{\log \beta}$$

Only half as many generations are needed for balancing the zygotic changes as in the equilization of the standard deviation in the series of gametic proportions.

Table 5 provides a general idea of how long an intermingling of foreign hereditary elements must have occurred for the differences in the gametic proportions of a simple Mendelian gene to be considered equalized, i.e., how soon various parts of the population attain the same composition so that the population is considered homogeneous.

As has been suggested, later interbreeding and the offspring resulting from it should best be separated and studied by themselves; if this is not possible, their effect should be investigated in a different way. In evaluating the periods of Table 5 one must remember the assumptions upon which it was built. We assumed that the process of equalization proceeded in the same manner— that each isolate received from the others a representative contribution of hereditary traits. As was indicated, this applies only if there is no connection between the deviations of the gametic proportions of isolates from the cross section of the population and the strength of the isolation (of the isolation factors). However, this frequently will not be the case in practice. That an isolate should deviate from the average can be taken as an indication of relatively strong isolation at that point. Furthermore, migration is greatest between areas of greater proximity, from which one can expect the same primary strains of hereditary traits.

Because of these conditions, equalization in the separated isolates is slower than indicated in formulas (6a) and (6b). An approach to homogeneity is usually slower under the assumptions that form the basis for Table 5. For this reason the values appear small. In practice, the methods of calculation may be modified according to the material at hand.

EFFECT OF ISOLATES ON THE CORRELATION BETWEEN GENES

It is obvious that genetic nonhomogeneity within a material in relation to two investigated traits causes correlation. We shall designate the mean value between two examined traits in the t isolates that were assumed to be in a material as A_1, A_2, A_3, $\ldots$, A_t, and B_1, B_2, B_3, $\ldots$, B_t. A and B, respectively, will be the mean values for the entire material; σ_A and σ_B will be the standard deviation.

We assume that the two traits are not conditioned by the same genes and, furthermore, that at least the one environment is not conditioned by the other.* In addition, we assume for simplicity that the changes, in the gene pools of the various isolates after interbreeding with foreign elements, have proceeded for so long that correlation caused by polymeria is no longer present. No correlation occurs in the isolates. There is, nevertheless, correlation present in the total material because the distributions of both traits were typical of hereditary distributions of which populations are composed and, furthermore, partially held together.

The correlation coefficient of the total population becomes

$$\frac{i_1(A_1 - A)(B_1 - B) + i_2(A_2 - A)(B_2 - B) + \ldots + i_s(A_s - A)(B_s - B)}{\sigma_A \sigma_B}$$

Meanwhile, as the different inequalities of the isolate are eventually equalized within the population through migration, the mean values of the isolates evolve toward the mean value of the population, even if not always in the same direction. The expressions in the numerator of the above formula for the correlation coefficients move away from 0, and joining them is the correlation coefficient of the total population.

In this way it can be determined that when foreign hereditary elements are bred into a population, a correlation results that endures— although even if not constant— in part because of polymeria and in part through the occurrence of isolates in the population over time. The correlation decreases eventually, and must disappear after a relatively few generations, at least when the hereditary factors of the characteristics are not too numerous and, above all, if the isolation is not too great.

CONCLUSIONS

Finally, some definitions are necessary for the practical application of the above results. It is evident from the preceding material that the effects of breeding foreign hereditary elements into a population break down into two phases.

---

* More is not necessary for our consideration. This does not mean that the environment plays no role in the correlation of two traits, one of which is restricted by the environment. That this is the case should be clear from the equation below (this page).

S. Wahlund

Primary phase: the direct change in the population, i.e., the shift which the interbred elements cause directly.

Secondary phase: the changes in following generations toward those values expected if the genes had randomly mixed. This latter process depends upon

a. A combination of genes in polyhybrid heredity. The genes can eventually be split only into randomly determined combinations.

b. The migration between the isolates in the population in which the foreign hereditary element has been divided unequally. Random crossing is not present in the population, thus explaining why only chance combinations of different hereditary elements result.

The primary effect of interbreeding can be determined by combining the immigrated elements into a special group. The secondary effect of the intermingling can be determined in a genealogical way, but only for the type of mixing which took place after that time from which the population can be traced genealogically. The offspring of the interbred elements can then be divided into groups which are sufficiently homogeneous. After a demographic study of the population (especially its migratory relationships), we can examine the extent to which the secondary effect of earlier mixing can be ignored, by using the preceding results.

We can thus determine to what extent and in what way a material can be divided into groups that are sufficiently homogeneous hereditarily. Within these groups there are no hereditary connections that are not determined by common genes or linkage. Hereditary differences in the material must lie between these homogeneous groups.

The results of our observations of correlation yield the possibility, in addition to the genealogical-demographic one, of examining the homogeneity of a material. Two traits that are not joined by common genes or linkage nor influenced by differences of environment cannot be correlated in a hereditarily homogeneous material. Such trait combinations appear to be, for example, eye color and body height or eye color and the length-breadth index of the head.

The following must be noted, however: that no correlation exists between two traits in a population, such as those mentioned, signifies only that this population is genetically homogeneous for these traits; it is not known if the population is homogeneous in relationship to other characteristics. Concerning these other characteristics, it is possible for primary mixing to have taken place later. Furthermore, a greater primary deviation from the randomly expected (limiting) value could have occurred. In addition, it is possible that traits other than those observed can be determined by a variety of hereditary factors. But even if the results of correlation investigations are applied to other traits and allow only a probable conclusion, they can nevertheless be used in practice. One need only remember the relationships discussed here—especially because disturbances take place in the hereditary balance during immigration into a population and changes thus occur to various traits simultaneously, even if in different degrees. We shall not now go into further detail on the application of the preceding results to anthropology and population statistics. It is obvious that such theoretical results attain their total value only on the basis of experience in practical application.

Anthropological material divided into geographical groups that are as homogeneous as possible should easily provide an idea of the hereditary composition of an area. As a result of the difficulty of gathering a sufficiently large amount of material, the intricate details are continually lost.

The preceding results point out another way in which one can proceed: the correlation-statistical method. The method of obtaining an idea of the characteristics of the primary hereditary components through the nature of their correlations is, however, possible only in heterogeneous populations, i.e., in populations in which the equalization of hereditary traits has not yet been completed.

REFERENCES

1. Dahlberg, G. 1926. Twins and twin births from a hereditary point of view. Stockholm.

2. Hultkrantz, J. V. and Dahlberg, G. 1927. Die Verbrietung eines monohybriden Erbmerkmals in einer Population und in der Verwandtschaft von Merkmalstragern. Arch. f. Rassen- und Ges.-biol., Bd. XIX.

3. Philiptschenko, J. 1924. Uber Spaltungsprozesse innerhalb einer Population bei Panmixie. Zeitschr. F. indukt. Abst.-und Vererbungslehre, Bd. XXXV.

4. Weinberg, W. 1909. Uber Verergungsgesetze beim Menshen. Zeitschr. f. indukt. Abst.-und Vererbrungslehre, Bd. I-II.

19

Reprinted from *Genetics*, **28**, 114–121, 136–138 (Mar. 1943)

ISOLATION BY DISTANCE*

SEWALL WRIGHT

The University of Chicago[1]

Received November 9, 1942

STUDY of statistical differences among local populations is an important line of attack on the evolutionary problem. While such differences can only rarely represent first steps toward speciation in the sense of the splitting of the species, they are important for the evolution of the species as a whole. They provide a possible basis for intergroup selection of genetic systems, a process that provides a more effective mechanism for adaptive advance of the species as a whole than does the mass selection which is all that can occur under panmixia.

RANDOM DIFFERENTIATION UNDER THE ISLAND MODEL

Mathematical consideration requires the use of simple models of population structure. The simplest model is that in which the total population is assumed to be divided into subgroups, each breeding at random within itself, except for a certain proportion of migrants drawn at random from the whole. Since this situation is likely to be approximated in a group of islands, we shall refer to it as the island model.

The gene frequency (q) of a subgroup tends to vary about a certain equilibrium point ($\hat{q}$) in a distribution curve ($\phi(q)$) determined by the net systematic pressure (measured by Δq, the net rate of change of gene frequency per generation from recurrent mutation, immigration, and selection) in conjunction with the cumulative effects of accidents of sampling (random deviation δq, variance per generation $\sigma_{\delta q}^2$) (WRIGHT 1929, 1931, 1942).

$$(1) \qquad \phi(q) = (C/\sigma_{\delta q}^2) \exp\left[2\int (\Delta q/\sigma_{\delta q}^2)dq\right].$$

Let N be the effective size of the subgroup, m the effective proportion of its population replaced in each generation by migrants, and q_t the gene frequency in the total population. The rate of change of gene frequency per generation in a subgroup, taking account only of immigration pressure, is $\Delta q = -m(q-q_t)$. In a random breeding population $\sigma_{\delta q}^2 = q(1-q)/2N$. Substitution in (1) gives the following, choosing C so that $\int_0^1 \phi(q)dq = 1$ (WRIGHT 1931, 1942).

$$(2) \qquad \phi(q) = \frac{\Gamma(4Nm)}{\Gamma(4Nmq_t)\Gamma[4Nm(1-q_t)]} q^{4Nmq_t-1}(1-q)^{4Nm(1-q_t)-1}$$

$$(3) \qquad \bar{q} = \int_0^1 q\phi(q)dq = q_t$$

* A portion of the cost of composing the mathematical formulae is borne by the Galton and Mendel Memorial Fund.

[1] Acknowledgment is made to the DR. WALLACE C. and CLARA A. ABBOTT MEMORIAL FUND of the UNIVERSITY OF CHICAGO for assistance in connection with the calculations.

(4) $$\sigma_q^2 = \int_0^1 (q - \bar{q})^2 \phi(q) dq = q_t(1 - q_t)/(4Nm + 1).$$

In the derivation of (1), it was assumed that Δq is sufficiently small that terms involving $(\Delta q)^2$ might be ignored. A more accurate value of σ_q^2 may be obtained directly. The deviation of a local gene frequency from the average, $(q - q_t)$, tends to be reduced to $(1 - m)(q - q_t)$ in the next generation. The mean sampling variance of $(q + \Delta q)$ is

(5) $$\frac{1}{2N} \int_0^1 [q - m(q - q_t)][1 - q + m(q - q_t)] \phi(q) dq$$
$$= [q_t(1 - q_t) - (1 - m)^2 \sigma_q^2]/2N.$$

Thus with a steady balance between the effects of immigration and of the accidents of sampling

(6) $$\sigma_q^2 = (1 - m)^2 \sigma_q^2 + [q_t(1 - q_t) - (1 - m)^2 \sigma_q^2]/2N$$

(7) $$\sigma_q^2 = q_t(1 - q_t)/[2N - (2N - 1)(1 - m)^2].$$

This is approximately the same as (4) for small values of m but becomes $q_t(1 - q_t)/2N$, the sampling variance, in the limiting case of no isolation whatever $(m = 1)$. This is about twice as great as given by (4) in this extreme case.

The variance, excluding the immediate sampling variance may be obtained by multiplying (7) by $(1 - m)^2$ as indicated in (6). Formula (4) lies between the values with and without the immediate sampling variance.

Under exclusive uniparental reproduction, whether vegetative or by self-fertilization, the distribution of alternative genotypes may be treated by the same theory except for replacement of $2N$ by N. Immigration pressure is the same but the sampling variance is $q(1 - q)/N$.

THE INBREEDING COEFFICIENT

Departures from panmixia may be expressed in terms of the average inbreeding coefficient of individuals, relative to the total population under consideration. This coefficient has been defined as the correlation between uniting gametes with respect to the gene complex as an additive system. It has been shown that its value can be found for any pedigree by finding all paths by which one may trace back from the egg to a common ancestor (A) and thence forward to the sperm along a wholly different path. According to the theory of path coefficients, the correlation between uniting gametes is the sum of contributions from all such paths (WRIGHT 1921, 1922b).

(8) $$F = \sum [(1/2)^{n_S + n_D + 1}(1 + F_A)]$$

where F and F_A are the inbreeding coefficients of the individual and of a common ancestor of sire and dam, respectively, and n_S and n_D are the numbers of generations from sire and dam, respectively, to this common ancestor. In a population in which the average inbreeding coefficient is F, the frequencies of genotypes (one pair of alleles) are as follows (WRIGHT 1921, 1922a).

	Genotype	Frequency
(9)	AA	$x_t = q_t^2(1 - F) + q_t F$
	Aa	$y_t = 2q_t(1 - q_t)(1 - F)$
	aa	$z_t = (1 - q_t)^2(1 - F) + (1 - q_t)F$

'The inbreeding, measured by F, may be of either of two extreme sorts: sporadic mating of close relatives with no tendency to break the population into subgroups, and division into partially isolated subgroups, within each of which there is random mating. The latter is the case in which we are primarily interested here. Assume that there are K subgroups each of size N. The proportion of heterozygotes within a subgroup is $2q'(1-q')$ where q' is the gene frequency in the parental generation, including immigrants.

$$(10) \qquad y_t = 2\sum_1^K q'(1 - q')/K = 2q_t - 2(\sum q'^2)/K.$$

The variance of the gene frequencies of the subgroups, not allowing for accidents of sampling in the last generation, is

$$(11) \qquad \sigma_{q'}^2 = \sum_1^K (q' - q_t)^2/K = (\sum q'^2)/K - q_t^2$$

$$(12) \qquad y_t = 2q_t(1 - q_t) - 2\sigma_{q'}^2 \quad \text{from (10) and (11)}$$

$$(13) \qquad \sigma_{q'}^2 = q_t(1 - q_t)F \quad \text{from (9) and (12).}$$

This formula does not allow for the contribution to variance due to accidents of sampling in the last generation. Thus it gives $\sigma_{q'}^2 = 0$ instead of $\sigma_q^2 = q_t(1-q_t)/2N$ for F=0. To compare with (7) it must be divided by $(1-m)^2$.

$$(14) \qquad \sigma_q^2 = q_t(1 - q_t)F/(1 - m)^2$$

$$(15) \qquad F = (1 - m)^2/[2N - (2N - 1)(1 - m)^2] \quad \text{from (14) and (7)}$$

$$(16) \qquad m = 1 - \sqrt{2NF/[(2N - 1)F + 1]}$$

$$(17) \qquad \sigma_q^2 = q_t(1 - q_t)[(2N - 1)F + 1]/2N.$$

The formula $F = 1/[4Nm+1]$ given in a preceding paper (DOBZHANSKY and WRIGHT 1941) is a satisfactory approximation if m is small.

This island model is not likely to be exactly realized in nature. In most cases, the actual immigrants to a population come from immediately surrounding localities in excess and thus are not a random sample of the species. This can be remedied to some extent by multiplying the proportion of replacement by an appropriate factor to obtain the effective immigration index. If q_m is the gene frequency in the actual immigrants (varying from group to group) the appropriate factor would be $(q-q_m)/(q-q_t)$. Unfortunately the values for effective m for different loci may be very different.

LOCAL INBREEDING IN A CONTINUOUS AREA

At the opposite extreme from the island model is that in which there is complete continuity of distribution, but interbreeding is restricted to small distances by the occurrence of only short range means of dispersal. Remote populations may become differentiated merely from *isolation by distance* (WRIGHT 1938, 1940).

Each individual has its origin at a particular place. Assume that its parents originated at distances from this place with a certain variance both in longitude

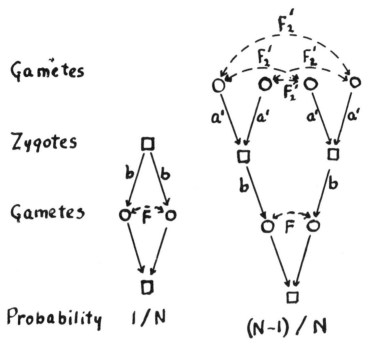

and in latitude. If the same condition held in preceding generations, the grandparents originated at distances with twice this variance in longitude and in latitude and the ancestors of generation K originated at distances with K times this variance in both directions. The parents may be considered as if drawn at random from a territory with a certain radius R and effective population size N. The ancestors of generation K may then be considered as drawn similarly from a territory of radius $\sqrt{K}\,R$ and effective population size KN.

We shall use the term parental group for the population (effective size N) from which the parents of an individual may be considered to be drawn; the term random breeding or panmictic unit will be used for any local population of the same effective size as the parental group.

The assumption of random union of gametes, including self fertilization (probability $1/N$) can be made with sufficient accuracy even though there is actually no self fertilization. It has been shown that such unions in a population of constant size N lead to fixation at the rate $1/2N$ in comparison with the rate $[(N+1)-\sqrt{(N^2+1)}]/2N$ either in a population of size N equally divided between males and females or in a population of N monoecious individuals in which self fertilization does not occur. As the latter formula may be written $[1-(1/2N)\cdots]/2N$ the difference is ordinarily negligible (WRIGHT 1931).

The inbreeding coefficient of individuals in such a population can be calculated from its definition as the correlation between uniting gametes. Let F_x be the correlation between random gametes drawn from a population of size xN and use primes to indicate preceding generations as in the text figure (p. 117). The inbreeding coefficient itself would be F_1 in this terminology. The values of these coefficients can be expressed in terms of coefficients for preceding generations by tracing all connecting paths and noting that the path coefficient b, relating gamete to parental zygote, has the value $\sqrt{(1+F')/2}$ and that the path coefficient, a, relating offspring zygote to one of the gametes that produced it, has the value $\sqrt{1/[2(1+F)]}$. The compound coefficient $ba'=\frac{1}{2}$ (WRIGHT 1921). It may easily be seen that (8) can be deduced at once from these considerations.

In the case of continuity

(18)
$$
\begin{cases}
F = \dfrac{1}{N} b_2 + \dfrac{N-1}{N} 4b^2 a'^2 F_2' = \dfrac{1}{N}\left(\dfrac{1+F'}{2}\right) + \dfrac{N-1}{N} F_2' \\[2ex]
F_2' = \dfrac{1}{2N}\left(\dfrac{1+F''}{2}\right) + \dfrac{2N-1}{2N} F_3'' \\[2ex]
F_3'' = \dfrac{1}{3N}\left(\dfrac{1+F'''}{2}\right) + \dfrac{3N-1}{3N} F_4''' \text{ etc.}
\end{cases}
$$

(19) Thus

$$
F = \frac{1+F'}{2N} + \frac{N-1}{N}\left\{ \frac{1}{2N}\left(\frac{1+F'}{2}\right) \right.
$$
$$
\left. + \frac{2N-1}{2N}\left[\frac{1}{3N}\left(\frac{1+F'''}{2}\right)+\cdots\right]\right\}.
$$

If the same population structure has continued indefinitely, primes may be dropped.

(20)
$$
F = \left(\frac{1+F}{2N}\right)\left[1 + \frac{1}{2}\left(\frac{N-1}{N}\right) + \frac{1}{3}\left(\frac{N-1}{N}\right)\left(\frac{2N-1}{2N}\right)\right.
$$
$$
\left. + \frac{1}{4}\left(\frac{N-1}{N}\right)\left(\frac{2N-1}{2N}\right)\left(\frac{3N-1}{3N}\right)\cdots\right].
$$

This is an infinite series, but in practice the value of F that is of interest is that relative to some finite population. The correlation between random gametes in a population of size KN is F_K which may be taken as zero, thereby stopping the series at $(K-1)$ terms. Let t_x be the xth term in the series in brackets and $\sum_1^{K-1} t$ the sum of first $(K-1)$ such terms

$$(21) \qquad F = \sum_{1}^{K-1} t \Big/ \left[2N - \sum_{1}^{K-1} t \right]$$

$$(22) \qquad t_x = \frac{(x-1)N - 1}{xN} t_{(x-1)}.$$

Let $t_{(x-0.5)} = (t_x + t_{(x-1)})/2$ and $\Delta t_{(x-0.5)} = t_x - t_{(x-1)}$

$$(23) \qquad \frac{\Delta t_{(x-0.5)}}{t_{(x-0.5)}} = - \frac{2(N+1)}{N(2x-1) - 1}.$$

If the values of t are treated as ordinates of a curve with abscissas x, we may write t and x in place of $t_{(x-0.5)}$ and $(x-0.5)$, respectively. The following then hold approximately

$$(24) \qquad \frac{dt}{tdx} = - \frac{2(N+1)}{2Nx - 1}$$

$$(25) \qquad t = C \left(x - \frac{1}{2N} \right)^{-(N+1)/N}$$

$$(26) \qquad \sum_{K_1}^{K_2-1} t = \int_{K_1-0.5}^{K_2-0.5} tdx \quad \text{approximately}$$

$$(27) \qquad \sum_{K_1}^{K_2-1} t = CN \left[\left(K_1 - \frac{1}{2} - \frac{1}{2N} \right)^{-1/N} - \left(K_2 - \frac{1}{2} - \frac{1}{2N} \right)^{-1/N} \right].$$

The value of the constant C can be obtained by equating actual and estimated values of t. Estimates for all but the first few terms in the series are in close agreement. Thus if $N = 10$

Actual series $[1 + .45 + .285 + .206625 + \cdots]$
Estimated series $C[1.05805 + .47969 + .30423 + .22067 + \cdots]$

The estimated value of C from the first term is .9451, from the second term .9381, from the third term .9363. The limiting value is .935774. The value of C approaches 1 as N increases. Thus for $N = 100$, $C = .994157$.

Estimates of $\sum_{1}^{K-1} t$ directly from (27) are not good approximations, but most of the error is in the first few terms. Good estimates can be made by using the actual values from (22) for these terms and the estimates from (27) for the later terms. For $N = 10$

	Actual (22)	Estimate (27)	Error of Estimate
$\sum_{1}^{3} t$	1.73500	1.86782	+.13282
$\sum_{4}^{9} t$	.79002	.79250	+.00248
$\sum_{10}^{39} t$	.99511	.99541	+.00030
$\sum_{40}^{99} t$	.57228	.57228	+.00000

269

A priori, one would expect F to approach i as a limit as the size of population is increased without limit. This requires that $\sum_{1}^{\infty} t$ approach N. Trial for values of N from 10 to 10,000 indicates that this is actually the case and thus gives a good check on the theory. Following are examples:

	N = 10	N = 20	N = 50		N = 100
$\sum_{1}^{39} t$ from (22)	3.52013	3.86519	4.09266	$\sum_{1}^{9} t$	2.797
$\sum_{40}^{\infty} t$ from (27)	6.47987	16.13481	45.90734	$\sum_{10}^{\infty} t$	97.203
	10.00000	20.00000	50.00000		100.000

LOCAL INBREEDING ALONG A LINEAR RANGE

In a species with an essentially one dimensional range (parents drawn from the whole width) the extent along the range from which the ancestors of generation K are drawn is proportional to $\sqrt{K}$ as with area continuity, but the effective size of the corresponding population is $\sqrt{K}\,N$ instead of KN. By analogous reasoning

$$(28) \quad F = \sum t/(2N - \sum t)$$

where

$$\sum t = \left[1 + \frac{1}{\sqrt{2}}\left(\frac{N-1}{N}\right) + \frac{1}{\sqrt{3}}\left(\frac{N-1}{N}\right)\left(\frac{\sqrt{2}N-1}{\sqrt{2}N}\right) \cdots \right]$$

$$(29) \quad t_x = \frac{N\sqrt{(x-1)} - 1}{N\sqrt{x}} t_{(x-1)}$$

$$(30) \quad \frac{\Delta t_{(x-0.5)}}{t_{(x-0.5)}} = \frac{2N(\sqrt{x-1} - \sqrt{x}) - 2}{N(\sqrt{x-1} + \sqrt{x}) - 1}.$$

Treating this expression as the slope at the mid-interval and replacing $(x-0.5)$ by x

$$(31) \quad \frac{dt}{tdx} = \frac{2N(\sqrt{x-0.5} - \sqrt{x+0.5}) - 2}{N(\sqrt{x-0.5} + \sqrt{x+0.5}) - 1}$$

$$= -\frac{N[1 + 1/(32x^2) + \cdots] + 2\sqrt{x}}{2Nx[1 - 1/(32x^2) + \cdots] - \sqrt{x}}.$$

Ignoring $1/(32x^2)$ and smaller terms in the brackets, this yields

$$(32) \quad t = Ce^{-2\sqrt{x}/N}[\sqrt{x} - (1/2N)]^{-[1 + (1/N)^2]}.$$

This seems to be as accurate an approximation as is warranted after replacement of $\Delta t/t$ by dt/tdx.

Comparisons of actual and calculated values of t indicate that estimates of C approach stability after a few terms. For N = 10, C = 1.1529 (from 30th to

40th terms). For $N = 100$, $C = 1.01465$ (from 9th and 10th terms). For larger values of N. especially if x is 10 or more, it may be sufficiently accurate to take dt/tdx as $-(N + 2\sqrt{x})/2Nx$, $C = 1$

(33) $t = e^{-2\sqrt{x}/N}/\sqrt{x}$ approximately.

In this case

(34) $\sum_{K_1}^{K_2-1} t = N(e^{-2\sqrt{\overline{K_1}}/N} - e^{-2\sqrt{\overline{K_2}}/N})$

The value of $\sum_1^{K-1} t$ can be approximated by finding actual $\sum_1^9 t$ from (29), estimating $\sum_{10}^{K-1}$ from (34) and multiplying the latter by the mean ratio of t from (32) to that from (33). Calculation of $\sum_1^\infty t$, $N = 10$, by this method (by steps) gave 10.008 (instead of theoretical 10) and for $N = 100$ gave 100.07 instead of theoretical 100. These theoretical values are on the assumption that the limiting value of F is 1 which again is seen to be verified.

[*Editors' Note:* Material has been omitted at this point.]

SUMMARY

Formulae are derived relating the variance of the gene frequencies of subgroups (σ_q^2) to the effective population number of these (N), the effective proportion of replacement per generation by immigrants (m), the inbreeding coefficient of individuals relative to the total population (F), and the mean gene frequency in the latter (q_t). Thus $\sigma_q^2 = q_t(1 - q_t)/[2N - (2N - 1)(1 - m^2)]$ $= q_t(1 - q_t)F/(1 - m)^2$ including the immediate sampling variance, but $\sigma_q^2 = q_t(1 - q_t)F$ excluding this.

The effect of isolation by distance in a continuous population in which there is only short range dispersal in each generation is worked out on the hypothesis that the parents of any individual may be treated as if they were taken at random from a group of a certain size (N). It is shown that the inbreeding coefficient of individuals in such a population relative to a population of size KN can be expressed in the form $F = \sum_1^{K-1} t/[2N - \sum_1^{K-1} t]$ where $\sum t$ is the sum of a series of terms in which $t_1 = 1$ and $t_x = t_{(x-1)}[(x - 1)N - 1]/xN$ or approximately $C[x - (1/2N)]^{-(N+1)/N}$ where C is a constant close to 1. The value of $\sum_1^{K-1} t$ can be obtained sufficiently accurately by actual calculation of the first few terms, supplemented by the approximate formula

$$\sum_{K_1}^{K_2-1} t = CN\left[\left(K_1 - \frac{1}{2} - \frac{1}{2N}\right)^{-1/N} - \left(K_2 - \frac{1}{2} - \frac{1}{2N}\right)^{-1/N}\right]$$

for later terms. The limiting value $\sum_1^\infty t$ is N. Thus F approaches 1 in an indefinitely large continuous population.

The preceding results apply to area continuity. With continuity in a linear range (for example, shore line), $F = \sum t / [2N - \sum t]$ as above, $t_1 = 1$ but $t_x = t_{(x-1)}[N\sqrt{x-1} - 1]/N\sqrt{x}$ or approximately $Ce^{-2\sqrt{x}/N}[\sqrt{x} - (1/2N)]^{-(N^2+1)/N^2}$.

In a continuous population with exclusive uniparental reproduction, the correlation between adjacent individuals is of the form $E = \sum t/N$ where $\sum t$ is the same as above for area or for linear continuity as the case may be.

The variance of gene frequencies in subdivisions of any size, N_i, within a more comprehensive population N_t is given by the formula $\sigma^2_{i \cdot t} = q_t(1 - q_t)$ $[F_t - F_i]/[1 - F_i]$ where F_i and F_t are the inbreeding coefficients relative to the populations of size N_i and N_t, respectively.

It is shown that in the absence of disturbing factors, short range dispersal (N less than 100 in the case of area continuity) leads to considerable differentiation not only among small subdivisions but also of large ones. Values of N greater than 10,000 give results substantially equivalent to panmixia throughout a range of any conceivable size. With linear continuity, there is enormously more differentiation than with area continuity. There is somewhat more differentiation under uniparental than under biparental reproduction.

Recurrent mutation, long range dispersal and selection are factors that restrict greatly the amount of random differentiation of large (but not small) subdivisions of a continuous population. A term $(1 - m_1)^{2x}$ under biparental, $(1 - m_1)^x$ under uniparental, reproduction is introduced into the expressions for t referred to above. In this $m_1 = [-\Delta q/(q - q_t)]$ where Δq is the rate of change of gene frequency (q) which such factors tend to bring about.

The effective size of a population characterized by the inbreeding coefficient F depends on whether F is due to a tendency toward mating of relatives not associated with territorial subdivision, or to such subdivision. In the former case the sampling variance is $\sigma^2_{\delta q} = q(1 - q)(1 + F)/2N$, in the latter, $q(1 - q)$ $(1 - F)/2N$, in contrast with $q(1 - q)/2N$ in a random bred population.

If different regions are subject to different conditions of selection, the amounts of both adaptive and nonadaptive differentiation depend on the smallness of m (if subdivision into partially isolated "islands") or of N, size of the random breeding unit (if a continuous distribution). If these are sufficiently large there is no appreciable differentiation of either sort; if sufficiently small there is predominantly adaptive differentiation of the larger subdivisions with predominantly nonadaptive differentiation of smaller subdivisions superimposed on this. Even under uniform environmental conditions, random differentiation tends to create different adaptive trends in different regions and a process of intergroup selection, based on gene systems as wholes, that presents the most favorable conditions for adaptive advance of the species.

LITERATURE CITED

DOBZHANSKY, TH., and S. WRIGHT, 1941 Genetics of natural populations. V. Relations between mutation rate and accumulation of lethals in populations of *Drosophila pseudoobscura*. Genetics **26**: 23–51.

THOMPSON, D. H., 1931 Variation in fishes as a function of distance. Trans. Illinois Acad. Sci. **23**: 276–281.

WRIGHT, S., 1921   Systems of mating. Genetics 6: 111–178.

    ̷1922a   The effects of inbreeding and crossbreeding on guinea pigs. III. Crosses between
highly inbred families. Bull. U. S. Dept. Agric. No. 1121.

    1922b   Coefficients of inbreeding and relationship. Amer. Nat. 56: 330–338.

    1929   The evolution of dominance. Amer. Nat. 58: 1–5.

    1931   Evolution in mendelian populations. Genetics 16: 97–159.

    1938   Size of population and breeding structure in relation to evolution. Science 87: 430–431.

    1940   Breeding structure of populations in relation to speciation. Amer. Nat. 74: 232–248.

    1942   Statistical genetics and evolution. Bull. Amer. Math. Soc. 48: 223–246.

WRIGHT, S., TH. DOBZHANSKY, and W. HOVANITZ, 1942   Genetics of natural populations. VII.
The allelism of lethals in the third chromosome of *Drosophila pseudoobscura*. Genetics 27:
363–394.

20

Reprinted from *Heredity*, **1**, 303–311, 315–321, 324–326, 333–336 (1947)

CONTAMINATION IN SEED CROPS
III. Relation with Isolation Distance

A. J. BATEMAN

John Innes Horticultural Institution, Merton

Received 7.vi.47

CONTENTS

1. Introduction
2. Behaviour of pollinating insects
   Observations
   Closeness of foraging
   Theoretical
   Fitting the data
   The effect of several flights
3. Insect-pollinated crops
   Theoretical
   Empirical
   Fitting the data

4. Wind-pollinated crops
   (*a*) Pollen dispersal
       Theoretical
       Fitting the data
   (*b*) Contamination
5. Conclusions
6. Summary
   References

I. INTRODUCTION

A NUMBER of results are now available from systematic experiments on natural cross-pollination between varieties of crops grown for seed, *e.g.* on radish by Crane and Mather (1943) ; on radish and turnip, using various planting arrangements by Bateman (1947*a*) ; and on the wind-pollinated crops, beet and maize, by Bateman (1947*b*). These data are in a consistent form which makes comparison and generalisation possible.

The conclusion that may be drawn from these experiments is that whatever the absolute level of contamination or the range of distance involved in any experiment, the shape of the curve relating contamination to isolation distance is the same. It is such that the rate of decrease of contamination per unit increase of isolation distance itself decreases with that increase. At first, increases in isolation distance rapidly reduce contamination, but at greater distances the contamination, though small, becomes persistent and in one instance (Bateman, 1947*a*) there was no detectable decrease in contamination when the isolation distance was increased from 50 yards to 200 yards.

The consistency of these results justifies the present attempt to derive a general mathematical expression which, by the substitution of appropriate values for the constants, could express the variation of contamination with distance under any conditions and for any crop.

In the following sections (the first of which contains new observations on the pollinating methods of bees) the argument is roughly divisible into three parts :—

(i) The discussion of the causation of the particular phenomena from which are derived one or more possible fundamental equations.

(ii) The choice of suitable simple equations which give an approximation to the results anticipated from the fundamental equations.

(iii) The testing of these empirical equations against experimental data.

To the pure mathematician this may be no more satisfactory as a method than a purely empirical one. There are, however, many possible empirical formulæ to fit a given set of data. Some of these would immediately prove unsuitable when applied to another set of data using a different range of factors. There would still be a large residue whose unsuitability would only slowly be exposed by accumulating evidence. The method herein adopted does, if the fundamental equation is satisfactory, enable one to make at the start a good choice of empirical equations, which must pass the severe test of agreement with the fundamental equation.

Since an attempt is made to discuss each phenomenon according to its causation, insect-pollination and wind-pollination have to be considered apart at first and, only at the end, taken together.

For the sake of consistency the same symbols have been used throughout for the same variables. This has meant transposing the formulæ of other authors.

The main symbols used are as follows :—

$F =$ the proportion of contamination
$D =$ the distance
$x =$ the power of D
$n =$ the insect density
$p =$ the pollen or spore density

2. BEHAVIOUR OF POLLINATING INSECTS

In insect-pollinated crops the effect of distance on cross-pollination is intimately connected with the behaviour of the pollinating insects. The connection is simplest in self-incompatible crops since with them all the functioning pollen has to be brought by insects. The proportion of contamination in the seed will then be equal to that in the available pollen on the insect visitor. There will of course be pollen in the baskets on the bee's legs and on some other parts of its body which will not be available for pollination. In self-compatible crops the contamination of the seed will be diluted to a varying extent according to the amount of automatic self-pollination in the particular crop.

The proportion of contaminant pollen on an insect must depend on the number of visits made after it has left the contaminating variety. The effect of distance on this depends on two factors : the flight habits of pollinating insects when at work, and (more difficult to observe) the rate of replacement per visit of contaminant by non-contaminant pollen.

Observations

The first feature observed about the flight of a pollinating insect is that there are two distinct kinds of flight. This is most noticeable in bees, which are the most important pollinators. Most of the insects' movement takes the form of slow flights which are easy for the observer to follow, and during which the insect hardly moves above the level of the flowers of the crop being visited. But every now and then an insect will soar up suddenly from the crop and disappear from sight. In hive bees these soaring flights probably mean the end of the forage and a return to the hive. By analogy, with bumble bees and solitary bees (such as Andræna), these flights would mean a return to the nest. In hover flies there is more doubt as to whether the two kinds of flight have distinct functions, but these insects are relatively unimportant in pollinating most crops.

In general, it would seem that pollen carried forward from one forage to the next is not likely to be responsible for much of the seed set because a single forage includes a large number of flights ; and although there does not appear to be any evidence as yet on the amount of pollen carried forward from one forage to the next, after the bee has cleaned itself in the hive, this carry forward is unlikely to be great. Most of the effective pollen will be collected and deposited during a single forage.

If information is to be obtained which will assist in understanding the process of contamination it is therefore the activity of the bee during the forage which needs most attention.

For this purpose observations were made on hive bees, solitary bees and hover flies visiting a turnip plot in bloom. The plants were spaced 6 in. apart each way. Individual insects appeared to move at random over the plot as judged by the paths traced for single insects during a forage. There was apparent disorderliness, the same plant often being revisited at irregular intervals.

Other observations made on bumble bees on a radish plot enable us to make a statistical test of randomness. Here the path of each bee was projected on to a line at right angles to the rows. Each flight was recorded in terms of its component parallel to this line. When bees did not cross from one row to the other the flight length was recorded as zero. Other flights were recorded as so many rows to the left or right. Three forages were analysed with respect to the direction of consecutive directed flights, *i.e.* those whose direction was recorded. The results are summarised on the following page.

Like follows like in 115 instances out of a total of 199. If the flights were completely random the expectation would be 99·5. There is therefore an excess of cases where consecutive directed flights were in the same direction. $\chi^2 = 4\cdot829$ for one degree of freedom with a probability of 0·02. In none of the three forages was there an excess of flights in one direction over the other. This means that, though over longer periods the bees did not appear to move in any particular direction, over shorter periods the bees tended to have an over-all

| | | Subsequent flight | |
		Left	Right
First flight	Left	55	42
	Right	42	60

direction first one way and then the other. The deviation from complete randomness, however, though statistically significant, is not very great.

The flight lengths of insects visiting the turnip plot were recorded in the manner illustrated in fig. 1. As the plants were 6 in. apart

Fig. 1.—The method of recording the flight lengths of pollinating insects in two dimensions. For explanation see text.

each way the unit of distance was taken as 6 in. Any plant just visited is taken as being at position 0 in fig. 1. The next plant visited can be taken as being on a square with plant 0 as the centre. The figures 1, 2, 3 and 4 denote progressively larger concentric squares whose sides are distances of 1; 2, 3 and 4 respectively from the centre. If the insect lands on a plant on square 2 the flight will be recorded as of length 2. Fig. 2 shows the distribution of flight lengths on turnip, of hive bees, solitary bees and hover flies. The distributions in the graphs appear similar for all the species. No record was kept of the length of the forages, but it can be stated that hive bees foraged longest of the three insect types, and hover flies shortest.

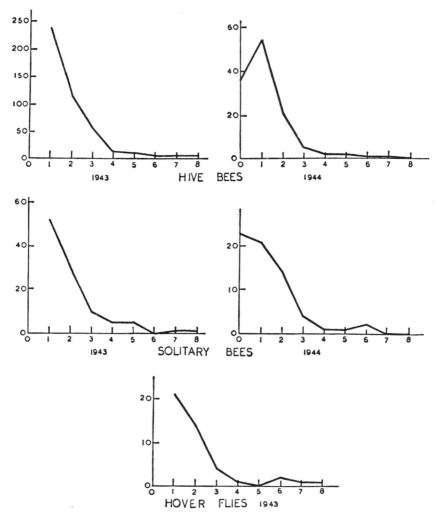

FIG. 2.--The frequency distribution of the two dimensional flight lengths of various insects on turnip. *D* is expressed in units of 6 inches.

Bumble bees do not visit turnip, possibly because the slender pedicels will not take their weight, but it would appear from observations on the radish that bumble bees forage longer than hive bees.

Closeness of foraging

The closeness with which insects forage is reflected in the distribution of the points which they reach by single flights, about the plant from which the flights start. The closer the foraging, the narrower this frequency distribution will be ; and the wider the foraging, the wider the distribution. The variance of the distribution can be taken, therefore, as a measure of foraging behaviour. An adjustment must, however, be made to allow for the two dimensional nature of flights, because the variance is measured relative to one dimension only.

This is done by dividing the frequency of flights, T, by a number proportional to the area of the zone into which the flight takes the insect. In Fig. 1 the boundaries of the zones are shown by dotted lines. The area of zone 4 is the area of the square enclosed by the outer zone boundary minus the area enclosed by the inner zone boundary. This is $9^2 - 7^2 = 32$. In general the area of a zone is given by $(2D+1)^2 - (2D-1)^2 = 8D$ where D is the number of the zone (*i.e.* the distance of the plant from plant o). This formula applies for all zones except zone o whose area is 1. If for all the other zones we use D as the correction factor for area, then the factor for zone o is $\frac{1}{8}$. The corrected value for flight distribution is termed n which is the density of insects round a source after one flight. The value of n for $D = 0$ is halved for estimation of variance, since the other values of D represent only half the distribution.

The bumble bee data from the radish plot with plants one foot apart gives the distribution in one dimension and therefore needs no correction. The observed frequencies give directly the distribution of n.

The mean squares for the distribution of insects after one flight are calculated as $\dfrac{\Sigma(nD^2)}{\Sigma n}$. Except in the bumble bee data, there is no means of estimating the mean, which is therefore taken as zero. D is therefore the deviation from the mean and no correction for the mean is required. As the mean is assumed, no degree of freedom is utilised in its estimation and the number of degrees of freedom is Σn.

In 1943 the frequency of flights within zone $D = 0$ was not recorded. The " mean square " calculated from these observations is thus not a true mean square but is inflated by the omission of the class closest to the assumed mean. A comparison of these false mean squares will show, however, any differences in the degree of dispersal of the flights.

Table 1 shows the values for T and n for six sets of observations with the calculations of the mean squares, true and false. The first three columns are strictly comparable, being based on the same plots at the same time. The next two differ from the first three only in pertaining to the following year. The true mean square for the bumble bee is six times that of the insects in columns 4 and 5. Here not only the years were different but also the crop, the spacing, and the method of recording, so little significance can be attached to the comparison.

The first five sets of observations suggest that if all cross-pollination is carried out during forages and none between forages, the dispersal of contaminating pollen on a given crop will be independent of the type of visitor. It is possible, however, that some pollen is carried forward from one forage to the next. If at the same time one forage of a visitor does not commence where its previous forage finished, conditions will be such that the shorter the forage the more wide-

TABLE 1

The dispersal of insect flights

D=Flight length.
T=Frequency of flights according to length in two dimensions.
n=Frequency of flights according to length in one dimension.
The mean squares have been corrected for grouping.
The frequencies actually recorded are in *italic* type : derived frequencies are in ordinary type.

D	Hive bees 1943		Solitary bees 1943		Hover flies 1943		Hive bees 1944		Solitary bees 1944		Bumble bees 1945
	T	n	T	n	T	n	T	n	T	n	n
0	...	...	...	...	...	...	36	144·00	23	92·00	506
1	238	238·00	52	52·00	21	21·00	54	54·00	21	21·00	...
2	118	59·00	30	15·00	14	7·00	21	10·50	14	7·00	169
3	56	18·67	10	3·33	4	1·33	5	1·67	4	1·33	...
4	16	4·00	5	1·25	1	0·25	2	0·50	1	0·25	20
5	12	2·40	5	1·00		0·00	2	0·40	1	0·20	...
6	5	0·83		0·00	2	0·33	1	0·17	2	0·33	7
7	5	0·71	1	0·14	1	0·14	1	0·14	...		...
8	5	0·63	1	0·13	1	0·13		0·00	...		5
9	1	0·11		0·00	1	0·11		0·00	...		...
10		0·00		0·00		0·00	1	0·10	...		...
11		0·00	1	0·09		0·00	...		...		...
12		0·00	...			0·00	...		...		1
13	1	0·08	...			0·00	...		...		...
14	1	0·07	...		1	0·07	...		...		...
15		0·00	...		1	0·07	...		...		...
16	2	0·13	...		...		...		...		...
·											
22	...		...		...		...		...		1
Total for all D	...		...		...		123	211·48	66	122·11	709
Total excluding D=0	460	324·63	105	72·94	47	30·43	87	67·48	43	30·11	...
True— Sum of squares	...		...		...			152·01		81·85	2196·0
Mean square	...		...		...			0·6355		0·5870	2·9306
False— Sum of squares		940·45		213·04		130·41		152·01		81·85	...
Mean square		2·8137		2·8375		4·2023		2·1694		2·6351	...

spread will be the contamination. As noted above, the average duration of a forage varies greatly with the type of visitor. For example, under these conditions, hover flies would produce more widespread contamination than hive bees.

Theoretical

The frequencies in table 1 may be used to test the validity of various formulæ which might express the flight of foraging insects. It is first necessary, however, to decide which formulæ are sufficiently appropriate to warrant consideration. Various formulæ have been previously proposed for similar phenomena, some on empirical grounds, some on theoretical grounds and some, after being derived theoretically, have been tested against actual data.

An example of the purely empirical approach is that of Wadley and Wolfenbarger (1944) who observed the dispersal of the Smaller European Bark Beetle from a centre. They found that the regression equation $n = a + b_1 \log D + \dfrac{b_2}{D}$ gave a good fit with the data.

In contrast, is the theoretical method of Pearson and Blakeman (1906) who considered the distribution of mosquitoes round a point source after r flights, on the assumption that all flights were of equal length (l) though random in direction. They found that as r increased the distribution approached a normal distribution. When r was greater than 7 the theoretical formula would be $n = \dfrac{Q}{\pi r l^2} \cdot e^{\frac{-D^2}{r l^2}}$ where Q is the total number of insects.

The same relationship was assumed by Frampton, Linn and Hansing (1942) who were concerned with the spread of viruses of the yellows type by means of leaf-hoppers. They likened the movement of leaf-hoppers over a crop to the two dimensional diffusion of a gas according to kinetic theory. The distribution of leaf-hoppers round a point source would then be a normal distribution the variance of which increased with time. No direct test of the validity of this hypothesis was made, but they derived from it a formula for the spread of virus which showed good fit with their observations.

Brownlee (1911) offered various criticisms of Pearson and Blakeman's theoretical formula for insect flight. On the basis of these he proposed a formula of his own $n = ge^{-cD}$. This formula was tested against data for the distribution of water-fleas and winkles at intervals after their liberation from a point source, and also for the distribution of epidemics both in space and time (substituting t for D). All these divers phenomena appeared to agree with his formula.

From the above, it would seem that two basic formulæ merit serious consideration to decide their suitabilities for accounting for the above observations on the flight of foraging insects. These can be expressed in a common form, viz. $n = ge^{-cD^x}$. According to Pearson and Blakeman, and Frampton et al., $x = 2$; according to Brownlee, $x = 1$. Other values of x might also be considered. On empirical grounds it was felt desirable to test the fit when $x = \frac{1}{2}$.

These formulæ can be tested against the data by the method of regression. This is done by taking logs throughout.

$$\text{Thus } \log_{10} n = \log_{10} g - c \log_{10} e . D^x$$
$$\text{or } \log n = a + b D^x.$$

We can now estimate the regression of log n on D^x for various values of x. The goodness of fit of several regressions can be compared by subtracting the sum of squares due to the regression from the total sum of squares for log n. The smaller the remainder sum of squares (which have the same degrees of freedom) the better the fit of the regression. The ratios between the remainders are treated as variance ratios and their significance estimated by reference to the usual tables.

[*Editors' Note:* Material has been omitted at this point.]

3. INSECT-POLLINATED CROPS

Theoretical

Formulæ have been derived which appear to express adequately the distribution of foraging insects. It is now necessary to see how these can be applied to the distribution of contamination.

The general formulæ for the distribution of insects after r flights from the source can be taken also to express the relative frequency with which insects visiting a plant at distance D from a contaminating plant have taken r flights to cover the distance. If the symbol for this frequency is N_r, then ΣN_r for all visits to a given plant will equal unity and it is necessary to adjust the factor g accordingly in the expression for n_r given above. If there are many contaminant plants there will be an increase in the magnitude of N_r for small values of r at the expense of the larger values of r. Thus g will be increased and c decreased in such a manner as to keep ΣN_r equal to unity.

At each visit an insect will deposit some pollen already on its body and replace it with a fresh supply. In this way the contaminant portion of pollen carried by an insect will decrease with every visit subsequent to its leaving the contaminant plot. If v is the proportion of contaminant pollen on an insect leaving the contaminant plot, we can consider it to decrease to a fraction w at every subsequent visit. The proportion of contaminant pollen on an insect r flights away from the plot will therefore be vw^{r-1}. In self-incompatible crops the effective r will be the number of flights up to the first visit to the plant now being visited, for successive visits to flowers on the same plant, though producing a decrease in the amount of compatible pollen carried, will not alter the proportion of that pollen which is contaminant. In self-compatible crops, on the other hand, r will be the total number of flights. It will include or exclude the last, according to whether the insect picks up pollen from a flower before or after touching the stigma.

The expressions for the frequency of flights and the proportion of contaminant pollen on an insect can be combined to give the following expression :—

$$p_r = \frac{vgw^{r-1}}{\frac{1}{r^x}} e^{\frac{-cD^x}{r}} .$$

p_r is the contaminant portion of the pollen on a stigma deposited by insects which have travelled in r flights from the source of contamination.

The contaminant portion of all insect-deposited pollen will be Σp_r for all values of r from one to infinity. There is frequently, however, some pollen deposited on the stigma of the same flower without first being picked up on to an insect's body. The proportion of contamination in the seed, F, will therefore be $(1-q)\Sigma p_r$ where q is the proportion of compatible self-pollen deposited without the aid of insects. In self-incompatible species and many self-compatible ones with an out-breeding mechanism such as protandry, $q = 0$. In crops such as the French bean q is almost equal to unity. The expression for contamination then becomes, by removing all factors, independent of r to outside the Σ :

$$F = vg(1-q)\Sigma \left(\frac{w^{r-1}}{\frac{1}{r^x}} e^{\frac{-cD^x}{r}} \right)$$

$$r = 1 \to \infty$$

The part of this expression included within the Σ sign cannot be simplified, but for any given set of conditions the expression $vg(1-q)$ can be treated as a single constant. The manner in which F varies with D can be observed if we attribute arbitrary values to the constant, calculate the contributions to F by various values of r, and sum them.

Let $\quad\quad\quad v = g = (1-q) = c = x = 1$
Let $\quad\quad\quad w = 0.5$

Then $\quad\quad\quad F = \Sigma \frac{e^{\frac{-D}{r}}}{2^{r-1}r}$

Table 5 gives the values of p_r and log p_r for the range of r from 1 to 8 and of D from 0 to 25. From these are calculated Σp_r and log Σp_r for r ranging from 1 to 8. The values of log p_r and log Σp_r which is equivalent to log F, are shown graphically in fig. 4.

It will be seen that though the relation between any log p_r and D gives a straight line, the line for log Σp_r is a continuous curve becoming parallel successively to each log p_r curve.

So far all the formulæ have been obtained on the assumption that the distributions of insects and contaminant pollen can be expressed as continuous curves of infinite extent. In other words it has been assumed that in practice there are no maximum values of r and D. Are these assumptions true, and if not do the derived formulæ lose their validity?

TABLE 5

Expected values for pollen distribution according to the formula $p_r = \dfrac{1}{2^{r-1}r}\,e^{\frac{-D}{r}}$ *and*

its equivalent $\log p_r = -\left[(r-1)\,0\cdot3010 + \log r + \dfrac{D}{r}\,0\cdot4343\right]$

	D	0	5	10	15	20	25
	r						
Log p_r	1	0	$\bar{3}\cdot829$	$\bar{5}\cdot657$	$\bar{7}\cdot486$	$\bar{9}\cdot314$	$\bar{11}\cdot143$
	2	$\bar{1}\cdot398$	$\bar{2}\cdot312$	$\bar{3}\cdot227$	$\bar{4}\cdot141$	$\bar{5}\cdot055$	$\bar{7}\cdot969$
	3	$\bar{2}\cdot921$	$\bar{2}\cdot196$	$\bar{3}\cdot475$	$\bar{4}\cdot749$	$\bar{4}\cdot024$	$\bar{5}\cdot303$
	4	$\bar{2}\cdot495$	$\bar{3}\cdot952$	$\bar{3}\cdot409$	$\bar{4}\cdot866$	$\bar{4}\cdot324$	$\bar{5}\cdot781$
	5	$\bar{2}\cdot097$	$\bar{3}\cdot663$	$\bar{3}\cdot228$	$\bar{4}\cdot794$	$\bar{4}\cdot360$	$\bar{5}\cdot926$
	6	$\bar{3}\cdot717$	$\bar{3}\cdot356$	$\bar{4}\cdot992$	$\bar{4}\cdot631$	$\bar{4}\cdot271$	$\bar{5}\cdot906$
	7	$\bar{3}\cdot349$	$\bar{3}\cdot041$	$\bar{4}\cdot728$	$\bar{4}\cdot420$	$\bar{4}\cdot107$	$\bar{5}\cdot798$
	8	$\bar{4}\cdot990$	$\bar{4}\cdot716$	$\bar{4}\cdot447$	$\bar{4}\cdot173$	$\bar{5}\cdot904$	$\bar{5}\cdot631$
p_r	1	1	0·0067	$0\cdot0^2005$	$0\cdot0^3000$	$0\cdot0^4000$	$0\cdot0^5000$
	2	0·2500	0·0205	$0\cdot0^2169$	$0\cdot0^3138$	$0\cdot0^4114$	$0\cdot0^6093$
	3	0·0834	0·0157	$0\cdot0^2298$	$0\cdot0^3562$	$0\cdot0^3106$	$0\cdot0^4201$
	4	0·0313	0·0090	$0\cdot0^2257$	$0\cdot0^3735$	$0\cdot0^3211$	$0\cdot0^4603$
	5	0·0125	0·0046	$0\cdot0^2169$	$0\cdot0^6622$	$0\cdot0^3229$	$0\cdot0^4842$
	6	0·0052	0·0023	$0\cdot0^2098$	$0\cdot0^3428$	$0\cdot0^3187$	$0\cdot0^4805$
	7	0·0022	0·0011	$0\cdot0^2053$	$0\cdot0^3263$	$0\cdot0^3128$	$0\cdot0^4629$
	8	0·0010	0·0005	$0\cdot0^2028$	$0\cdot0^3149$	$0\cdot0^3080$	$0\cdot0^4427$
	range of r up to						
Σp	2	1·250	0·0273	$0\cdot0^2173$	$0\cdot0^3139$	$0\cdot0^4114$	$0\cdot0^5093$
	4	1·366	0·0519	$0\cdot0^2728$	$0\cdot0^2144$	$0\cdot0^3328$	$0\cdot0^4814$
	6	1·383	0·0588	$0\cdot0^2995$	$0\cdot0^2249$	$0\cdot0^3743$	$0\cdot0^3246$
	8	1·386	0·0604	0·0108	$0\cdot0^2290$	$0\cdot0^3951$	$0\cdot0^3352$
Log Σp	2	0·097	$\bar{2}\cdot436$	$\bar{3}\cdot238$	$\bar{4}\cdot142$	$\bar{5}\cdot055$	$\bar{7}\cdot969$
	4	0·135	$\bar{2}\cdot715$	$\bar{3}\cdot862$	$\bar{3}\cdot157$	$\bar{4}\cdot515$	$\bar{5}\cdot911$
	6	0·141	$\bar{2}\cdot769$	$\bar{3}\cdot998$	$\bar{3}\cdot395$	$\bar{4}\cdot871$	$\bar{4}\cdot391$
	8	0·142	$\bar{2}\cdot781$	$\bar{2}\cdot032$	$\bar{3}\cdot462$	$\bar{4}\cdot978$	$\bar{4}\cdot546$
	∞	0·15	$\bar{2}\cdot80$	$\bar{2}\cdot05$	$\bar{3}\cdot53$	$\bar{3}\cdot07$	$\bar{4}\cdot65$

The values for log Σp for r up to infinity are obtained by visual extrapolation of graphs showing the effect of increases in r on log Σp. The index numbers to the o's signify the number of times they are to be repeated.

It is possible that there is a maximum length to single flights, *i.e.* that there is a maximum range of D when $r = 1$. When, as in some wild species, plants are very widely spaced, this maximum is likely to limit the distance over which cross-pollination takes place. In a seed crop, however, the closeness of foraging is so great, that the frequency of such a maximum flight length being exceeded, if the distribution were continuous, is insignificant. Where plants are growing densely, therefore, the discrepancy between expectation and fact due to the existence of a maximum flight length would be negligible.

The existence of a maximum value of r beyond which contaminant pollen would not be carried depends on whether pollen is carried forward from one forage to the next. If pollen is not carried forward then the maximum value of r will be set by the number of flights per forage. This is already known to vary according to the species of

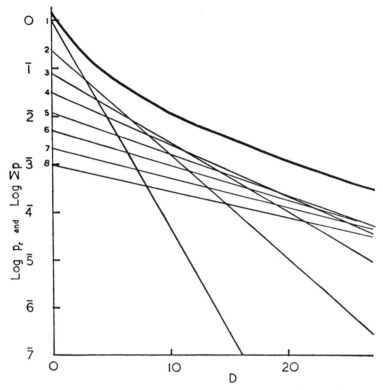

FIG. 4.—Variation of $\log p$ and $\log F$ with D according to the equation

$$p_r = \frac{1}{2^r - 1_r} e^{\frac{-D}{r}}.$$

Curves for p are shown for values of r from 1 to 8. The $\log F$ curve (thick line) is that estimated as $\log \Sigma p$ for r ranging from 1 to ∞.

insect and is probably affected by other factors such as the amount of nectar obtained per visit. The effect of a maximum r would be that the curve relating $\log F$ to D would become parallel to the p_r curve for the maximum r. If successive forages by an insect continue from where its previous ones finished, and there is a complete carry forward of pollen from one forage to the next there will be no maximum r. If there is an incomplete carry forward of pollen the contribution due to the higher values of r will be correspondingly reduced. Even if there should be a maximum r it is likely to be relatively large for most pollinating insects.

A further possibility is that consecutive forages are incompletely correlated in position or even uncorrelated. This could only have an effect if pollen were carried forward from one forage to the next. The result would be a flattening of the curve relating log F and D for the higher values of D, the degree of flattening being increased by decrease in the spatial correlation between forages and increase in the amount of pollen carried forward. If there were no correlation between forages the log F curve would eventually become parallel to the D axis resulting in a minimum contamination. This, however, is an impossible extreme as there must be some correlation between forages if only because bees will remain within a certain distance of their hive. The presence of wandering insects (not foraging closely) as postulated by Butler *et al.* (1943) would also flatten the curve at greater values of D.

It has been implied by Butler (*ibid.*) that there is a further limitation of insect flight—that foraging tends to be restricted to a given area. This would involve a maximum D, equal to the diameter of the foraging area, irrespective of the number of flights, r. There does not appear to be any evidence of this phenomenon and it need not therefore be considered further.

There is a possibility of secondary contamination arising through insects picking up contaminant pollen from a contaminated flower. The effect would merely be to reduce the slope of the log F curve at all distances, and the appropriate formula would be the same.

Even though all the above considerations involve speculation about circumstances not yet understood, none necessitate any radical changes in the formula for F as originally proposed. This can be restated in a simpler form, using K as a constant for any particular pair of contaminant and contaminated plots.

$$F = K\Sigma \frac{w^{r-1}}{r^{\frac{1}{x}}} e^{\frac{-cD^x}{r}}$$

$$r = 1 \to \infty$$

The evidence of the insect flight data is that the most appropriate value for x is 1 or $\frac{1}{2}$.

Empirical

The next problem is to decide on an empirical formula which will closely parallel the effects of the above fundamental formula. As can be seen from fig. 4, when the theoretical log F is plotted against D a curve is obtained which deviates by upward curvature from a straight line, the deviation being the effect of successively higher values of r. Fig. 3 shows a very similar curve corresponding to the regression of log r on $D^{\frac{1}{2}}$ which is derived from the equation $n = ge^{-cD^{\frac{1}{2}}}$.

The corresponding expression for F is $ye^{-kD^{\frac{1}{2}}}$. A similar upward curvature would be a feature of the regression equation

$$\log F = a - bD - \log D. \quad \text{This corresponds to } F = \frac{ye^{-kD}}{D}.$$

These empirical formulæ can be tested for conformity with the theoretical formula in the following way. From the values in table 5 one can calculate the values of $\log \Sigma p$ for the following ranges of r: 1—2; 1—4; 1—6; and 1—8 for values of D from 5 to 25. ($D=0$ is an impossible value for actual cross-pollination and the corresponding value of $\log F$ is in another sense unreal, being positive, which would make $F>1$.) It is then possible to construct graphs for each value of D showing the effect of increasing the range of r on $\log \Sigma p$. By extrapolation of these graphs we obtain the values of $\log \Sigma p$ to be expected when the range of r is complete (i.e. $1 \to \infty$). This method is obviously subject to personal error in the extrapolation. However, it is now possible to treat these estimates of $\log \Sigma p$ as " observed " $\log F$ and to test the goodness of fit of the two empirical equations by the method of regression.

For the equation, $F = ye^{-kD^{\frac{1}{2}}}$ the appropriate regression is that of $\log F$ on $D^{\frac{1}{2}}$, for the values of D from 5 to 25. The total sum of squares for $\log F$ is 2·8288 whilst the sum of squares due to the regression is 2·8279, leaving a remainder sum of squares for 3 degrees of freedom of only 0·0009. The fit between the fundamental and empirical equations is thus very good, though the extremely small remainder may be deemed a coincidence.

Similarly, for the equation $F = \frac{ye^{-kD}}{D}$ the appropriate regression is that of ($\log F + \log D$) on D. The total sum of squares of ($\log F + \log D$) is 1·2875, whilst the regression sum of squares is 1·2816, leaving a remainder of 0·0059 for 3 degrees of freedom. This still represents a good fit. The difference between the two remainders is not significant. It must be concluded that both empirical equations give a good fit with the fundamental equation and no choice can be made between them at this stage.

Comparison of these empirical formulæ with the original shows that the coefficient y corresponds to the part outside the Σ sign, namely $vg(1-q)$. These are respectively the proportion of contaminant pollen on an insect leaving the contaminant plot; a constant derived from the insect density formula which decreases with increasing dispersal of the insect flights; and the proportion of compatible pollen which is deposited by insects. The compound coefficient y expresses the contamination at zero isolation distance.

The coefficient k corresponds to those parts inside the Σ sign, namely w, c and x. These are respectively the diminution of contaminant pollen with each successive forage; the inverse of the degree of dispersal of insects (i.e. the closeness of foraging); and the power of D which expresses most closely the effect of D on insect foraging. The degree of dispersal of the insects, $\frac{1}{c}$, is determined largely by the way in which the plants are arranged, whether densely or sparsely, for example. The compound coefficient k expresses the rate of decrease of contamination with distance.

Before proceeding further to test these formulæ by comparison with the data on contamination, attention should be paid to any formulæ hitherto proposed, which might have a bearing on the question. From their formula for the distribution of leaf-hoppers, Frampton *et al.* derived a formula for the spread of virus disease. The assumption was that the infection I was directly proportional to time t, and to the number of insects feeding. On this basis the following formula was obtained :—

$$I = yte^{k't - kD}$$

This situation must inevitably differ from cross-pollination, for an insect vector can continue to produce a new infection with each fresh bite, whereas in the process of contamination the pollinating insect steadily loses its ability to contaminate. The time factor is therefore of no importance (provided, of course, that the contaminant and contaminated varieties are synchronous in flowering period). If we are to apply the formula for virus infection to contamination by cross-pollination we must adjust the expression to $F = ye^{-kD}$.

This formula is supported by the experimental evidence of Currence and Jenkins (1942), who observed the effects of distance on contamination in a cross-pollinating variety of tomato. They found that a good fit with the results was obtained by using the regression equation log $F = a + bD$ where b was negative. This is merely the logarithmic form of the above modification of Frampton's formula.

There appear, therefore, to be three equations which merit testing for their agreement with observation :

$$F = ye^{-kD}$$
$$F = ye^{-kD^{\frac{1}{2}}}$$
$$F = \frac{ye^{-kD}}{D}$$

[*Editors' Note:* Material has been omitted at this point.]

4. WIND-POLLINATED CROPS

(a) Pollen dispersal

Theoretical

A detailed discussion of the mechanics of the dispersal of air-borne fungal spores has recently been published (Gregory, 1945). Basing the arguments on formulæ proposed by Sutton (1932) for the dispersal of a smoke cloud from a point source, a formula 'has been obtained for spore dispersal. The main feature of the discussion is the importance attached to atmospheric turbulence as a dispersing agent. The following formulæ are given :—

(a) for deposition when the rate of deposition has a negligible effect on the amount of spores in suspension.

$$p = \frac{zQ}{\pi^{\frac{3}{2}}CD^{\frac{1}{2}(m+2)}}$$

Where p = the mean number of spores deposited per unit area in all directions from a point source.

z = the proportion of suspended spores deposited as a spore cloud passes over unit area.

Q = the total number of spores.

D = the distance from the point of liberation of the cloud.

C = an atmospheric constant.

m = a factor varying between 1·24 when turbulence is at a minimum and 1·76 when turbulence is at a maximum.

(b) for the change of total suspended spores as deposition proceeds

$$Q_D = Q_o e^{\left[\frac{-2zD^{(1-\frac{1}{2}m)}}{\pi^{\frac{1}{2}}C(1-\frac{1}{2}m)}\right]}$$

Combining the two expressions we obtain the complete formula for the dispersal of air-borne spores.

Modification of this formula are deduced to account for deposition down wind from a point source (p_w) and deposition down wind from a line source (p_{lw}).

These are $$p_w = \frac{2zQ_D}{\pi C^2 D^m}$$

and $$p_{lw} = \frac{2zQ_D}{\pi^{\frac{1}{2}}CD^{\frac{1}{2}m}}$$

where Q_D is derived from Q_o as above.

Doubt has been thrown by Bosanquet and Pearson (1936) on the practical importance of eddy diffusion or atmospheric turbulence. They assert that when there is a continuous source and observations are made over a period of time which allows for considerable variation in wind velocity, wind variation predominates over eddy diffusion in determining the resultant dispersal. If attention is paid to dispersal in all directions no account need be taken even of wind direction. Over a given period, then, during which Q pollen grains are liberated, ignoring loss through deposition, the grains will pass outwards in two dimensions under the influence of wind so that the number of grains passing over unit circumference at distance D from the source will be

$\dfrac{Q}{2\pi D}$. Allowance can be made for deposition by assuming that the amount of suspended grains is reduced by a certain proportion per unit distance travelled from the source.

Then $p = \dfrac{ke^{-cD}}{2\pi D}$ or in general form $\dfrac{ge^{-cD}}{D}$.

This formula is the same as is obtained from Gregory's expression by making m equal to zero. Taking logs this equation becomes of the form $\log p = a - b_1 D - \log D$. This means that when $\log p$ is plotted against D a line will be obtained curving upward from a straight line of negative slope.

Another formula for eddy diffusion in a stationary cloud has been proposed by Schmidt (1925). This is only the normal distribution in another form.

$$p = \dfrac{Q}{2\sqrt{\dfrac{A\pi t}{\rho}}} e^{\left(\dfrac{-D^2}{\dfrac{4At}{\rho}}\right)}$$

where p is the spore density at distance z from the source at time t, Q is the total number of spores, and A and ρ are atmospheric constants. It may be written as $p = ge^{-c_2 D^2}$.

If Schmidt's expression is used and allowance is also made for deposition rate

$$p = ge^{-c_1 D - c_2 D^2}$$

If $\log p$ is plotted against D, this equation will give a line curving downward from a straight line with negative slope ($\log p = a - b_1 D - b_2 D^2$). Reference to fig. 5 shows that where definite curvature from a straight line is discernible it is always in the upward direction. This equation may therefore be dismissed.

We are left then with the formulæ proposed by Gregory and the modification which ignores the effects of turbulence (assumes $m = 0$).

Gregory's formula can be simplified in the following manner :—

$$p = \dfrac{ge^{\dfrac{-cD^{(1-\frac{1}{2}m)}}{(1-\frac{1}{2}m)}}}{D^{(1+\frac{1}{2}m)}}$$

where g varies with the amount of pollen liberated and the rate of deposition, c varies with rate of deposition and m varies with the turbulence.

[*Editors' Note:* Material has been omitted at this point.]

CONCLUSIONS

It is now possible to compare the formulæ derived for insect- and wind-pollination. In insect-pollinated crops two formulæ appear equally suitable :

$$F = ye^{-kD^{\frac{1}{2}}} \quad \text{and} \quad F = \frac{ye^{-kD}}{D}$$

In wind-pollinated crops the corresponding formula is

$$\frac{F}{1-F} = \frac{ye^{-kD}}{D}$$

In spite of the very different ways in which these formulæ have been derived, they all give similar results in practice. If F is small so that $F \simeq \dfrac{F}{1-F}$ one formula can be used in all cases, viz. :

$$F = \frac{ye^{-kD}}{D}.$$

If this formula is expressed as a graph in which log F is plotted against D (see fig. 4) a curve of negative slope is obtained, its steepness decreasing as D increases. If the curve relating log F to D were a *straight line* of negative slope the *proportionate* decrease in contamination with unit increase in distance would be constant. The proposed formula means, therefore, that even the proportionate decrease in contamination becomes less with each increase in distance. In practical terms successive increases in isolation distance become less and less effective in improving the isolation. If, for example, contamination is complete for no isolation distance, and 100 yards reduces it to one in a hundred, the contamination at 200 yards will be greater than one in ten thousand.

The practical importance of the proposed formula to seed growing would be in the prediction of the contamination at any isolation distance provided the contamination at two distances were known with sufficient accuracy and that other conditions were constant. Under given conditions it would be possible to predict the distance at which contamination of one in a hundred or one in a thousand would be obtained.

The formula also has a bearing on population genetics. It gives the distribution in a continuous population of the pollen parents of the progeny of an individual seed parent. The nature of this distribution suggests that if the seed were not widely dispersed, there would be a correlation between the proximity of two plants and the closeness of their relationship. Even in an infinite and continuous population, therefore, gene combinations would spread slowly and there would be slight inbreeding. In a population of finite size more genetic variability and greater inbreeding would occur than would be expected on the basis of population size alone. In the past, the occurrence of greater variability than was anticipated from the population size has been explained by the assumption of an internal discontinuity in the population (Sewall Wright, 1939). A limited dispersal of pollen or seed would effectively explain the same situation.

It is worth noting that in a seed crop which is harvested in bulk, mixed mechanically and resown, there is no correlation between the proximity of plants in the field and their relationship. Crop plants are therefore probably the only examples in which the total number of plants in a population really corresponds to Sewall Wright's N statistic.

SUMMARY

Insects and air produce pollination by very different means which can be contrasted as follows :—

1. Insects move independently of one another, but the experiments show that they do so in a statistically predictable manner. Air, on the other hand, moves in large masses broken up by turbulence, the effects of which are also predictable ; but the movement is modified by variation in wind direction and velocity which is unpredictable.

2. Insects carry pollen systematically from flower to flower of the same species : under normal wind conditions, if the species is evenly dispersed, the pollen is distributed equally in all directions. Air-borne

pollen, on the other hand, is distributed down-wind and alights on a stigma without regard to its species.

3. An insect can carry only a limited amount of pollen available for pollination. Consequently the amount of pollen of one variety on an insect can increase only at the expense of other varieties. In the air, on the other hand, the amount is almost unlimited. The atmospheric concentration of pollen of one variety, therefore, has no direct influence on the concentration of another.

Bearing in mind these distinctions, formulæ are derived for the effect of distance on contamination in insect- and wind-pollinated crops. These agree well with the experimental results now described. The formulæ for the two classes of pollination are unexpectedly similar ; so similar indeed that one can derive a common formula for the two.

With this formula one can use the contamination observed at two distances to predict what will be found at a third. The formula also throws some light on the breeding behaviour of natural populations.

Acknowledgments.—This work has been carried out under the auspices and financial assistance of the Agricultural Research Council. The author also wishes to express his gratitude to Dr K. Mather for his guidance during the work.

REFERENCES

BATEMAN, A. J. 1947*a*.
Contamination in Seed Crops. I. Insect Pollination.
J. Genet. 48, 257.

BATEMAN, A. J. 1947*b*.
Contamination in Seed Crops. II. Wind Pollination.
Heredity 1, 235.

BOSANQUET, C. H., and PEARSON, J. L. 1936.
The spread of smoke and gases from chimneys.
Trans. Faraday Soc. 32, 1249.

BROWNLEE, J. 1911.
The mathematical theory of random migration and epidemic distribution.
Proc. roy. Soc. Edinb. 31, 262.

CRANE, M. B., and MATHER, K. 1943.
The natural cross-pollination of crop plants with particular reference to the radish.
Ann. appl. Biol. 30, 301.

CURRENCE, T. M., and JENKINS, J. M. 1942.
Natural crossing in tomatoes in relation to distance and direction.
Proc. Amer. Soc. hort. Sci. 41, 273.

FRAMPTON, V. L., LINN, M. B., and HANSING, E. D. 1942.
The spread of virus diseases of the yellow type under field conditions.
Phytopathology 32, 799.

GREGORY, P. H. 1945.
The dispersion of air-borne spores. *Trans. Brit. mycol. Soc. 28*, 26.

JENSEN, I., and BØGH, H. 1941.
On conditions influencing the danger of crossing in the case of wind-pollinated cultivated plants.
Tidsskr. Planteavl. 46, 238.

PEARSON, K., and BLAKEMAN, J. 1906.
Mathematical contributions to the theory of evolution. XV. A mathematical theory of random migration.
Drap. Co. Mem. biom. Ser. 3, 1.

SCHMIDT, W. 1925.
Der Massenaustausch in freier Luft und verwandte Erscheinungen.
Probl. kosm. Phys. 7, 1.

SUTTON, O. G. 1932.
A theory of eddy diffusion in the atmosphere.
Proc. roy. Soc. A, 135, 143.

WADLEY, F. M., and WOLFENBARGER, D. O. 1944.
Regression of insect density on distance from centre of dispersion as shown by a study of the Smaller European Bark Beetle.
J. agric. Res. 69, 299.

WRIGHT, S. 1939.
The distribution of self-sterility alleles in populations.
Genetics 24, 538.

Editors' Comments
on Papers 21 Through 24

21 SUTTER and TRAN-NGOC-TOAN
The Problem of the Structure of Isolates and of Their Evolution Among Human Populations

22 CAVALLI-SFORZA
Some Data on the Genetic Structure of Human Populations

23 HAJNAL
Concepts of Random Mating and the Frequency of Consanguineous Marriages

24 SCHULL and LEVIN
Monte Carlo Simulation: Some Uses in the Genetic Study of Primitive Man

EMPIRICAL STUDIES OF HUMAN POPULATIONS

From a practical point of view, most populations are not isolates in any real sense, and yet the number of potential mates of any one individual is clearly circumscribed. The populations in which demographic vital rates are sufficiently well known for the computation of N_e are generally so large that the value so computed bears little resemblance to the actual mating behavior of living individuals. It is therefore of importance and interest to develop means, direct or indirect, by which one can determine the effective size of breeding groups that really exist. Although populations are not really isolated, mating behavior restricts contact to the extent that we can estimate the *effective size of the isolate*, that is, the effective size of the existing mating groups.

This concept of isolate size was first developed by Dahlberg in 1928 and elaborated in his book-length treatise in 1948. Dahlberg notes that in complete panmixia in a finite population there will be a certain probability that a person will marry a cousin (or any other relative, for that matter). If we assume that each couple produced exactly two offspring which survive to adulthood, as well as complete panmixia, then if there are N individuals in the reproductive pool each generation (the effective size), the frequency of cousins mating in panmixia will be known; by observing this frequency one can infer an estimate of N to the degree that the assumptions are valid. This can be generalized to include other types of relatives and other assumptions; in fact, Dahlberg's was a very important paper even though his work was based on restrictive assumptions, since it stimulated many subsequent efforts. It should be noted that Dahlberg was only interested in general estimates of isolate size and was well aware of the limitations of his assumptions. Since his paper appeared in English and in accessible sources, we have not reprinted it in this volume.

Dahlberg's work has been criticized by several workers, and most severely by Morton (1955), who showed that, for various populations (Japan, Brazil, Europe), isolate size estimates varied up to 10 times depending on the type of "cousin" marriage used in making the estimate. Of several other critical papers, at least two notable ones provide analysis of Dahlberg's formulas for specific populations and a discussion of the error introduced by his simplifying assumptions. Frota-Pessoa (1957) uses Brazilian data on consanguineous marriages gathered mainly by Freire-Maia (1957) to show that variance in family size is considerable and the mean greater than 2, and he provides an improved formula for this case. See also Jacquard (1974). A different treatment is given by Ballonoff (1974c).

In another paper, Sutter and Goux (1961) use data from French Departments considered as isolates and based on the data of Sutter and Tabah (e.g., 1948, 1951) to show the effect of real demographic characteristics on estimates of isolate size based on consanguinity; they demonstrate that time depth is required in the data if accurate assessments are to be made.

Paper 21 is an excellent short review of the problem of isolate structure and its study among human populations. This paper summarizes the approach of Dahlberg and Wahlund toward the effect of isolation and notes the importance of considering sociocultural factors in dealing with human beings. The authors point out that Wright's continuous model of isolation by distance requires knowledge of σ, the standard deviation of migration distance distribution, from which follows the neighborhood size; hence f, the local inbreeding coefficient, is dependent ultimately only upon the migration distribution in this model, and knowledge of the distribution gives all the essential information. They go on to present evidence from French data on this, showing that the distribution is leptokurtic (a high proportion of short-range distribution, with a rapid decline as distance increases). Commenting on Bateman's work, the authors note that many natural dispersion distributions can be fitted by the same functional form, a natural harmony that they find appealing.

Paper 22 applies data from the Parma diocese of Italy to the question of the basic genetic population structure in human populations. This paper represents another early effort to determine the applicability of available models for population structure. Using the Parma Valley data, the author first shows that the population is clustered at various levels; using the village or parish as the basic population unit, he finds that the effective size of these populations is about 88 percent of the total census size. Now whether an island or continuous model is a better fit to this data is determined by the migration patterns between villages. Migration distance distributions of a negative exponential, normal, and gravitational form are tried by Cavalli-Sforza, with none being fully satisfactory; he then uses the fraction of parish-endogamous marriages as a single parameter to gauge migratory patterns for these particular data.

Dahlberg's formulas are applied to estimate isolate size from consanguinity data; Cavalli-Sforza finds little relationship of consanguinity to population size but an inverse relationship to density. Difficulties with these estimates are discussed. He then uses variation in gene frequencies and Wright's island model to get other estimates of effective size, and ends by discussing the discrepancies among these various measures.

Cavalli-Sforza and his colleagues have followed this paper with several studies, generally of theoretical developments, designed to analyze the problem of migration, isolation, and population effective size from a more sophisticated point of view (Cavalli-Sforza, 1962). In particular, Cavalli-Sforza et al. (1966) explore various approaches to migration between villages in Parma with the goal of predicting the probabilities of consanguineous marriages; they find only fair agreement with the data, but the exploration of the theoretical and empirical difficulties is important.

The concept of a *migration matrix* was given full theoretical development by Bodmer and Cavalli-Sforza (1968) (see also Smith, 1969; Hiorns et al., 1969). This model is developed primarily to analyze the generation of genetic variability in a stochastic model of interconnected demes and is essentially a representation of Malecot's original work (1959). The model ties together concepts from Wright, Malecot, Kimura and Weiss, and others, who worked on various approaches to isolation by distance and the genetic variation produced.

The examination of the relationships among the demographic characteristics of a population, including socially relevant aspects, and the frequencies of consanguineous matings has been made in rather full detail in two papers. The first, by J. Hajnal (1963), is reprinted here; the other, by Cavalli-Sforza et al. (1966), is a fuller development including more factors (see Paper 19 in Ballonoff, 1974a).

Paper 23, a long but excellent and highly readable paper, begins with a discussion of Dahlberg's estimations of isolate size from cousin marriage rates; he notes two special difficulties not considered in detail by other critics of Dahlberg. These are the complications introduced by overlapping generations and the fact that social behavior affects the probabilities that persons of any particular genetic relationship will marry. Hajnal is not interested in predicting isolate size from consanguinity but rather in predicting the number of consanguineous unions.

Hajnal constructs a hypothetical population with seasonal births to illustrate his model. He notes that marriage is more likely for small age differences than for large ones, and hence that consanguineous marriages are more likely to take place between some pairs of related individuals than between others of equal genetic relationship, merely because of age differences between them (these are related to sex–age patterns as well). Hajnal argues that truly random mating in a demographically structured population of humans should more realistically be defined as that which occurs when the chance of marriage between two individuals is a function only of their age difference.

On the assumptions that a population is closed and stationary, that it practices random mating as he defines it, and that there is no correlation between the birth and death rates of relatives, Hajnal shows how to compute the number of marriages of any degree of consanguinity x that will occur in a given year. He also shows that this number is constant if the population is large enough so that stochastic forces can be ignored, and that it is independent of population size. He shows that this is true for stable populations $(r \neq 0)$ and discusses means to deal with migration and, briefly, the effects of small populations.

Hajnal uses many examples to show that random mating, in his definition, seems to occur in many populations. His elegant treatment is true demographic genetics, considering biological as well as social factors in demography. It is an effort at a rather comprehensive treatment of the genetics of the reproductive process in demographically interesting populations. The study by Cavalli-Sforza et al. (1966) extends Hajnal's work to include the details of the migration pattern and considers assortative mating effects as well.

The use of family names has a place in studies of consanguinity rates and should be mentioned at this point. Genetic applications of surname studies go back at least to the nineteenth century and Galton, but the explicit use of surnames to determine inbreeding coefficients was first made by Crow and Mange (1965). They study *isonymy,* or marriages between individuals of the same surname. If such individuals can be assumed to have common ancestry as reflected in their identical surnames, the frequency of isonymous marriages can be related to overall consanguinity rates. Crow and Mange show how consanguinity studied in this way can be partitioned into that due to finite population size and that due to nonrandom mating, which was discussed earlier. Studies of isonymy have subsequently been made by many authors, and this approach to human populations has thus had an important role in empirical assessment. However, serious problems are the assumptions, which must be made, that surnames all have single, unique origins and that all surnames now observed are genealogically descended through males from that origin. Great care must be exercised, therefore, before applying this very appealing method (see Cavalli-Sforza and Bodmer, 1971). A good theoretical treatment of the use of isonymy is in Yasuda et al. (1974).

By now we have covered much of the development of studies relating to the mating structure as a demographic genetic problem. The basic considerations were the size and degree of isolation of populations, the migration between then, and the use of inbreeding (or similar) measures to gauge the behavior of the population.

SIMULATION STUDIES

Thus far we have discussed theoretical studies of demographic genetics and attempts to verify or employ them empirically. Often, realistic situations are far too complex to be analyzed adequately with available theory; just as often, empirical data are insufficient to reflect theoretical expectations. In such cases the advent of high-speed computing techniques allows us to simulate real conditions better and for a long enough "time" period to derive answers to some of the questions we ask but cannot answer otherwise. The challenge to the simulator is to decide what population parameters are of importance and how to mimic their behavior on the computer or, in particular, what probability distributions to use in treating stochastic aspects.

Computer simulation relevant to population genetics began in the late 1950s and early 1960s. Paper 24 was the first to deal explicitly with demographic genetics in small human populations. Schull and Levin discuss the general problems with computer simulation, and go on to argue that the major input parameters for a simulation of human populations with overlapping generations should include population size, age distribution, a formula for mating, and the pattern of selection which is to be applied. Of course, the problem being studied will dictate which parameters are important in any given case. Schull and Levin are particularly interested in the analysis of gene frequency changes under a variety of conditions by Monte Carlo simulation of stochastic aspects of the genetic process.

This same computer simulation was further developed by MacCluer (1967), who has applied it to South American Indian genetics problems (MacCluer et al., 1971) to investigate questions of consanguineous matings in tribal populations. Cavalli-Sforza and Zei (1967) have used a similar model for the Parma Valley populations. Cannings and Skolnick (1975; Skolnick and Cannings 1974) have used a different approach to the life history of individuals and obtained somewhat differing results, which are at present difficult to interpret. Fraser has done considerable work on simulation as well (e.g., Fraser, 1962). In 1970, MacCluer and Schull used their simulation program to analyze frequencies of consanguineous mating and inbreeding in a Japanese population (MacCluer and Schull, 1970); they compare their results to the theoretical predictions from Hajnal's formulations and those of Cavalli-Sforza and Zei, and demonstrate the usefulness of simulation methods in these problems. In particular, they note that small fluctuations in demographic characteristics can produce significant differences in results.

Computer simulation is clearly a future tool of population genetics, especially since the detail and sophistication of questions now being asked is rapidly exceeding the power of analytic methods to provide easily interpretable answers. A recent symposium on computer simulation in this context is highly recommended as a presentation of this growing field (Dyke and MacCluer, 1974).

However, many aspects of simulation studies must be worked out in better detail before their use can become profitable. Simulation results are often very difficult to interpret because one is never sure whether the specific parameter values used, or their stochasticization, have produced the results or whether there is theoretical generality in them. Furthermore, there are usually so many parameters or variables interacting in cases for which one is likely to apply simulation that it is difficult to isolate the effects of each individual variable. It is certain that many runs of any given program must be made with carefully controlled parameter values before one can have any confidence in the distributions, or other results, that are found. It will always be preferable to find analytic solutions to properly simplified models. This is illustrated well by the fact that several of the papers in Dyke and MacCluer find completely different and conflicting results when analyzing the same problem (e.g., whether the imposition of an incest rule increases or decreases the rate of genetic drift in small populations). The dilemma is that one is likely to use simulation in those cases which are too difficult to allow analytic solution, and these are just the cases for which the results are very difficult to interpret.

21

Reprinted from *Cold Spring Harbor Symp. Quant. Biol.*, **22**, 379–383 (1957)

The Problem of the Structure of Isolates and of Their Evolution Among Human Populations

JEAN SUTTER AND TRAN-NGOC-TOAN

Institut national d'études démographiques, Paris, France

The notion of isolates dates back to the Swedish scientist S. Wahlund (1928). While investigating the genetical composition of a population and the correlations between different characters, he considered the case where two closely related populations, equally panmictic and bearing the same demographic aspects, exchange part of their individuals.

He then acknowledged that a big human population is only formed of smaller populations, with their extension restricted by factors of different natures; geographical, social, religious, professional, and so on. Wahlund has given the name of isolate to each of these smaller populations. The isolate may still be correctly represented by the population within which each individual may find his partner. This notion of restricted population has shown itself to have the same importance in animal biology or botany or in human biology.

The following year, another Swedish scientist, Dahlberg (1929), proposed to evaluate, within the bounds of panmixia, the size of the isolate, that is the number of its inhabitants, by taking into account the number of intermarriages between first cousins which usually take place. On this occasion, Dahlberg introduced an important demographic concept by pointing out that the number of consanguineous marriages was closely related to the size of the population and that of the families.

The most important fact which, stemming from this concept, has been brought into light in the populations of western Europe is what is called the breaking up of the isolates. For Dahlberg (1943) has drawn attention to the rapid disappearance of intermarriages within these populations, a phenomenon whose origin he arbitrarily situates in the middle of the 19th century. We have shown (Sutter and Tabah, 1948), for the period extending from 1926 to 1945 how this phenomenon was perceptible at the level of each of the 90 departments dividing France into administrative units. The breaking up of the isolates reveals the scope of the genetic homogeneity which has taken place in Europe during the last decades. The blending of innumerable restricted populations has brought about a transformation of individual structures, the importance of which cannot fail to appear if one bears in mind that, for instance in numerous villages of Brittany, the rate of intermarriages (to the 6th degree of kinship, that is between second cousins), has in a few years dropped from 20 to less than one.

The evaluation of the rapidity and degree of genetic homogeneity of the great European populations is one of the most interesting problems which up to now has been offered to the genetics of populations. We shall examine the difficulties inherent in its solution.

The Swedish school propounds two formulae which may be simultaneously used for the evaluation of the increase in size of the isolates and its genetic effects. The Dahlberg formulae which make use of the rate of intermarriages, and Wahlund's formulae which help to estimate how fast the homogeneity is settled after human exchanges have been established between two neighbouring isolates. Both ways are only acceptable on the hypothesis of panmixia. Now if panmixia is easily acknowledgeable and even obvious as far as concerns animal or plant populations, it is very difficult to observe among human populations. As a matter of a fact, on this theory the population is supposed to be closed, marriages to take place at random, and fertility to be identical for all couples. Facts of observation showing that human populations depart from it, can be presented thus:

1) Number and limits of the population the number of individuals it encloses and their geographical position;

2) migrations;

3) intermarriages within the population exist intermarriages due mainly to interest or to the structures of kinship belonging to certain cultures;

4) mutations one or many genes suddenly acquire a new quality of manifestation which can be transmitted;

5) selection there is a differential fertility;

6) positive or negative assortative mating individuals carrying the same features marry more willingly (or not) than by mere accident.

It is only by disregarding demographic and cultural phenomena that one may propound that Dahlberg's formula or its derivatives on the one hand, Wahlund's formulae, on the other, make possible a minute outlook on the mode of development of genetic homogeneity. Unfortunately it is extremely difficult to disregard purely human phenomena in this research upon human populations. Looking at things in this manner, there have been many doubts expressed about the value of Dahlberg's test (Sutter and Tabah, 1951; Morton, 1955). The latter has tried to prove that intermarriages at different degrees of kinship could in no way insure a true evaluation of the size of the isolate since this size varied (for instance passing from one to ten times its former size), when one considered marriages between uncle and niece or between second cousins. One may answer this by saying that it is not logical to compare on the sociological plane the significance of intermarriages between uncle and niece (which are exceptional cases) and between first cousins or second cousins. We have personally shown that the two first categories were open to criticism as concerns evaluation, whereas marriages between second cousins had, strictly on the demographical plane, a much greater significance in the panmictic sense. They are much more closely related to the theory than marriages of another degree of kinship.

Yet the hectic appearance of intermarriages constitutes only a small part of the difficulties arising. The greater the development that research on the isolates undergoes, the more obvious it is that one must take into account the cultural differences between populations. Social structures imposed by the different institutions varying from one culture to another control the structure of isolates and impose methodological adaptations.

For some years ethnology has taught us how different the structures of kinship are, and how they can vary from one tribe to another, from one group to another. Preferential marriages, which belong to numerous forms of civilisation, for instance the rule, which seems to be general enough (Lévi-Strauss, 1949) of matri-lateral marriage (that is, the preferential marriage with the daughter of the mother's brother as opposed to the marriage with the daughter of the father's sister) seems to bring difficulties in the formal methodology of the isolate.

The taboos attached to certain categories of inter-marriages have the same result. In our study on the isolate of polar Eskimos (Sutter and Tabah, 1956), we have shown that the taboo resting on the intermarriage between cousins (up to second cousins), has made it impossible to evaluate endogamy with the aid of consanguinity. Even in the populations of a western type one meets with such phenomena. Thus, in the Isle of Sein (Finistère) where intermarriages between second cousins are frequent we have observed the total absence of marriages between first cousins during the period 1919 to 1930, for a rate of 17 per cent concerning intermarriages between second cousins. The fishermen of this island believe it unnatural that their brothers and sisters should encourage marriages between their respective children. We must then consider that the classical methodology of the isolate cannot have applications in all these cases, and that it must be made more flexible so as to have significance in all the innumerable circumstances which may crop up in the bounds of the structures of kinship, this being a polymorphic cultural phenomenon.

This criticism does not only apply to limited populations, isolated or tribal, counting a small number of inhabitants and having between themselves but few human exchanges. We meet the same oppositeness if we consider what goes on in a population which is numerically large, and living in a very wide area like western populations. A further difficulty arises in this case, since the determination of the isolates has to be made artificially. For these populations are divided into numerous administrative sections: the parishes or communities which differ one from the other according to their demographical composition and the density of their population. In all countries with a registry office such subdivisions must be taken into account for the determination of the state of endogamy for instance, by means of the number of intermarriages within each community.

Our research, which has been carried through in two French departments (Sutter and Tabah, 1955), has pointed out, for instance, that the evolution of consanguinity taken in the sense of the breaking up of the isolates, has not followed the same rhythm throughout the course of time, if one considers the destiny of each parish by arranging the parishes in order of the number of inhabitants. In the department of Loir-et-Cher, one can see that not only the rate of intermarriages varies from one parish to the other, but the scarcity of intermarriages has grown more rapidly from 1870 to 1954 in the parishes with the highest population; a phenomenon which may not be systematized since the parishes of the department of Finistére counting 1,000 to 1,100 inhabitants have undergone a slower evolution in this field than in the parishes of 2,000 to 2,500 inhabitants or those of 700 to 800 inhabitants.

This very accentuated polymorphism which one can meet within human populations proves that, when dealing with human problems, one comes up against social structures and cultural phenomena which may not be avoided in the genetic problems we are dealing with.

One can easily believe that under such working conditions, it is very difficult to set up, upon the basis of observation, and after the formulae of the Swedish school, numbers accounting very accurately for the mode of evolution of genetic homogeneity in the western populations. The least one can say is that the valuation of the breaking up of isolates and its genetic consequences can only come up against almost insuperable difficulties. It has thus become necessary to go on with this research but with other concepts in mind. In this way Sewall Wright's theories may have very fruitful applications.

In Wright's theory, one does not deal (as in the Swedish system) with restricted populations whose evolution can be studied taking two at a time, but with populations offering a uniform distribution in a wide area. In these populations marriages between individuals are only possible in a neighbourhood circumscribed within restricted distances, so that the individuals who live at the farthest distance have almost no chance of ever getting married. It is this mode of dividing continuous populations in successive neighbourhoods, that Wright has called isolation by distance (Wright, 1943, 1946). The words "isolate" and "neighbourhood" are closely connected, and the two theories should not be opposed so systematically (Morton, 1955). The importance of distance was not missed by either Wahlund or Dahlberg. Yet from the strictly mathematical point of view, the models which have been provided by Wright have an undeniable superiority, for their general scope as well as for the depth of their insight. Their essential characteristics have been well expounded by Li (1948).

If one calls N the size of a neighbourhood, this means that the parents of an individual belong to a neighbourhood formed of N individuals. On the other hand, one supposes that the distribution of individuals· is uniform within the geographical area considered, in such a way that N should be in direct proportion to the area of the neighbourhood. The four grandparents of each individual may be considered as being derived, at random, from a surface $2 N$, and the ancestors of generations K of an area measuring KN. Supposing a neighbourhood of $2 N$ dimensions to be a circular surface with a radius R, the surface of the circle $2N$ would be $R \sqrt{2}$, the surface of KN dimensions would be inscribed within a circle of $R \sqrt{K}$ radius, which means that when the surface of the neighbourhood increases by nine times, the radius increases three times only. With such considerations one may embark upon genetic correlation. Wright published tables on the subject. Since the genetic variance σ^2 of different groups of neighbourhoods gives the measure of the degree of differentiation (heterogeneity, variability) of a wider population, and since F, the inbreeding coefficient, is directly proportional to the genetic variance, F may be used to measure the differentiation in the big continuous populations.

Yet Wright's computations prove that F depends essentially on the size of the neighbourhood, which in its turn is proportional to the variance of the distribution of the distances separating, in one direction, the birth places of the children and those of the parents; which shows the importance of the determination of this distribution.

One can then understand how the distance which separates the birth places of children and of parents expresses all the characteristics which concern us, and enables the appreciation of the course of genetic homogeneity in western populations. It finally makes it possible to neglect the influence of social, economical and cultural factors which are capable of causing prejudice to research based on too rigid models. As N.E. Morton (1955) has said: "Application of Wright's theory to human data, particularly to relatively sedentary rural communities, would permit the comparison of inbreeding rates calculated independently from consanguinity and migration".

Unfortunately, in practice, all the documents of the Registrary Office give no indication whatsoever on the birth places of the bride's and bridegroom's parents. The only information which may be given rapidly on a large scale is the addresses of the consorts at the time of the wedding. If one reflects on this one can see that this distance may make up very satisfactorily for that which separates the birth places of different generations. In the first case, one has an indication on the preceding generation and the present one, whereas in the second case one has indications on the present generation and the future one. We have given ourselves to research upon the evolution of the distance separating the dwelling-place of consorts in a French department (Loir-et-Cher) of 275,000 inhabitants distributed over 300 parishes. The parishes have been divided into 25 groups according to the number of inhabitants (from 0 to 8,000 or more), arranged as follows: From 0 to 200, from 200 to 300, from 300 to 400 and so on. The distance has been studied for the three following periods: 1870–77, 1919–25, 1951–53, on the basis of a sample comprising 53,000 marriages. For each period, the marriages have been classified in three groups: The two consorts living: 1) in the same parish; 2) in two different parishes of the same department; 3) in a parish of the department and in a parish outside the department. This study has shown that from 1870 to 1953, the rate of the marriages contracted in the same parish of the department had gone down regularly, that the marriages contracted in two different parishes had remained almost stationary whereas the rate of marriages contracted in a parish of the department and a parish outside the department had increased regularly, thus materializing the effects of migrations. We have measured the distance separating the dwelling-place of the consorts on a map to the scale of $^1\!/_{200,000}$, every centimeter representing a distance of $2\,Km$, and we have determined for each group of parishes the average distance from 1870 to 1953.

Until now we have studied the distribution of the distance of the consorts' dwelling-places during the periods 1870 to 77 and 1919 to 24, in the case of consorts living in two different parishes of the department. The observations were made on 9,600 marriages for 1870 to 77, and 7,700 marriages for 1919 to 24. Two interesting facts may be inferred concerning the variance and the aspect of the distribution. The variance and the average distance have increased with time. It has passed from 146 ± 6.6 in 1870 to 77 to 184 ± 10.8 in 1919 to 24, and the classical statistical tests immediately reveal that the acknowledged variability is not due to a fortuitous choice of sampling but does actually exist. This means that the size of the neighbourhood has increased, in other words, that genetic homogeneity has been spreading.

As for the distribution curve, it presents the same shape for both periods. We have endeavoured to fit a theoretical curve, and more particularly to verify the hypothesis of its normality. Wright and Haldane (1948) had assumed, in fact, that gene dispersion follows a normal law—a natural and reasonable hypothesis whilst lacking data to the contrary. New work having cancelled this assumption, Wright observed (1951) that the results of his theory are for the most part, independent of the trend of the distribution curve. It is important nevertheless for future researches to know precisely the trend of the distribution.

It is necessary to bring about some precision with respect to an important fact, to avoid any confusion in the comparison with other studies on neighbouring subjects. There is no question here of the distribution in one direction, as was obtained by Bateman (1947), for instance in his study on the dispersion of the flights of fertilizing insects, by, measuring, not the length of the flights, but only their projections on a predetermined axis. In numerous cases, among which is our own, the data are not given in an adequate form for the immediate determination of the distribution according to a direction. Bateman (1947) had to face this problem while studying the dispersion of the flights of insects, which brought him to make an adjustment on the data, "An adjustment must, however, be made to allow for the two dimensional nature of flights, because the variance is measured relative to one dimension only. This is done by dividing the frequency of flights, by a number proportional to the area of the zone into which the flight takes the insect.... In general the area of a zone is given by $(2D + 1)^2 - (2D - 1)^2 = 8D$ where D is the number of the zone (i.e. the distance of the plant from plant 0)." In fact the solution of this problem is not so simple, though it presents no serious mathematical difficulty. We shall not go further into this discussion. We shall merely underline an essential point. If the dispersion around a point is determined by a law of the type Cde^{-ad^2}, the uni-dimensional distribution, taken in the sense made explicit above, is normal, (supposing that the dispersion is always the same in all directions), which in a way justifies Bateman's theory. Bearing in mind these considerations, we have examined the different formulae which up to now have been propounded for the representation of phenomena comparable to the one we are dealing with (Bateman, 1950). Some are based upon theoretical considerations, others have been suggested by experience. The formulae which seem best to correspond to our curves are of two types:

$$f = ade^{-bd^x} \quad \text{and} \quad f = ae^{-bd^x}.$$

The values to be taken in consideration for the exponent X are: $\frac{1}{2}$, 1 and 2.

$X = 2$ corresponds to the normal distribution around a center for the first type, and to the normal distribution in a direction for the sec—

type. The fact that X should be smaller than 2 will mean that the distribution in either case is leptokurtic. The comparison of the formulae can be made by the method of regression by rising logarithms, and evaluating the constants a and b by the classical method of least squares. The smaller the sum of the residual squares

$$S = \bar{\Sigma}(\log f_{obs.} - \log f_{theor.})^2$$

the better the adjustment will be. The comparison of the S is done by regarding them as variances and applying Fisher's tests. We have confined ourselves to distances below 30 Km (which already corresponds to 98% of the distribution). The frequencies of distances over 30 Km are weak though they be much superior to theoretical frequencies in the case of a normal distribution, and the variance of the corresponding logarithms is extremely important. In spite of these limits, the superiority of the regressions in X and $X^{\frac{1}{2}}$ over X^2 appears very clearly. The formulae which give the best adjustments are $f = ae^{-bd}$ and $f = dae^{-b\sqrt{d}}$.

The present state of these calculations does not give us further precision. But already, we may assert that the distribution of the distance is certainly not normal, but that it presents a very pronounced leptokurtic character.

This distribution accounts for the scattering of the genes within the department observed, starting from each parish which may be considered as a center of dispersion. Through this observation for human populations, one meets with the general problem of the scattering of genes, of which the formal and quantitative estimation has already been the subject of important pieces of work. In 1950, Bateman observed: "The following types of gene dispersion have thus been shown to be leptokurtic: Automatic movement of airborne insects (Dobzhansky and Wright, 1943, 1947; Bateman, 1947, 1950); water fleas (Brownlee, 1911); passive movement of airborne organisms (Gregory, 1945; Bateman, 1947); dispersal of pollen by insects (Bateman, 1947). There remain a number of important types of gene dispersion for which the necessary information is not yet available—the movement of terrestrial animals and birds; the movement of plankton; and the dispersal of seeds by explosive and other automatic mechanisms. When such a varied collection of known dispersal types has been shown to be consistently leptokurtic there is no reason to doubt that the balance would be similarly non-normal. It is possible, therefore, that gene dispersion is as a general rule, leptokurtic." We hope that the research we are carrying on with at the present time on human populations of two different demographic types, will soon prove definitely that in this field, as in others, man follows the destiny of all living creatures, but certainly not for the same causes.

REFERENCES

BATEMAN, A. J., 1947, Contamination in seed crops. III. Relation with isolation distance. Heredity *1:* 303–336.

——— 1950, Is gene dispersion normal? Heredity *4:* 353–363.

BROWNLEE, J., 1911, The mathematical theory of random migration and epidemic distribution. Proc. Roy. Soc. Edinburgh *31:* 262–289.

DAHLBERG, G., 1929, Inbreeding in man. Genetics *14:* 421–454.

——— 1943, Matematische Erblichkeitsanalyse von Populationen. Supp. 148, Acta Med. Scand., 219 pp.; English Trans. 1948: Mathematical Methods for Population Genetics. Basle, Karger; London, New York, Interscience Pub.

DOBZHANSKY, TH., and WRIGHT, S., 1943, Genetics of natural populations. X. Dispersion rates in *Drosophila pseudoobscura.* Genetics *28:* 304–340.

——— 1947, Genetics of natural populations. XV. Rate of diffusion of a mutant gene through a population of *D. pseudoobscura.* Genetics *32:* 303–324.

GREGORY, P. H., 1945, The dispersion of airborne spores. Trans. Brit. Mycol. Soc. *28:* 26–72.

HALDANE, J. B. S., 1948, The theory of a cline. J. Genet. *48:* 277–284.

LÉVI-STRAUSS, C., 1949, Les Structures élémentaires de la Parenté. Paris, P. U. F.

LI, C. C., 1948, An Introduction to Population Genetics. Peiping, National Peking Univ. Press; 2nd ed., 1955, Population Genetics. Chicago, Univ. of Chicago Press.

MORTON, N. E., 1955, Non-randomness in consanguineous marriages. Ann. Human Gen. *20:* 116–124.

SUTTER, J., and TABAH, L., 1948, Fréquence et répartition des mariages consanguins en France. Population *3:* 607–630.

——— 1951, Les notions d'isolat et de population minimum. Population *6:* 481–498.

——— 1955, L'évolution des isolats de deux departements francais: Loir et Cher, Finistère. Population *10:* 645–674.

——— 1956, Méthode mécanographique pour établir la généalogie d'une population. Application à l'étude des esquimaux polaires (Recens. J. Malaurie) Population *11:* 507–530.

WAHLUND, S., 1928, Zusammensetzung von Populationen und Korrelationserscheinungen vom Standpunkt der Vererbungslehre aus betrachtet. Hereditas *11:* 65–106.

WRIGHT, S., 1943, Isolation by distance. Genetics *28:* 114–138.

——— 1946, Isolation by distance under diverse systems of mating. Genetics *31:* 39–59.

——— 1951, The genetical structure of population. Ann. Eugen. *15:* 323–354.

[*Editors' Note:* The discussion has been omitted.]

22

Reprinted from *Proc. 10th Intern. Congr. Genetics 1958,* Vol. 1, University of
Toronto Press, Toronto, 1959, pp. 389–407

SOME DATA ON THE GENETIC STRUCTURE OF HUMAN POPULATIONS

L. L. Cavalli-Sforza[1]

IT HAS BEEN for some time a commonplace of human genetics to consider man a poor object of genetic study in spite of the fact that since geneticists themselves belong to this species a study of man becomes as desirable or more desirable than that of any other species. But in recent years it has been increasingly recognized that man can be a good, if not optimal, organism for certain types of genetic problems, in spite of the disadvantages caused by the fact that the life span of the research worker is no longer than the generation time of the species under analysis. There are, of course, problems for which man has been a favorite object of study in fields other than that of genetics, but which supply information interesting from a genetic point of view. They range from biochemistry and medicine to history. One of them is the study of the genetic structure of populations; demography of the human species is more developed than that of any other species and can offer real help to an adequate description of population structure and an analysis of some of its problems.

Population structure can perhaps be defined best in a negative way, i.e. as the ensemble of known factors governing changes in gene or genotype frequencies other than mutation or selection. It represents the limits set to the action of these two primary evolutionary forces by the fact that any population is not infinite in size, and only approximates the simplifying conditions of random mating, which is the basis of the simplest evolutionary predictions.

In the theory of population structure, which has been presented by Wright (1943, 1946, 1951), we have the choice between two models which represent two somewhat extreme population structures. In one (the island model) a number of small population units or clusters, each of effective size N, all exchange genes with any other cluster, m being the effective proportion of the gametes of a cluster which is replaced in each generation by immigrants. Here we are required to estimate the two parameters N and m in order to define our population.

According to the alternative model (continuous distribution), the population is not grouped in clusters but is diffuse and mobile over an area or along a line (a strip of land or water). It is then considered as a population of units moving at random with given mobility and density; and essentially one parameter is required to describe it from a genetic point of view. Such a parameter has been called "neighborhood," and is defined mathematically as the inverse of the probability of self-fertilization of an individual. In the linear case, the neighborhood is calculated from $N = 2\sqrt{(\pi\sigma)}d$ where d is population density and σ is the standard deviation of the distance between birth places of parent and offspring. Individual mobility is assumed to operate according to the laws of normal diffusion. A range such as that occupied by a number of individuals equal to N includes 92.4 per cent of the actual parents of the individuals at the center. In the area model the neighborhood is $N = 4\pi\sigma^2 d$, and is equivalent to the number of individuals in a circle of radius 2σ. It includes 86.5 per cent of the parents of the individuals at the center.

[1]Laboratorio di Genetica (Istituto di Zoologia), Università di Parma, Parma, Italy.

If we want to make a decision in favor of one of these models in order to describe a real population, we must somehow conduct an analysis of the tendency to a formation of clusters in the population under study and of the geography and stability of these clusters.

When we think of a human population, one distinction of importance is that between its urban and rural parts. For urban populations the continuous model may seem appropriate inside the town, not forgetting, however, all the intricacies resulting from social stratification. But for some rural populations it seems that the clustering in villages is sufficiently pronounced that it has to be taken into account.

I have had an opportunity, in the last few years, to investigate a human population in an area of northern Italy, the Parma Diocese. I shall draw largely from the results attained in this investigation for the purpose of illustrating the topics under discussion. The work has proceeded along several interconnected lines, and would have been impossible without the help of a number of collaborators. The results, of which only a very preliminary and partial analysis has been published (Cavalli-Sforza, 1956), will appear in full in due time. Demographic records kept independently and in somewhat different form, by both governmental and religious authorities, are available in some cases for more than three centuries. The Parma Diocese includes about three-quarters of the Parma province with a total population of about 300,000. It covers a wide variety of physical conditions from mountains, which are scarcely inhabited, to plains with a high population density. Except for the immediate surroundings of towns where industry is prevalent, its economy is essentially agricultural. Changes in social and hygienic conditions characteristic of our contemporary world seem to have affected breeding patterns only in the last few decades and to a moderate extent.

VILLAGE SIZE

An idea of the spatial distribution of this essentially rural population and its tendency to cluster can be obtained from an analysis of the distribution of the sizes of "villages" (Fig. 1, based on the population census of 1951). We can distinguish at least three levels in the clusters of houses. The first, the smallest identifiable physical clusters, are usually not endowed with community centres ("nuclei abitati"). A fairly large proportion (more than one-quarter) of the total population lives in isolated houses, while the rest (about one-half, excluding the town) live in the "nuclei e centri abitati," which have a median size of forty-five people. All these scattered houses and small clusters, however, are attached to centers, called "frazioni," which do not have administrative independence but which form religious or social centers (second level). The smallest administrative units (third level) are the "comuni," with a median number of inhabitants approximately one hundred times that of a "nucleo abitato" and ten times that of a "frazione."

The religious unit, the parish, is almost identical with the "frazione." It is thus a relatively small unit; advantageously so from our point of view, because religious records, which take the parish as the geographical unit, are as a consequence more detailed from the point of view of spatial relations than are governmental records, which take the "comuni" as the smallest geographical unit. The social importance of the parish is such that it forms a "village," although in some cases one would, of course, like to be able to break it down further.

If we take the parish as the cluster, then enumeration of the people living in it and multiplication by simple factors might supply a direct estimate of "effective

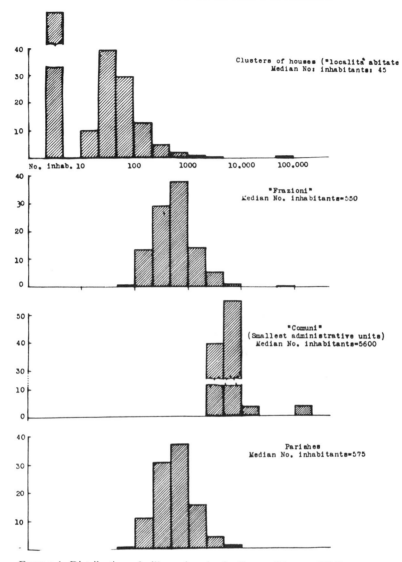

FIGURE 1. Distribution of village sizes in the Parma Diocese (1951 census).

size." It is the easiest to obtain and naturally the least accurate because, however strong the *a priori* considerations are for considering the parish as the population unit, its choice for this purpose is clearly arbitrary. Moreover, a parish is often socially heterogeneous and social heterogeneity will tend to reduce the effective size calculated from the parish total. In a similar calculation Böök (1956) had evidence that the "effective units" calculated from Dahlberg's formula were smaller than a parish. Swedish parishes are, however, probably larger on an average than Italian ones and are more similar to our "comuni."

If, as a first approximation, we take the parish as the cluster or population subgroup for use in Wright's island model, then the calculation of "effective size" N

demands only the knowledge of which fraction of the population is genetically active, or more exactly the number of gametes used per generation. This can be obtained from the number of fertile marriages per generation multiplied by four, or by similar related methods (Böök, 1956; Glass et al., 1952). When only census data are available, a convenient approximation is to calculate the number of marriages per generation by multiplying village size $\times$ generation time in years $\times$ 0.007. This figure is derived from the ratio of marriages per year to number of inhabitants, and is a demographical constant which has a remarkable stability, even over long periods, but may vary somewhat from country to country.

In the present instance, N turns out to be equal to about 88 per cent of the village size.

MIGRATION

Estimation of migration can be easily made from parish data because marriage books, at least those available to us, indicate the parish of residence and sometimes of birth of the spouse. Actually, one would like to know the birth places of parent and offspring to be able to cover migration over a full generation cycle, and this can in some instances be done. In the population which I have analyzed a substantial part of the total amount of migration taking place over one generation is covered by using the distance between the residences of the spouses.

More than 99 per cent of the marriages are performed in the Catholic Church and are, therefore, recorded in the parish books. According to the local rules, marriage almost always takes place in the bride's parish. After marriage the couple usually reside either in the bridegroom's parish or in the bride's parish, a little oftener in the former.

One can analyze migration by the fraction of marriages where both partners are from the same parish; various coefficients (for example, hexogamy coefficient, Freire-Maia, 1952, 1957) are simple functions of this fraction. Here we shall refer to E and "endogamy" as the fraction or percentage of marriages in which both spouses reside in the same parish. Or, alternatively, one can examine the whole distribution of distances between the residences of the spouses, as has been done repeatedly by several demographers, sociologists, and geneticists (Sutter and Tran-Ngoc-Toan, Schwidetzky, quoted by Freire-Maia, 1957a; Bossard et al., 1932).

It should be noted that the two modes of analysis may be completely unified, and in a very easy way if the whole distribution can be expressed by a single parameter. Actually, it turns out that we are close to this ideal.

Let us, therefore, examine first the distribution of distances. Our present data represent a stratified sample of twenty-one out of the ninety parishes of the Parma valley (to be described later), in which the marriages contracted from 1800 to 1950 were analyzed from this point of view.

The observed distribution is given in Figure 2. Classes were chosen according to a quadratic proportion of distances, to account for rapid falling-off of frequencies with distances. The theoretical curve for normal diffusion, corresponding to the equation:

$$f(x, y) \frac{1}{\pi v} e^{-(x^2+y^2)/v}$$

where v is the mean square distance from the origin, x and y the distances in rectangular co-ordinates from the origin, is given in Figure 4, in a form suitable for

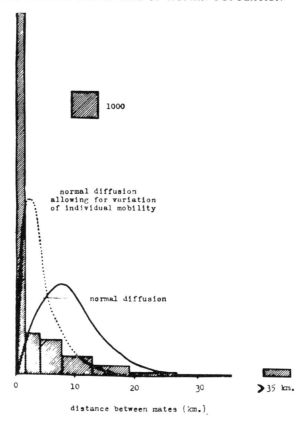

1000

normal diffusion
allowing for variation
of individual mobility

normal diffusion

0   10   20   30   > 35 km.

distance between mates (km.)

FIGURE 2. Distribution of 8,665 marriages according
to distance of place of residence between mates.

graphical representation, that is, as the distribution of distances from the origin
r $(r = \sqrt{(x^2 + y^2)})$

$$f(r) = \frac{2r}{v}e^{-r^2/v}$$

where v is the mean value of r^2. Clearly this theoretical curve is very far from
describing matrimonial migration. In statistical terminology one can say that the
actual distribution deviates from the normal in being highly leptokurtic.

The leptokurtic aspect of distributions of biological diffusion has been established
before, and also for other organisms, for example, insects (Bateman, 1950). An interesting suggestion by Skellam (1951) is pertinent here. This author has considered
the effect of variation in individual mobility or migratory tendencies. The result is a
leptokurtic distribution which, by assuming a certain distribution of individual
mobility (more precisely a gamma distribution of $1/v$), turns out to give a good fit,
for example, for the wanderings of millipedes. It is likely that human individuals
differ one from the other even more than millipedes in the degree of their sedentary
or nomadic tendencies, in their initiative, allegiance to or rejection of tradition, taste
for the new, etc., and it was anticipated that some such variation would be found.

However, even assuming an extreme variability of individual mobility (such as a J-like, or almost J-like, distribution), according to Skellam's model we are not yet able to cope with the extreme leptokurtosis of the data (Fig. 2, dotted curve).

This is not too surprising, considering how far the actual situation is from that of ordinary diffusion. If the standard model of diffusion were correct, an individual would be assumed to wander like a molecule, and every movement would start from the place where the individual arrived last without any memory of previous whereabouts. In reality, everyone goes home almost every night and explores a limited territory during his daily activities. Usually he has a chance of meeting his or her future mate in places which are most of the time at a distance from home of one day's walk or travel. He has, therefore, a good chance of exploring the immediate neighborhood fairly well. Diffusion would come only if he changed residence repeatedly, but this is an infrequent event in rural populations.

In the lack of a good model to represent this type of migration, we may look at least for an adequate description of the data. Previous experience with insects (Bateman, 1950) and partially with man (Sutter and Tran-Ngoc-Toan, 1957; Fraccaro, 1958) showed a good fit of an empirical function

$$F = ce^{-k\sqrt{r}}.$$

The constant c is equal to $k^2/2$ and this function depends on one parameter only. Sutter and Tran-Ngoc-Toan (1957) have found that some distributions are to be fitted instead with the function e^{-kr}.

If we try to apply this empirical formula to our data, especially if we try to fit the data from individual parishes, we have to take account somehow of the vagaries of the geographical distribution of the population, which greatly distorts the probability of marriage as a function of distance alone ($f(r)$). Instead of estimating this we can take the number of marriages at a given distance divided by the number of people living at that distance,[2] and then obtain a probability $p(r)$ of marrying any given individual at a given distance. In other words while $f(r)$ expresses a probability per kilometer, $p(r)$ expresses a probability per kilometer person. If, however, we prefer the former type of probability, $f(r)$, we can account for the vagaries of geographical distribution by dividing $f(r)$ by population density at distance r. The two values $f(r)$ and $p(r)$ are easily transformable one into the other, but we should keep in mind that in a homogeneously distributed population the number of inhabitants grows linearly with r.

By this method of correction it was found that irregularities are considerably smoothed down. The only exceptions are rural villages near towns, where the effect of the town is smaller than that calculated from its number of inhabitants, presumably because there is no great admixture between rural and urban poplations.

A graphic representation (Fig. 3) shows a rather good fit of $p(r)$ with the formula $e^{-k\sqrt{r}}$ for all points except, unfortunately, the most important one, namely, the marriages of the parish villagers between themselves. As this is a major disadvantage unless it can be further corrected, there is nothing to recommend this representation. It was, however, useful here in pointing out that when the individual parishes are analyzed individually, the rather remarkable result is observed that all the parishes show practically the same slope of the curve. Thus, although some of the parishes are from mountainous regions, others from perfectly flat country,

---

[2]"Distance" here means the average distance from the center of the points in the area comprised between two circles, whose radii are the class limits of the distance considered. Road distances are used throughout.

mobility as expressed per kilometer person is almost the same for every parish. This is indirectly shown in Figure 3 by the ranges of individual parish values, which are relatively small and almost constant throughout. It will be noticed that the range of the first point (for $r = 0$) falls off the line; the range of the second point

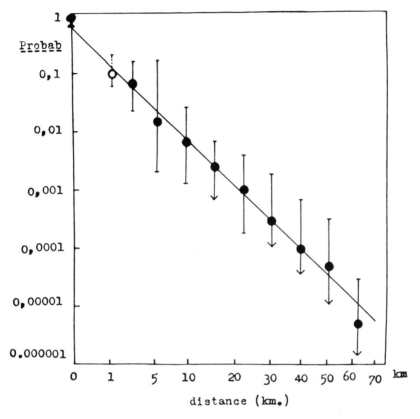

FIGURE 3. Matrimonial migration. Fitting : $e^{-k\sqrt{\text{distance}}}$

is smaller because only five parishes out of the twenty-one here examined have neighboring parishes situated at such a small distance (1–2 km.). For this reason this datum is given as a white circle in Figures 4 and 5. In fact, the error of this point is larger than the error of the others although the (rather fallible) measure of variation employed in the graph, the range, is smaller.

A more interesting representation stems from a straight application of the formula used to describe gravitation, or the attraction between electric charges, as well as the effect of distance on some social phenomena. According to this formula the "attraction force" of an individual of a village situated at distance r, and hence the probability of marriage is directly proportional to the number of people living in the village, and inversely to the square of the distance.

On the Parma data the fit of such a hypothesis is good for the shorter distances (Fig. 4), but poor for longer ones (which, however, contribute less than 5 per cent of the total marriages).

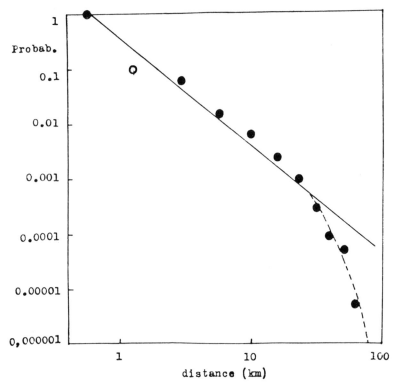

FIGURE 4. Matrimonial migration: the gravitational model applied to the data of Figure 2.

The data on the Swedish parish of Ovanåker (collected by Fraccaro and communicated at this Congress), which I have been able to examine, fit the curve in a remarkable way, and so do short-range data collected in the city of Philadelphia (Bossard, 1932).

In formulae, the attraction of an individual and, therefore, the frequency of marriage at distance r will be

$$f = \frac{k'A}{r^2};$$

therefore $\frac{f}{A} = p = \frac{k'}{r^2}$

where A is the number of people living at distance r. For a population homogeneously distributed over an area, $A = 2\pi rd$ where d is population density and therefore f is proportional to d/r.

One practical difficulty which is encountered in fitting a law of "Newtonian" attraction is that of determining the distance of an individual from his own village, in order to estimate endogamy. I have tentatively used half the radius of the parish taken as a circle, as is sometimes done for the calculation of population potentials in demography (Stewart, 1947). This last concept stems in fact from essentially the same considerations as those used in testing the Newtonian type of attraction.

Actually, the Newtonian model is interesting because it is somewhat opposite to that of diffusion. The gravitational model is based, in fact, on a dynamic equilibrium between villages. With diffusion, on the other hand, the effect of distance terminates at equilibrium, at which any person can mate with any other person with constant probability irrespective of distance. If a bachelor allows enough time to elapse before marrying, or if mobility is very high, diffusion can proceed a long way before marriage takes place, and eventually complete panmixia is reached at the limit of infinite time or mobility.

Actually, some hints are available that diffusion does play some role. If we examine the correlation between age at marriage and distance between the place of residence of the spouses, we find an effect which suggests that the older the spouses, the greater the distances between their birthplaces (or residences) since they have had more time to move around. Here the analysis has been carried out in a preliminary way, only to the extent of camparing the average age of brides or bridegrooms in endogamous and hexogamous marriages (Table I). The effect is significant but small. It is possible that a better model should be a compromise between a gravitation theory and a diffusion theory.

TABLE I

MEAN AGE AT MARRIAGE
(20 villages, 1800-1950)

	Wife c_c	Husband c_c	No. of marriages
Both from the village	24.72	28.71	1,077
Husband from elsewhere	25.09	30.11	972
Wife from elsewhere	27.19	30.26	31
Both from elsewhere	26.05	30.37	95
Resident in the parish	24.89	28.75	
Resident elsewhere	26.33	30.13	
Difference	1.44±0.55	1.38±0.30	

However, coming back to our primary interest, that of estimating migration, we do not have a perfectly satisfactory formula, but a single parameter formula similar to $e^{-k\sqrt{r}}$ or kA/r^2 can probably represent the situation, or at least extract most of the interesting information. If a single parameter formula is eventually found to be valid for representing the data, any particular estimate of migration will be a simple function of this parameter and, if properly estimated, any particular parameter chosen to indicate migration will be a function of all the others. Of the estimates presently available, at least for the purpose of making internal comparisons in our set of data, "endogamy" E, or the frequency of marriages in which both partners belong to the same parish, is convenient. Besides extracting a fair amount of the information, it also has the advantage that m is easily calculated from it for the island model, taking the parish as the unit cluster ($m = 1 - \frac{1}{2}E$). This will be an underestimate of m if the contribution to migration of the changes of residence, other than those connected with marriage alone, are frequent, a fact which demands a closer scrutiny in every case. From a preliminary exploration it is not important in our case.

When E values (x) for twenty-one parishes of the Parma valley are correlated with three ecological variables: village size (y), local population density (w), and altitude (z), some interesting results are obtained (Table II). E is directly correlated with village size when the effects of density and altitude are eliminated $(r_{xy.wz} = +.66)$. There is a negative correlation with density which persists after elimination of village size $(r_{xw.y})$ though it is reduced to a non-significant amount after elimination of density. However, density and altitude are here highly correlated, in an inverse way, in this sample $(r_{wz} = \sqrt{.84})$ and in the area in general, so that almost all that is stated of one is inevitably found to be true of the other variable as well. One exception is, however, the correlation of E with altitude which persists after elimination of density effects $(r_{xz.y} = +.82$ and $r_{xz.yw} = +.57)$.

TABLE II

ENDOGAMY (PERCENTAGE OF MARRIAGES IN WHICH BOTH PARTNERS RESIDE IN THE SAME PARISH) IN 21 PARISHES AND ITS CORRELATION WITH DEMOGRAPHIC AND ECOLOGICAL DATA

Variables	Mean	Range
x Endogamy	54.5%	34–77%
y Village size	700	159–2,495
w Pop. density	86.5	37–180
z Altitude (m.)	478	60–950

Correlation coefficients

Zero order	First order	Second order
$r_{xy} = +.25$	$\begin{cases} r_{xy.w} = +.52^* \\ r_{xy.z} = .67^{**} \end{cases}$	$r_{xy.wz} = +.66^{**}$
$r_{xw} = -.67^{**}$	$\begin{cases} r_{xw.y} = -.76^{**} \\ r_{xw.z} = -.28 \end{cases}$	$r_{xw.yz} = -.34$
$r_{xz} = +.66^{**}$	$\begin{cases} r_{xz.y} = +.82^{**} \\ r_{xz.w} = +.25 \end{cases}$	$r_{xz.wy} = +.57^*$
$r_{yw} = +.20$		
$r_{yz} = -.34$		
$r_{wz} = -.84^{**}$		

*Significant at 5% level.
**Significant at 1% level.

While the conclusions thus reached deserve to be further tested and amplified by an increase in the size of the sample of parishes, it is interesting that the conclusion on correlation of E with village size is in agreement with the expectations from the gravitational model. The model taken in its simplest form does not directly give information on the possible effects of altitude and density, unless these environmental parameters are further analyzed, as is done below.

The effect of altitude is worth observing on the whole distribution of distances. Subdividing the twenty-one parishes into two groups, those located above 400 meters and those below, and taking the frequency of marriages $f(r)$ per kilometer in the two groups (Table III), one finds that the mountain parishes have a lower mobility at short and intermediate ranges though a slightly higher one at far distances. Lower mobility of the mountain population is a fairly obvious result as one mile on a mountain path takes much more time and effort to travel than a mile on a regular road in a hilly or flat country. However, the difference in mobility is not as striking as one might have expected. Moreover, as already noted when analyzing Figure 3, there is no important variation between parishes when mobility is given per kilometer person rather than per kilometer. This point can be confirmed in particular for the comparison between averages for mountain and plains parishes.

Another factor likely to play a role is the fact that in the mountains the population is more often concentrated in villages. The fraction of population living in "isolated houses" as opposed to clustering of houses and minor or major centers increases strongly with population density (and, therefore, decreases strongly with altitude) (Table IV).

TABLE III

MATRIMONIAL MIGRATION IN A GROUP OF 10
PARISHES AT LOW ALTITUDE (P) AND A GROUP
OF 11 PARISHES AT HIGH ALTITUDE (M).

Distance (1 unit = 625 km.)	P (%)	M (%)
0–2.5	51.3	64.2
2.5–6.5	9.9	9.9
6.5–12.5	15.2	6.6
12.5–20.5	9.9	6.3
20.5–30.5	6.0	3.9
30.5–42.5	3.7	1.8
42.5–56.5	1.0	1.8
56.5–72.5	0.7	1.3
> 72.5	2.0	3.6
TOTAL NUMBER OF MARRIAGES	5,604	3,060

$\chi^2[8] = 287.6$.

TABLE IV

DISTRIBUTION OF POPULATION IN CLUSTERS AS AFFECTED BY DENSITY

Density class	Average density	No. of villages ("frazioni")	Median size of villages	Village size/density	Percentage of population living in isolated houses
0–50	44	70	300	6.8	11.1%
51–95	66	62	475	7.2	25.5%
96–140	122	74	680	5.6	47.4%
141–180	157	56	915	5.8	65.8%

It was also tested whether the spacing between villages is increased in the mountains; a possibility which would be important if the gravitational model is correct and could help explain the higher rate of endogamy observed there. In fact, in this case, if two villages, each of 1,000 inhabitants, were at a distance of 10 km. they would, with the law of the square of distance, be more endogamous than four villages each of 500 distributed regularly over a total distance of 10 km., in such a fashion as to give the same population density. However, it is found that the spacing of villages is almost independent of population density; the lower population density being almost exactly compensated by the smaller average size of the village and, therefore, their higher number per unit of area (Table IV). Moreover, a fraction of the mountainous area is totally uninhabited, and villages may even turn out to be a little closer one to the other in the mountains than in the plains. An exact comparison has to take account, however, of the different spatial distribution in the two cases (see Table VII), a fact which by itself has a relevance in the present connection.

Consanguinity Data

The classical approach to the study of population structure in man has been through consanguinity studies (for a recent review, see Freire-Maia, 1957b). The files of dispensations existing in the archives of Catholic bishoprics have formed a favorite material for this kind of investigation.

One caution that is perhaps not widely appreciated is that dispensations thus collected form a list of planned marriages, not of marriages actually performed. The extent of the discrepancy is now being evaluated and may be relatively large, perhaps in view of the delay, especially in earlier generations, between requesting and obtaining a dispensation.

The main difficulty attached to drawing conclusions from consanguinity data as regards population structure is that of non-randomness in the sense of a more or less strong bar against consanguineous marriages, which is posed by tradition and civil or religious customs. An attempt is being made to evaluate the effect on the closer consanguineous marriages.

Although objections can be raised against it, the study of consanguinity in man offers too tempting an approach to be neglected, in view of the availability of data.

Figure 5 shows a synthesis of data from consanguinity dispensations from 1851 to 1950 existing in the archives of the Parma Diocese. The 295 parishes belonging to the diocese have been divided into sixteen categories, according to four classes of population density (of the comuni to which each parish belongs), and four classes of village sizes. The bars in Figure 5 represent the average value of F (Haldane and Moshinski, 1939). The fractions of which each bar is composed show the contributions to F of the various types of consanguineous marriages, or more precisely the marriages with given coefficient of inbreeding, as shown by the illustrative bar on the right of the figure.

It is clear that population density has a high inverse effect on F, as has been shown before for Brazil by Freire-Maia (1957a). Village size has also had some effect, though smaller than in the partial collection of these data previously published (Cavalli-Sforza, 1956), and relating to the last quarter of the century, the examination of which has not been completed.

It is also clear that the contribution of marriages, other than those of first cousins, is different in the various classes of population density, being about a half of the total in the zones of low population density and one-fifth in those of high density, with lower F value.

One use which can be made of consanguinity data in order to evaluate properties of populations is to calculate the so-called "isolate size" of Dahlberg (N_D). This is the number of people among whom a spouse is selected and is based on the observed frequency (c) of marriage between relatives of a given kind. Assuming that the choice between a consanguineous and a non-consanguineous mate is a random one, and knowing how many relatives of a given kind each individual has, or is expected to have on the average (b), the number of people among whom a spouse is selected (N_D) can be obtained from $c = b/N_D$. The original formulae by Dahlberg have been recently improved by Frota Pessoa (1957) so as to take into account the effect of the variability in number of fertile individuals per family. However, as these formulae require knowledge which is not easily available and the approximation obtainable is, for reasons to be discussed later, influenced proportionately more by other uncontrolled factors, I have used for calculation the standard formulae.

The estimates of "isolate sizes" obtained with the Dahlberg method from these

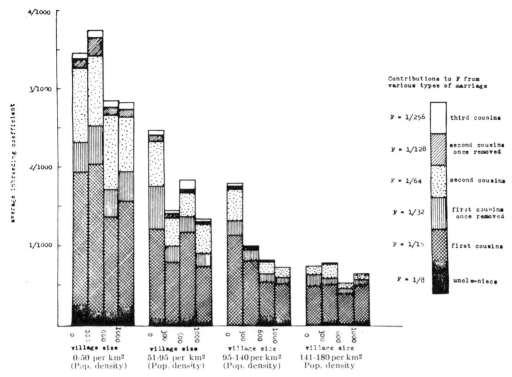

FIGURE 5. Average inbreeding coefficient as a function of population density and village size.

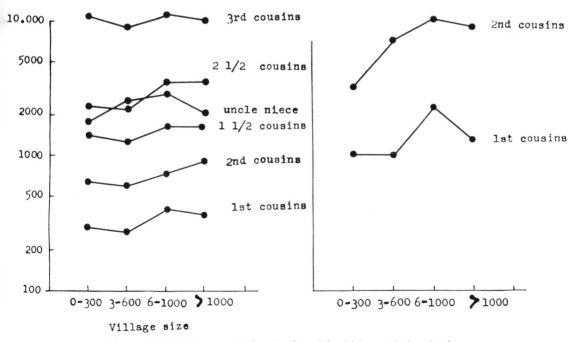

FIGURE 6. *Left*: Low population density; *right*: high population density.

data are given graphically in Figure 6 in such a way as to show how they vary with village size, type of marriage, and population density. Only the two extreme density classes are given, but the intermediate ones yield intermediate results. The values are actually calculated not from the frequencies of the indicated marriage types, but from the frequencies of consanguineous marriages with the corresponding degree of inbreeding; the difference, however, is trivial. Some discrepancy is noticeable between these calculations and the earlier ones (Cavalli-Sforza, 1956) which were based on the last quarter of the century here considered. These discrepancies are to be further analyzed in future work.

In general it can be concluded that: (1) the estimates of isolate sizes increase with population density; (2) the effect of village size, if any, is modest (the mode of graphic representation is such that if there were proportionality between isolate size and village size the curves should be straight lines at a 45° angle); (3) of the various types of consanguineous marriages, first cousins give the lowest estimates, and cousins of other degrees progressively lower estimates, probably because they are progressively more influenced by migration. In fact, the more generations there are between the common ancestors and the consanguineous individuals the greater the amount of migration, and therefore of diminution of the number of consanguineous people in the immediate neighborhood; (4) degrees of relationship involving an unequal number of generations, such as uncle-niece (or aunt-nephew) or cousins once removed, give higher estimates of isolate size because of their comparatively lower frequency. This is probably to a large extent the result of the unfavorable age relation between such relatives, and naturally this effect tends to cancel out with increasing remoteness of the relationship.

The concept of "isolate size," or at least its estimates, are therefore to be taken with caution if they are to be considered at all. But unquestionably these estimates can become more useful than they are now if it is possible to remove disturbing effects, such as those of age and of migration. A step in this direction can be provided by a more detailed examination of the data at the level of distinguishing classes of consanguineous marriages of the same general type, according to the sex of the ancestors intermediate between the common ones and the consanguineous mates. There are, for instance, four classes of first cousins and sixteen of one and a half or second cousins, and they have been shown in other situations to have discrepant frequencies (for example, Morton, 1955). This is true also of the present data, but the relative frequency of the various types differs with time and space, and the problem will require a special analysis.

More difficult to remove is the effect of non-randomness, which derives from traditional bars against consanguineous marriages whenever they are incomplete and ill-defined, but an attempt could perhaps be made at estimating their strength.

When the factors mentioned have been duly analyzed and their effects assessed it is likely that a clearer picture may emerge from consanguineous data.

GENE FREQUENCIES

The ultimate goal of population genetics is naturally that of being able to make predictions about frequencies of genes, genotypes and phenotypes, their distributions, trends and equilibria. Clearly the direct estimation of gene frequencies whenever possible gives more direct and more satisfactory information than any other method, but may, if taken without consideration of the demographic data, be insufficient and even misleading.

A sampling survey of blood-group gene frequencies has been undertaken in the area which has formed the object of this investigation, and exactly in that section of it which is defined by the valley of the Parma River, which occupies the eastern third of the area. Approximately an equal number of adults was sampled from each parish. Out of the ninety parishes only fifty-three have been sampled so far and the data subjected to a preliminary statistical analysis. While waiting for the more complete data, certain tentative conclusions are worth mentioning.

Two types of analysis can be used on the data that were collected. One is the examination of whether genotype frequencies are in agreement with the equilibrium values expected under random mating, or the particular mating conditions of the population. The other is the analysis of the variation in gene frequencies over the geographical area which is being examined.

Analysis according to the first approach has not been completed. However, the distribution of phenotype frequencies for the blood-group systems ABO (A_1A_2B) and MN was tested for the goodness of fit of the Hardy-Weinberg equilibria in individual parishes. When the results are pooled for all parishes, some significant departures from expectation are found, especially for the MN system, in that a slight excess of heterozygotes exists.

A fairly clear-cut result has been obtained by the second approach. For this analysis the available data were grouped in a hierarchy as follows. Adjacent parishes showing closer geographical relations were assigned to groups (called the first order), containing usually two or three parishes. These groups were further classified into seven major groups (of second order), which correspond to the four southern mountain "comuni," Corniglio, Monchio, Tizzano, Palanzano, the intermediate "comune" of Calestano,[3] and two areas forming respectively the hilly and flat parts of the country which descends north of the Apennines towards Parma town.

A hierarchical analysis of the variation between parishes for the frequencies of genes A, B, M, R_1, R_2, and r yields the results summarized in Table V. A fair amount of variation is encountered, and almost all the heterogeneity chi squares (between parishes within groups of first order) are significant, and so are a number of those for the higher categories.

However, the heterogeneity between groups of first order and between groups of second order cannot be considered proved by the significance of their relative chi squares, because of the existence of heterogeneity at the lowest level. Rather, chi-square analysis for the higher levels of hierarchy must, as a consequence of such heterogeneity, be transformed into an analysis of variance. When this is done (Table V, F values), none of the heterogeneity at the higher level remains significant. As a first approximation we can therefore conclude that in this area there is variation above that expected from random sampling, but it is mostly confined to differences between adjacent parishes and no clear gradients or pockets of gene frequencies are observed. Variation is, in a word, confined to a microgeographical level.

When the total variation is subdivided according to the groups of second order, it is found that significant heterogeneity is entirely confined to the four parishes in the mountains, with no special preference for any blood groups (except perhaps blood group A for which the variation is smaller), while the flatter part of the region seems perfectly homogeneous in regard to blood groups.

[3]Discarded in Table VI because of the paucity of data.

TABLE V

BLOOD-GROUP GENE DISTRIBUTION: VARIATION IN GENE FREQUENCIES

System	Gene	Variation between adjacent parishes within groups of first order	Variation between groups of first order within groups of second order		Variation between groups of second order	
ABO	A	$\chi^2[28] = 34.62$	$\chi^2[18] = 31.22^{*\ddagger}$	$F = 1.40$	$\chi^2[6] = 10.20\dagger\dagger$	$F = 1.20$
	B	51.68**	37.85 **	1.14	11.31	1/1.03
MN	M	66.43**	47.10 **	1.10	15.21*	1.04
	R_1	39.38	19.38	1/1.30	9.36	1.22
Rh	R_2	48.26**	22.01	1/1.38	15.23*	1.65
	r	43.60*	14.94	1/1.89	4.18	1/1.60

*$P < 5\%$ **$P < 1\%$ ‡df 18,28   ††df 6,46

Since the mountainous part of the area is the one in which lower density, smaller villages, less migration, etc. prevail, this result is far from surprising, but the magnitude of the variation encountered is somewhat astonishing. Although some of the variation, for example, that of the Rh genes, is probably partly due to the use of inefficient methods of estimation of gene frequency, to be improved in the later and more complete analyses, no cause has been found which might have artificially increased this variation. While these data can, therefore, be taken to constitute *prima facie* evidence of drift, no possibility exists so far of discarding local differences of selection, although the variation would require a microheterogeneity of selective conditions which may be difficult to imagine. There is a chance of testing this further by analyzing the heterogeneity of the Hardy-Weinberg equilibria among parishes.

However, instead of limiting the analysis to the picture given in Tables V and VI, one may try to carry it further and estimate effective sizes from the observed variation in gene frequencies, using the formulae given by Wright.

TABLE VI

BLOOD-GROUP DISTRIBUTION χ^2 BETWEEN PARISHES FOR GROUPS OF SECOND ORDER

	Mountain area				Low area		Sums	
	Corniglio (df 7)	Monchio (df 9)	Tizzano (df 5)	Palanzano (df 3)	Hills (df 12)	Plains (df 10)	Mountain (df 24)	Low area (df 22)
A	16.01*	10.32	21.04**	5.39	17.76	5.64	52.76**	23.40
B	4.79	41.31**	11.88*	4.40	10.04	18.11	62.38**	28.15
M	17.17*	46.58**	13.01*	3.37	19.45	13.95	80.13**	33.40
R_1	10.46	14.77	6.49	2.69	8.88	15.47	34.41**	24.35
R_2	17.67*	13.24	4.20	8.33*	12.93	13.10	43.44**	26.03

*5% significance level.
**1% significance level.

As only the mountains show heterogeneity between parishes, the analysis will be limited to them, subdividing the parishes into two groups, smaller and larger in size, to test the effect of village dimensions. The island model will be used, considering that: (1) in the mountain area the degree of clustering is very high; (2) the spatial distribution of people is somewhat intermediate between the two types offered by the continuous model, namely the linear, and the area distribution (Table VII), so that the choice between them would be difficult; and (3) there being no clear-cut macrogeographical variation in gene frequencies over the area, immigration from neighboring groups gives on an average almost equivalent results as immigration from more remote groups. This corresponds to the expectation of the island model.

TABLE VII

PATTERN OF GEOGRAPHICAL DISTRIBUTION OF PARISHES†

	Low altitude (<400 m.)			High altitude (>400 m.)		
		Expected			Expected	
Distance	No. of parishes	Area distribution	Linear distribution	No. of parishes	Area distribution	Linear distribution
1.5–2.5	1	3.1	17.0	4	2.8	15.3
2.5–6.5	32	27.8	68.0	47	25.0	61.3
6.5–12.5	101	88.1	102.0	73	79.3	91.9
12.5–20.5	189	204.0	136.0	167	183.8	122.5
	323	$\chi^2_{[3]} = 5.0$	$\chi^2_{[3]} = 54.8^{**}$	291	$\chi^2_{[3]} = 21.9^{**}$	$\chi^2_{[3]} = 31.7^{**}$

Difference between distributions at low and high altitudes   *P < 5%
$\chi^2_{[3]} = 8.82^*$ **P < 1%

The spatial distribution of villages is analyzed by counting, for each of the twenty-one parishes, how many parishes are found within circles at the distances given in the table. Here, as elsewhere, distances are road distances. The frequencies of parishes thus counted are cumulated for the various parishes distinguished into two groups, namely mountain and plains. The observed frequencies are then compared with those expected for a linear and an area distribution.

From the smaller parishes the variation in gene frequency $\sigma^2 P$, after elimination of the sampling component, gives rise to estimates of N which, for $m = .15$, range between 75 and 230 (for various blood-group genes). The average village size is 233 and from it a "demographic" effective size can be estimated as 205. This demographic estimate is higher than the genetic one and any estimate from consanguinity data would be even higher. It should be noted that any demographic estimate so far obtained is vitiated in the excess direction by failing to consider social differences within the population. All consanguinity estimates are vitiated usually in the same direction, and especially by non-randomness of marriages. The estimate of N from the variation in gene frequency may seem low, especially if one thinks that the existence of selection for blood-group markers must tend to raise it above the "true" unknown value.

Data from the larger mountain parishes give similar, and occasionally even lower estimates, but the magnitude of the error involved in estimation makes comparisons difficult at the present stage of the investigation.

However, it has been noted from Dahlberg's "isolate sizes" that these estimates do not increase proportionately with village size. Possibly the increase of village size is accompanied by an increase of social complexity which keeps "effective sizes" low. Also, m decreases somewhat with village size. The data of Table VII suggest, however, that the pattern of geographical distribution is likely to be an important factor of gene variation in this area. An assessment of its relative importance is left for future investigators.

In conclusion, the study of population structure in man is probably not as direct a task as it may seem, in spite of the availability of many demographic data. I am still convinced, however, that man is the organism of choice for many such problems. It would be especially desirable—an opinion which I believe I share with some other people—if these studies were extended to a variety of primitive populations, while they are still primitive, although clearly, the more primitive the subjects of study are, the less amenable they will be to demographic investigations.

SUMMARY

1. On the premise that man is a "good" organism for the study of population structure, an investigation has been started in an area chosen for the variety of physical environment it offers and the availability of good demographic records, the Parma Diocese in Italy. An analysis is being carried out of (a) clustering and geographical distribution of the population; (b) intermarriage, and genetic migration; (c) consanguinity; (d) distribution of blood group genes.

2. Of the various orders of clusters of people that can be considered as "villages," the "parish" (about 575 inhabitants on the average) is the most satisfactory one for a first approach. "Effective size" in Wright's sense is about 88 per cent of the village size, but this "demographic" estimate constitutes a very rough approximation.

3. Genetic migration is here estimated from the distance between the residence of the two spouses. This is only a fraction, but an important one, of the total migration which it is desired to estimate, that is, the distance between birthplaces of parent and offspring.

4. The distribution of matrimonial distances does not fit the normal diffusion model, but, as already shown for a number of other organisms including man, it shows a better fit with the empirical formula $e^{-k\sqrt{r}}$ (r = distance).
of villages of N individuals at distance r is proportional to N/r^2.

6. The "attraction" does not vary for villages from different environments such as mountain or plains. In more general words, when the probability of marrying at distance r is weighed for population density at that distance, it is constant in the whole area investigated.

7. Diffusion when measured per kilometre is slightly higher, at short distances, in the plains than in the mountains.

8. "Endogamy" or the fraction of people who marry in the same village increases with village size and with altitude, and is relatively independent of population density by partial correlation analysis.

9. Consanguinity has a high dependence on population density and altitude, but it is very little, if at all, dependent on village size. Not only the quantitative, but the qualitative picture of consanguinity, that is, the relative contributions from marriages of various consanguinity levels, change with density.

10. As stressed by so many other authors, the Dahlberg formulae do not supply easily interpreted values of "isolate sizes." They show, however, that isolate size is almost independent of village size.

11. An analysis of the presently available data on blood group gene distribution shows that some groups of adjacent parishes have significantly different frequencies for blood group genes, irrespective of the blood group system investigated (ABO, MN, Rh). There is no major gradient or variation over the area: the variation is entirely "microgeographical" and confined to the mountain region (where isolation is highest).

12. The pattern of geographical distribution of parishes in the mountains is neither linear nor two-dimensional, but somewhat intermediate.

13. Estimates of "effective sizes" from variation in gene frequencies with the "island model" are somewhat smaller than demographic estimates even for the group of smallest villages, but, for a more satisfactory analysis, further data will be necessary.

14. In the bigger villages "effective size," in Wright's sense, in parallel with "isolate size" in Dahlberg's sense, is no higher than in the smaller villages. This probably reflects the fact that social stratification increases with village size, and tends to keep effective sizes low and almost independent of village size.

REFERENCES

BATEMAN, A. J. 1950. Is gene dispersion normal? Heredity *4*: 353–363.

BOSSARD, J. H. S. 1932–33. Residential propinquity as a factor in marriage selection. Am. J. Sociol. *38*: 219–224.

BÖÖK, J. A. 1956. Genetical investigations in a North-Swedish population. Ann. Human Genet., Lond. *20*: 239–250.

CAVALLI-SFORZA, L. L. 1956–7. Some notes on the breeding patterns of human populations. Acta genet., Basel *6*: 395–399.

DAHLBERG, G. 1929. Inbreeding in man. Genetics *14*: 421–454.

FRACCARO, M. 1958. Private communication.

FREIRE-MAIA, N. 1952. Frequencies of consanguineous marriages in Brazilian populations. Am. J. Human Genet. *4*: 194–203.

——— 1957a. Inbreeding in Brazil. Am. J. Human Genet. *9*: 284–298.

——— 1957b. Inbreeding levels in different countries. Eugen. Quart. *4*: 127–138.

FROTA-PESSOA, O. 1957. The estimation of the size of isolates based on census data. Am. J. Human Genet. *9*: 9–16.

GLASS, B.; M. S. SACKS; E. F. JAHN; and C. HESS. 1952 Genetic drift in a religious isolate: An analysis of the causes of variation in blood group and other gene frequencies in a small population. Am. Natur. *86*: 145–159.

HALDANE, J. B. S., and P. MOSHINSKY. 1939. Inbreeding in Mendelian populations with special reference to human cousin marriage. Ann. Eugen., Cambr. *9*: 321–340.

MORTON, N. E. 1955. Non-randomness in consanguineous marriage. Ann. Human Genet., Lond. *20*: 116–124.

SKELLAM, J. G. 1951. Gene dispersion in heterogeneous populations. Heredity *5*: 433–435.

STEWART, J. Q. 1947. Empirical mathematical rules concerning the distribution and equilibrium of population. Geogr. Rev. *37*: 461–485.

SUTTER, J., and TRAN-NGOC-TOAN. 1957. The problem of the structure of isolates and of their evolution among human populations. Sympos. Quant. Biol. *22*: 379–383.

WRIGHT, S. 1943. Isolation by distance. Genetics *28*: 114–138.

——— 1946. Isolation by distance under diverse systems of mating. Genetics *31*: 39–59.

——— 1951. The genetical structure of populations. Ann. Eugen., Cambr. *15*: 323–354.

23

Reprinted from *Proc. Roy. Soc.*, **B159**, 125–174 (1963)

Concepts of random mating and the frequency of consanguineous marriages

By J. Hajnal

London School of Economics†

Formulae due to Dahlberg for computing the numbers of consanguineous matings to be expected under random mating fail in the case of overlapping generations. The concept of random mating can be extended to cover overlapping generations by making the chance of mating between a male and female depend on the interval between their dates of birth. This concept of random mating is illustrated by a numerical example based on a simple hypothetical organism.

The implications of the suggested concept of random mating can be explored for species with complicated demographic characteristics, and man in particular, by means of a mathematical model in continuous time. Calculations of the frequencies of various types of consanguineous marriage to be expected under random mating are presented and compared with statistics of consanguineous marriages occurring in various countries. Discrepancies and similarities between observed and theoretical values are discussed.

1. The Dahlberg formulae

In the last few decades, human geneticists have become increasingly interested in consanguineous marriages. It is widely known that the offspring of marriages between relatives are more frequently affected by certain abnormalities. But the significance of the frequency of consanguineous marriages for geneticists goes considerably beyond this aspect; the subject has been involved, for example, in the study of the long-term genetic effects of radiation or the total load of deleterious mutations. As a result of this interest geneticists have collected a large volume of data on the frequency of consanguineous marriages.

In the analysis of such data one of the questions which arises is this: How many consanguineous marriages of each type would there be if such marriages occurred at random, i.e. if there were no special preferences for such marriages, or aversions to them, or circumstances which tended to bring cousins together, and so forth?

There is an apparently simple way of working out the amount of consanguinity which would occur under random mating. Consider an isolated population, and

† Part of this work was done at Dartmouth College, New Hampshire, U.S.A., during a visit supported under the National Science Foundation grant to the Dartmouth Mathematics Project. The research was also supported by the population Investigation Committee, London.

suppose that every one is equally likely to marry every potentially marriageable person.

Let us take the case of marriages between first cousins. Suppose every marriage produced two children who grow up to marry in their turn. We can then work out the number of cousins available for marriage to a man of marriageable age. His father had one sibling who married. Of this marriage there are two children, of whom on average one is a girl. Our hypothetical individual, therefore, has on average one marriageable girl cousin on his father's side. By the same argument he will have one marriageable girl cousin on his mother's side. So a marriageable man has on average two marriageable girl cousins. If there are 100 marriageable girls in the population, and if marriages occur at random, two out of 100 marriageable men will marry their cousins. A similar argument can obviously be used for relatives of any degree. It is of course an over-simplification that every marriage produces two children who grow up to marry in their turn, but it is possible to correct the argument and allow for the variation in the number of children of a marriage (Frota Pessoa 1957). This line of argument was originally developed by a Swedish geneticist, Dahlberg (1929, 1948), and the formulae deriving from it are usually associated with his name. It is a convenient formulation for many purposes. For example, the argument can be reversed. If 2 % of all marriages are known to be cousin marriages, then, if people marry at random, we may reason that the average individual has a choice of 100 partners.

The Dahlberg approach turns out to have a number of snags. It is assumed that there is a fixed universe of potential spouses any one of which a person is equally likely to marry. But who are these potential spouses? For a girl, is it the men aged 20 to 40, or all men between 16 and 90? Moreover, suppose we agree that for unmarried men potential spouses are the girls aged 16 to 40. The unmarried girls aged 16 to 40 available in 1955 to a man who is then aged 20 will be different from the unmarried girls of the same age from whom he can choose if he is still unmarried at the age of 25 in 1960. It is in fact impossible to define a fixed population of potential spouses.

It is equally impossible to say just how many of a man's cousins constitute potential spouses. The Dahlberg approach assumes that all who survive to marriageable age do; but suppose a man is born when his father is 55 (not a rare event) and that the father has a sister 15 years older than himself. This aunt will be 70 years older than the man we are considering. Suppose she had a daughter when she was 20. This girl would be unlikely to defer marriage until she is 70, when her young cousin is 20.

To take another difficulty, suppose that a pair of cousins both survive to be 20 and then die, for example in an epidemic; they clearly have some chance of marrying each other, but would have much more if they both survived to be 40. The difficulties arising out of mortality cannot be overcome by working in terms of the numbers who survive to marry. Two cousins may both survive to be 40 and yet one may be 30 years older and hence die when the other is only 10 years old.

Another weakness is the neglect of age preferences in the choice of a spouse. A woman has the same number of nephews, on average, as a man has nieces. By the Dahlberg formulae, uncle–niece and aunt–nephew marriages should be equally frequent. It scarcely needs data to confirm that they are not. Nieces are on average a good deal younger than uncles, and nephews younger than aunts, but marriages between older men and young women are much more frequent than marriages between young men and old women.

2. SPECIFYING THE AGES AT PARENTHOOD

To overcome these and other difficulties a basic reformulation is needed. The fundamental requirement is to take explicit account of the ages at which reproduction takes place. Dahlberg's formula for first-cousin marriages is logically satisfactory only for an organism which reproduces only once at a fixed point in its life cycle. For example, the formula will apply to an animal which has periodic breeding seasons if both males and females mature in one season and reproduce in the next one. This implies discrete, non-overlapping generations. In any breeding season a male can mate with all those females (and only those females) who were born in the previous breeding season. For such an animal random mating can be taken to mean that a male has an equal probability of mating with all females one season old. Under these assumptions mating between members of unequal generations such as uncle and niece would be impossible.

A satisfactory formulation for any organism with overlapping generations (even for one whose pattern of reproduction is much less complicated than for man) necessitates bringing into the calculation explicitly the ages at which reproduction takes place. This is particularly clear for mating between individuals who are not of the same generation. For example, consider an organism which matures for twenty years and is then capable of reproduction for two seasons. In such an organism, mating between parent and offspring or uncle and niece is impossible. Members of unequal generations can only mate if they are very distantly related. On the other hand, in an organism which matures for two years and is capable of reproduction for twenty years, mating between parent and offspring or uncle and niece is extremely frequent if mating occurs at random.

What is involved in a logically satisfactory method of calculating the frequencies of consanguineous matings under assumptions of randomness can be illustrated by a simple numerical example. Readers who do not care for such calculations can skip to the end of the section.

Imagine a species where each male and each female mates once only. There are distinct breeding seasons and males or females can mate either in the first or the second breeding season after that of their birth. 50 % of the males and 60 % of the females mate when one season old, 50 % the males and 40 % of the females mate when two seasons old. An individual only mates once in a lifetime. The result of mating is always a litter of two offspring, one male and one female. We consider a closed population within which mating takes place 'at random' (in some sense). For convenience, we take the population to be large.

Under the assumptions given, a male always has two female cousins. His female cousins are born either in the season of his own birth or in the one before or the one after. The probabilities that a male have female cousins who are younger, older or the same age as himself can easily be worked out from the assumptions. The result is shown in table 21.

What is the probability that two individuals mate in our imaginary species? Under random mating this probability does not depend on their relationship, but it will presumably depend on the age difference between them. If the female is born in the breeding season following that of the male's birth, she is not mature in the first breeding season in which he can mate. When the female is 2 years old the male has passed the reproductive period. On the other hand, a male and female born in the same breeding season will both be of reproductive age in two seasons and thus presumably have more chance of mating with each other than in the case where the male is born one season before the female.

TABLE 21. IMAGINARY SPECIES: FEMALE FIRST COUSINS OF MALE, BY AGE

ages of cousins relative to male	probability
both younger	0·06
one younger, one older	0·13
both older	0·06
one younger, one same age	0·25
one older, one same age	0·25
both same age	0·25

Assumptions. Each male and female mates, producing simultaneously one male and one female offspring. No mortality before mating. Both male and females mate either when 1 or 2 seasons old. 50 % of the males and 60 % of the females mate when 1 season old; the remainder when 2 seasons old. Large population.

TABLE 22. IMAGINARY SPECIES: PROBABILITY OF MATING BETWEEN A PARTI-
CULAR MALE m AND A PARTICULAR FEMALE f BORN IN SAME SEASON

(first calculation)

probability that m mates when 1 season old with any female	0·5
probability that f mates when 1 season old with any male	0·6
probability that both these events occur	$0·5 \times 0·6 = 0·3$
conditional probability that, if both these events occur, they mate with each other	$1/M$
probability that they mate with each other when 1 season old	$0·3/M$
probability that they mate with each other when 2 seasons old	$0·2/M$
total probability that they mate with each other	$0·5/M$

Note. M = number of marriages per year in the isolate. For characteristics of species, see note to table 21.

The assumptions stated so far are not sufficient to determine the probabilities of mating. We have specified only how likely mating is for males and females at each age (i.e. in successive seasons after their season of birth). The probabilities that two individuals of given age difference mate are determined if, for example, we make the additional assumption that among those who mate in any season any two are as likely to pair off as any other two. This involves thinking of the process of mating as made up of two stages, a decision to mate and the selection of a partner. Under the assumptions stated earlier, the number of individuals mating every season would be equal to the number born in a season. (We take the population to be large enough to render chance fluctuations negligible.) Suppose that M males and M females are born every season. There will thus be M males and M females mating in every season. The process of pairing one male with one female may be carried out in $M!$ ways. Of these there are $(M-1)!$ ways which involve the mating of a particular one of the M males with a particular one of the M females. The probability of this event is, therefore, $(M-1)!/M! = M^{-1}$. Table 22 shows how the assumptions stated may be used to compute the probability of mating between two individuals born in the same season.

If the male was born one season before the female the probability of their mating, computed by the same method, turns out to be $0 \cdot 3/M$. If the female was born one season before the male, the result is $0 \cdot 2/M$. These probabilities may be combined with the results of table 21 to compute the average number of first-cousin matings per male which will occur in an isolate with M matings per year, i.e. the probability that in an isolate of this size a male mates with a first cousin (table 23).

TABLE 23. IMAGINARY SPECIES: NUMBER OF FIRST-COUSIN MATINGS PER MALE IN ISOLATE

male older than female by	av. no. of female first cousins per male	probability of mating	av. no. of first-cousin matings occurring per male
1 season	$\frac{1}{2}$	$0 \cdot 3/M$	$0 \cdot 15/M$
0 season	1	$0 \cdot 5/M$	$0 \cdot 5/M$
− 1 season	$\frac{1}{2}$	$0 \cdot 2/M$	$0 \cdot 1/M$
total	2	—	$0 \cdot 75/M$

Notes. av. = average, M = number of matings per year in isolate. For characteristics of species, see note to table 21.

The average number per season of matings between first cousins may be computed as follows:

av. no. of first-cousin matings

$$= \text{no. of male births} \times \text{av. no. first-cousin matings per male birth}$$
$$= M \times 0 \cdot 75/M = 0 \cdot 75.$$

Under the assumptions made there will, on average, be three first-cousin matings in 4 years. This result is independent of M, which measures the size of the isolate. In a larger isolate there will be more pairs of first cousins than in a small one, but under random mating the probability that a pair of first cousins actually mate will be correspondingly less. The result is, of course, not restricted to first-cousin matings. A similar statement applies to consanguineous matings of any kind.

The method developed in this example for specifying what is meant by random mating is flexible and can cope with a good deal of biological reality. There is one aspect of reality (especially important for realistic calculations concerning the human species) which this model, as so far stated, cannot cope with, namely age preferences in the choice of a mate. The method of calculation used in table 22 implies that the age at which a male mates has no influence on the age of the mate he selects, i.e. that the ages of male and female in mating pairs are independent. In the example given, the distribution of 100 matings by the ages of the participants would be as follows:

	age of male		
age of female	1 season	2 seasons	all ages of males
1 season	30	30	60
2 seasons	20	20	40
all ages of females	50	50	100

The method of table 22 may be modified to accommodate dependence between the ages of males and the ages of their mates. Suppose, for example, it were specified that the distribution of the ages of mates in our imaginary species were as follows:

	age of male		
age of female	1 season	2 seasons	all ages of males
1 season	40	20	60
2 seasons	10	30	40
all ages of females	50	50	100

Given such a distribution, the probability of a mating between a particular male and a particular female with a given age difference can be worked out by a modification of the previous argument, as is shown in table 24. The randomness in 'random mating' now consists in the assumption that any of the pairs which can be formed of the mating males and females *with a given combination of ages* are equally likely to occur. For the special case of independence, this method gives the same results as that of table 22.

TABLE 24. IMAGINARY SPECIES: PROBABILITY OF MATING BETWEEN A PARTICULAR MALE m AND A PARTICULAR FEMALE f BORN IN SAME SEASON

(second calculation)

probability that m mates when 1 season old with any female 1 season old	0·4
probability that f mates when 1 season old with any male 1 season old	0·4
probability that both these events occur	$(0·4)^2 = 0·16$
conditional probability that if both these events occur they mate with each other	$1/0·4M$
probability that m and f mate when 1 season old	$0·16/0·4M = 0·4/M$
probability that m and f mate when 2 seasons old	$(0·3)^2/0·3M = 0·3/M$
total probability that m and f mate with each other	$0·7/M$

Notes. M = number of marriages per year in the isolate. For characteristics of species, see note to table 21.

If the method of table 24 is applied to compute also the probability of mating for the cases where the male is one season younger or one season older than the female, the results may be used to recalculate the average number of first-cousin marriages per male birth. It is found that with this amount of dependence between the ages of mating pairs, there would be 0·85 first-cousin matings per year in the isolate, not 0·75 as computed earlier.

In man there is no division of time into successive breeding seasons. We have to use an arbitrary subdivision of the continuous stream of time. In principle this causes no difficulty. Provided we use intervals sufficiently small, our model can approach reality more and more closely. Unfortunately, if we let the time intervals get smaller while keeping the population size finite, difficulties arise. The assumption, that among those mating in a time interval any male and any female are as likely to pair off as any other two, becomes unworkable. If we want to operate with continuous time in a random mating model of this kind, we must have an infinitely large population. Fortunately the problem can be side-stepped. (For details, see Appendix I.) All quantities concerned with the probabilities of consanguineous marriage can be calculated provided a probability can be assigned to the mating of a male and female separated by a given age difference.

For human populations we can determine numbers appropriate to represent the probabilities that a newborn boy and a newborn girl separated by some given age difference should eventually marry. Such probabilities can be determined from two conditions. First, the probabilities for different age differences will bear the same relationship to each other as the frequencies of marriage with those age differences in the population. For example, if there are three times as many marriages between spouses of the same age as marriages where the groom is 10 years older than the bride, the probability assigned to the eventual marriage of a boy and a girl born at the same time will be three times as great as for a boy and a girl born 10 years later. Secondly, the absolute level of such probabilities must be such as to ensure that the right proportion of people are married. To exaggerate, if the probabilities are such that only one in a hundred ever get married, they are clearly wrong and they are equally wrong if they result in five marriages per person. The precise level required can be worked out.

It is therefore appropriate to make this the defining characteristic of random mating, i.e. random mating will be said to hold if *the chance that two newborn individuals marry depends only on the interval between their dates of birth*. This definition avoids the unrealistic notion that there is a fixed group of women each of whom a man is equally likely to marry. Some such extension of the conventional concept of random mating is unavoidable if overlapping generations are to be dealt with adequately. From the point of view of the geneticist, it would be appropriate to substitute 'dates of conception' for 'dates of birth' in the definition given, but the statistical data are all in terms of birth.

3. The four types of first-cousin marriage

The previous section showed how one may deduce from the demographic characteristics of a simple, imaginary species the numbers of consanguineous matings which would occur under a system of random mating. Such calculations can be made even if the species under consideration has more complex demographic characteristics, but a different kind of mathematical formulation is required. It is not necessary to make any simplifying assumptions about ages at which reproduction takes place, variation between matings in numbers of offspring or their distribution by sex. For man the results of such calculations can be compared with data on the actual occurrence of consanguineous marriages.

The mathematical apparatus by which the required formulae are derived is the main contribution of the present work and is set out in Appendix I. The calculations themselves are very simple; they are explained in Appendix II.

The general character of such calculations and comparisons between the observed numbers of consanguineous marriages and those that would occur under random mating can be presented without entering into the mathematical background. In this section, the example of the four types of first-cousin marriage is described in detail to make clear the limitations of the method. Other comparisons are dealt with more briefly in §§ 4 and 5.

TABLE 25. PROPORTION (%) OF CONSANGUINEOUS MARRIAGES
AMONG ALL MARRIAGES

population	period	first cousins	1½ cousins*	second cousins
Belgium	1920–24	0·7	0·2	1·0
	1955–59	0·2	0·1	0·7
England	ca. 1870–1920	0·6	0·005	0·1
Italy	1903–23	3·5	0·7	1·8
Japan	ca. 1930–53	4·1	1·2	1·7
Brazil	1950–51	9·5	2·7	5·0
Senegal	ca. 1930–57	11·3	5·0	11·3

Belgium: all Catholic marriages. England: parents of hospital patients. Italy: Milan Archdiocese. Japan: cities of Hiroshima, Nagasaki and Kure. Brazil: Petrolina diocese in State of Pernambuco. Senegal: Fouta Valley.

* 1½ cousins = first cousins once removed.

Sources. Twiesselmann *et al.* (1962), Bell (1940), Serra & Soini (1959), Schull (1958), Freire-Maia (1957), Cantrelle (1960).

As a result of geneticists' interest in the topic, a considerable body of data on consanguineous marriages has been accumulated in the last few years. Some figures are shown in table 25.

The largest source of data has been the marriage dispensations of the Roman Catholic Church. By Canon Law marriages between close relatives require a special dispensation in each case. The records of these dispensations are carefully kept and have been made available to research workers in many countries. Another body of information is a consequence of the dropping of the atomic bombs on Hiroshima and Nagasaki. In investigating the effects of the radiation to which people in those areas were exposed, data have been collected on consanguinity among the parents of newborn children and these studies have been linked with information on this topic from other parts of Japan, the collection of data being facilitated by the family registration system which exists in Japan. Some data from special surveys are also shown in table 25, for example, those derived from a survey of hospital patients in England (these, the only available data for this country, are probably rather incomplete). The frequencies of consanguineous marriages show a great variety. For first-cousin marriages the frequencies range from well under 1 %, which is probably characteristic of much of present-day Europe and the United States, to over 10 %, perhaps to be found in much of Africa and many communities in other parts of the world, for example India. Great heterogeneity may be found within one country; for example, in large parts of Brazil consanguineity is at European levels while other regions, such as the area shown in table 25, display much higher frequencies.

We proceed to discussion of the four types of marriage between first cousins. The four types are distinguished by the sex of the siblings through whom the cousins are related, as shown in figure 9. The terms applied in this paper to the four types have been adapted from the usage of social anthropologists.

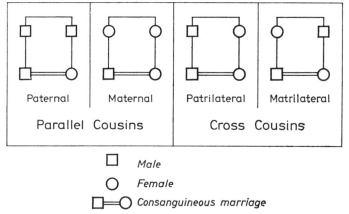

FIGURE 9. Types of first-cousin marriage.

One can work out how many of each type of cousin marriage there would be if the frequency of the various types were determined by random mating. To do this, one needs first to know the number of potential marriages and their distribution by the interval between the dates of birth of the spouses. For example, we may take the case of paternal parallel cousins. Consider a randomly selected newborn boy. We need to know on average how many girls are born who are his paternal parallel cousins and when they are born, i.e. how many are born in the year of his birth, how many the previous year, how many in the following year and so forth.

Estimates of these quantities can be obtained. The *number* of a boy's paternal parallel cousins is determined by the number of brothers which his father has and the numbers of daughters which these brothers have. From the distribution of families by size and other data calculations can be made on the numbers of cousins an individual is likely to have. (Unfortunately, for lack of information it is necessary to assume that there is no correlation between the vital rates of relatives.) The *interval* between the birth of a boy and that of a paternal parallel girl cousin is determined by three quantities: (1) the age of the boy's father at his birth; (2) the length of time between the birth of the father and that of his brother; and (3) the age of this brother at the birth of his daughter. There is a good deal of information on the ages of parents at the birth of their children. The distribution of the intervals between the births of brothers and sisters is a much more difficult matter and a fairly crude method (Appendix III) has been used to cope with this problem. (Another weakness is the need, in the absence of better information, to assume that the intervals between the births of brothers and those of brothers and their children are independent.) Calculations can thus be made concerning *potential* cousin marriages of each type.

The next step is to compute how many of the potential marriages would, under random mating, actually occur. We have taken random mating to mean that the probability for a new-born boy and girl to marry is a function of the interval between their dates of birth. How these probabilities can be obtained has already been explained.

All quantities entering into the calculations presented have been chosen so as to be broadly appropriate to populations of high fertility and mortality. The results will be applicable to most of the existing data on consanguineous marriages, even for European populations, because most of the data so far collected are only slightly affected by the rapid declines in mortality and fertility which in most of Europe set in during the last quarter of the nineteenth century. The calculations assume, in the first instance, that we are dealing with a closed population which will be called an 'isolate'. Adjustments can be made for migration in and out of the isolate (in view of the possible selective effects of migration, it is perhaps not a legitimate extension of the term 'random mating' to apply it even in this case). Also as a first approximation, we assume that the population is stationary in the technical demographic sense. Again, it is possible to generalize the results to cases of growth (or decline) at a constant rate and, with rather more trouble, to more complex situations of changing vital rates.

To return to the four types of first-cousin marriage, it may be supposed at first sight, and was at one time commonly supposed, that all four types should be equally frequent. In fact, it is readily seen that one would expect more matrilateral cross-cousin marriages than patrilateral cross-cousin marriages, for the same reason that uncle–niece marriages are more frequent than aunt–nephew marriages. For, since on average mothers are younger at the birth of their children than fathers, a man will on average have matrilateral cross-cousins who are slightly younger and therefore of a suitable age for marriage, whereas girls who are his patrilateral cross-cousins will on average be somewhat older.

Some actual data may be seen in table 26. It turns out that this expectation is fulfilled, i.e. that matrilateral cross-cousin marriages are, in fact, more frequent than patrilateral cross-cousin marriages. However, another feature appears regularly in the data. Maternal parallel marriages are considerably more frequent than paternal parallel ones; maternal parallel marriages are, indeed, often even more frequent than matrilateral cross-cousin ones. The high frequency of maternal parallel cousin marriages has been something of a puzzle. Very different explanations have been suggested. One view, a European one, maintains that because of the Oedipus complex men prefer to marry women who resemble their mothers. Another explanation, put forward in Japan, is based on the idea that many people think that heredity in some way has more influence through the father than through the mother. People who hold this view and also fear that marriages between relatives might produce defective children may be less reluctant about marriages between maternal parallel cousins than between paternal parallel cousins.

The predominance of maternal parallel cousin marriage is not universal; in a very extensive set of data from various parts of Brazil (Freire-Maia 1961) the paternal parallel is the most frequent type of first-cousin marriage. The Brazilian data have a number of other peculiarities and it is hardly possible to discuss them without full knowledge of the demography of Brazil. The data may be biased by selection. They are based on Catholic marriage dispensations; but it has been shown (Mortara 1948) that, though the population of Brazil is almost entirely Catholic,

TABLE 26. DISTRIBUTIONS OF FIRST-COUSIN MARRIAGES BY
TYPE (PERCENTAGES)

country	no. of marriages on which percentages are based	parallel		cross		all types
		pat.	mat.	pat.	mat.	
Germany	1327	21	29	23·	27	100
Austria	822	18	33	21	28	100
Italy	4384	22	28	21	29	100
Japan	689	22	33	18	27	100

Germany: dioceses of Mainz (1890–1935) and Trier (1901–35). Austria: archdiocese of Vienna (selected years during 1901–31). Italy: dioceses of Parma and Piacenza (1851–1957) and Reggio nel Emilia (1927–57). Japan: cities of Hiroshima, Nagasaki and Kure (pregnancy terminations of 1948–52).

Sources. Ludwig (1949), Orel (1932), Barrai *et al.* (1962), Morton (1955).

church marriages comprise fewer than two-thirds of the socially recognized unions, as revealed by the census. In addition to the Brazilian statistics, table 26 does not include the data for Israel, where the situation is clearly abnormal, and studies where a large proportion of first-cousin marriages could not be classified by type. With these exceptions, all substantial bodies of data on the distribution of first-cousin marriages by type are included. (At least 500 first-cousin marriages are needed for reasonably adequate determination of the frequencies discussed.)

The predominance of maternal parallel cousin marriage accords with theoretical expectation on our concept of random mating. How the differences in theoretical frequencies arise is illustrated in figure 10, which displays a number of normal curves. Four of these are designed to show the probability that a randomly selected newborn boy has a girl cousin of a particular type born at various periods before or after his own birth. These curves were not derived from actual data on pairs of cousins, but computed from information on the intervals between the births of sibs and on ages at paternity and maternity. These curves are inevitably to some extent speculative because of the imperfections of the available information. Their true shape is probably something like normal. For convenience they have been taken to be exactly normal.

Figure 10 also shows the probability that two newborn individuals should marry each other, expressed as a function of the interval between their dates of birth. The absolute value of these probabilities is inversely proportionate to the size of the isolate, which is reflected in the constant M. In a small isolate, marriage between two randomly selected individuals is more likely than in a large isolate. This curve has been given a normal shape by deliberate distortion. The distortion probably makes very little difference to the result; but if all the curves are normal the integrations required come out easily.

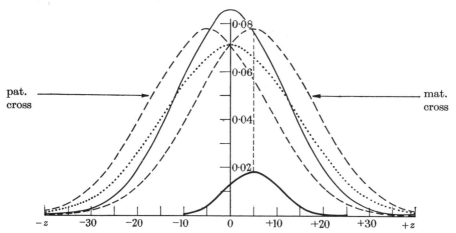

FIGURE 10. Distribution of intervals between births of first cousins (groom 5 years older).*
Probability (per year) of birth of girl cousin z years after birth of boy.

    . . . ., Paternal parallel
    ———, maternal parallel
    – – –, cross-cousin

——, $M \times$ probability that marriage takes place;
(M = no. of marriages per year)

If the two curves for parallel cousin marriages are compared, it is seen that the one for paternal parallel cousins is farther spread out. This is because the reproductive life of men is longer than that of women, i.e. there is a greater variation in age at paternity than age at maternity. The consequence is that paternal parallel cousins are likely to be born farther apart in time and hence less likely to marry each other.

It can be seen now why maternal parallel cousin marriages are very frequent. The reason is that, in this case, the potential spouses are likely to be born closer together than with other types. Matrilateral cross-cousin marriage is also common because in this case the age difference between the potential spouses corresponds most often to the age difference which maximizes the probability of marriage.

Figure 10 relates to a population where the bridegroom is on average five years older than the bride. The position of the various curves for different types of cousins in relation to each other depends upon the average age difference between the bride

* In the notation of Appendix I the four larger curves represent
$$qU\bar{n}[2\pi V_\xi(h)]^{-\frac{1}{2}} \exp\left[-\{z-\mu_\xi(h)\}^2/2V_\xi(h)\right].$$
The small curve represents
$$(pq)^{-1} U^2[2\pi V(a)]^{-\frac{1}{2}} \exp\left[-\{z-\mu(a)\}^2/2V(a)\right].$$
$\mu(a) = 5$, $U = 0{\cdot}25$, $\bar{n} = 20$. Variances as in Appendix II.

and groom, since this is the main factor in influencing the average difference between maternal and paternal ages at the birth of children.

Given the information embodied in the curves in figure 10, it is possible to work out how first-cousin marriages would be distributed between the different types under random mating. The results of such calculations for different levels of the difference between groom's age and bride's age may be seen in table 27. The main effect of a large age difference on the theoretical frequencies is a higher ratio of matrilateral to patrilateral cross-cousin marriage. (This point is further discussed below.)

TABLE 27. THEORETICAL DISTRIBUTIONS (%) OF FIRST-COUSIN
MARRIAGES BY TYPE

(closed stationary population)

mean excess of groom's age over bride's age (years)	parallel		cross		all types
	pat.	mat.	pat.	mat.	
0	23·2	27·1	24·9	24·9	100
2	23·3	27·1	24·3	25·3	100
5	24·1	27·5	21·1	27·3	100
10	25·6	27·6	12·3	34·4	100

For method of computation, see Appendix II below, p. 165.

In Japan the mean excess of groom's age over bride's age is of the order of 6 years. In Europe it is usually between 2 and 5 years. Of the countries in table 26, Italy is at the top end of this range, Germany near the bottom. (It must, however, be remembered that the data in table 26 cover only small areas within each country.) Comparison between tables 26 and 27 shows that the theoretical frequencies differ in the right direction, with maternal parallel and matrilateral cross-cousin marriages as the most common. Also the ratio of matrilaternal to patrilateral cross-cousin marriage increases as one passes from top to bottom of table 26 in a way which corresponds to the theoretical expectation revealed by table 27. However, the differences between the theoretical frequencies do not seem as big as in the actual data. In particular, the excess of maternal parallel over paternal parallel marriages in the theoretical frequencies is not big enough.

There are various possible explanations of this discrepancy. In the first place, there are defects in the data. Persons contracting marriage are much less likely to be ignorant about the identity of their mother than of their father. One reason is illegitimacy. Something of its effects may be seen in the data for Vienna analyzed in admirable detail by Orel (1932). He recorded thirty-one maternal parallel cousin marriages, but only two paternal parallel marriages, where one or other of the spouses was stated to have been an illegitimate child. Many of the latter may well be unrecorded. The total number of marriages recorded in Orel's data as involving an illegitimately born partner is obviously too low. Moreover, death of the father when a child is young (or indeed before birth) is far more frequent than the loss of the mother. Again, as a result of separation or desertion, a child is much more likely to lose contact early in life with the father than with the mother. For all these reasons it is probable that paternal parallel cousin marriages are most under-reported, while the reporting of maternal parallel cousin marriages is probably the most complete. These differences in completeness will probably be greater in urban mobile populations than in rural ones.

A second possible explanation for the discrepancy between the expected and the actual values lies in the various imperfections and unrealistic assumptions entering into the calculations. A third possible explanation is to be found in migration. If men are more likely than women to move out of the community of their birth, and especially to move far enough to lose contact with their relatives, maternal parallel cousin marriages will be more frequent than paternal ones. A correction (see Appendix II (a)) corresponding to the amount of overseas migration of men and women from Sweden in 1900–10, as it happens, yields a distribution of first-cousin marriages by type which comes very close to the observed values in table 26.

The correction works out as follows for the case where the groom is on average 5 years older than the bride:

| | parallel | | cross | | all |
	pat.	mat.	pat.	mat.	types
without migration	24·1	27·5	21·1	27·3	100·0
with migration	22·3	29·4	21·0	27·2	99·9

For various reasons the agreement between the facts and expected frequencies corrected for emigration must be regarded as largely coincidental. The effects of migration vary with time and place and are rather complex (e.g. local migration over short distances may have very different characteristics from overseas migration). What the calculation does show is that one cannot be sure from the existing data that any factors other than the availability of potential spouses substantially influence the relative frequencies of different types of first-cousin marriage.

The values of table 27 are for populations which are not only closed, but also stationary. Adjustments to take account of growth (or decline) make very little difference unless both the rate of population change and the mean excess of groom's age over bride's age are very great. Growth tends to increase the proportion of paternal parallel cousin marriages at the expense of the maternal parallel type.

As already mentioned, the mean excess of the age of the groom over the age of the bride in Europe is mostly of the order of 2 to 5 years. In many non-European populations the excess is considerably greater than this. When the excess is large, matrilateral cross-cousin marriage becomes the predominant type under random mating and patrilateral cross-cousin marriage becomes rare. Why this is so may be understood from figure 11.

As the age difference between the spouses is increased, the highest points of the curves representing the distribution of intervals between potential spouses will be pushed apart. The curve for matrilateral cross-cousin marriages will continue to correspond to its maximum point to the curve representing the chances that a potential marriage actually takes place. But the curves for other types of cousin marriage will now be much further removed from the curve of marriage probabilities.

In particular, in the case of patrilateral cross-cousins the girl is likely to be much older than the man, so that marriage between them is highly improbable.

There are apparently very few available statistics on the four types of cousin marriage among pre-industrial populations. It is not really possible to test whether the predominance of matrilateral cross-cousin marriage expected on theoretical grounds is confirmed by the facts. Some statistics are shown in table 28.†

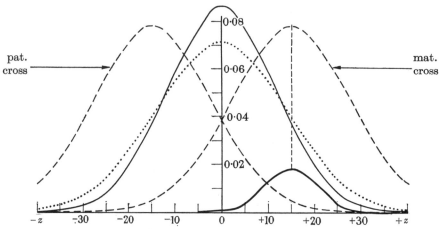

FIGURE 11. Distribution of intervals between births of first cousins (groom 15 years older). Annotations as for figure 10. Equations of curves same as for figure 10, except that $\mu(a) = 15$.

TABLE 28. DISTRIBUTION (%) OF FIRST-COUSIN MARRIAGES BY TYPE

(large excess of groom's age over bride's age)

area	no. of first-cousin marriages	parallel		cross		all types
		pat.	mat.	pat.	mat.	
Bombay	282	26	28	15	31	100
Senegal	46	33	17	9	41	100
Bechuanaland	89	15	21	11	53	100
theoretical distributions						
excess of groom's age { 8 years		25	28	16	31	100
12 years		26	27	9	38	100
16 years		25	24	5	47	100

Bombay: Moslems and Parsees. Senegal: Fouta Valley. Bechuanaland: Tswana. Calculation of theoretical distributions as in Appendix II (*a*).
Sources. Sanghvi *et al.* (1956). Cantrelle (1960). Schapera (1957).

The figures for India are part of an investigation of a number of endogamous groups (Sanghvi *et al.* 1956). Among the Hindu groups studied the overwhelming majority of first-cousin marriages were of the matrilateral cross-cousin type which is highly favoured by tradition. Paternal parallel cousin marriages are strictly forbidden. The Bombay figures in table 28 are for Moslems and Parsees, among

† Data covering 54 first-cousin marriages published by Kang & Cho (1959) are relevant here and in agreement with the theoretical frequencies. They have been omitted because the other demographic data in the paper are clearly highly unreliable.

whom, it seems, no special rules governing the various types of first-cousin marriage are observed. The lumping together of Moslems and Parsees to obtain adequate numbers is perhaps dubious, since the demographic characteristics of the Parsees are somewhat distinct from those of other Indian populations (Chandra Sekar 1948). However, in regard to the frequencies of consanguineous marriages of different types (not only first-cousin marriages), they appear similar to the Moslem groups studied.

From the Bechuanaland study (Schapera 1957), the figures for 'commoners' were selected for inclusion in table 28. The figures for 'nobles' relate to a population which is not closed, since nobles take wives from outside their own circle; Schapera remarks that the distribution of first-cousin marriages among nobles corresponds with that which would be expected as a result, i.e. there is a very high proportion of paternal parallel cousin marriage.

The frequencies in table 28 seem in good agreement with the theoretical expectations. In the Indian groups covered by the data, the mean excess of groom's age over bride's age is probably of the order of 8 years (Agarwala 1957) and for the African figures it may well be a good deal higher, say 12 years or more (e.g. Gibson (1958) and recent official studies† in the former French African colonies). Unfortunately the numbers of first-cousin marriages are very small and the agreement between theory and observation may be largely coincidental. It is perhaps of greater significance that anthropologists have observed that matrilateral cross-cousin marriage is the preferred type of cousin marriage in many preliterate societies. This preference fits in well with a system where the groom is normally a good deal older than the bride.

4. DISTRIBUTION BY PEDIGREE TYPE OF CONSANGUINEOUS MARRIAGES OTHER THAN THOSE BETWEEN FIRST COUSINS

The numbers of uncle–niece and aunt–nephew marriages are small. Table 29 compares observed numbers of the various types in two studies with theoretical frequencies under random mating. Theoretical frequencies were computed by distributing the combined total for both uncle–niece and aunt–nephew marriages according to the theoretical proportions. For Italy‡ these proportions were calculated using constants (for mean ages at parenthood) known to hold for the population under study. The figures show a very good agreement between the observed distribution and that expected under random mating. Data from Austria (Orel 1932) display much the same features as the Italian data.

For Brazil table 29 also shows a close correspondence between observed and theoretical values. However, the correspondence is close mainly because, in the

† *Présidence du Conseil, Rapport du Haut-Comité Consultatif de la Population et de la Famille*, Tome v, 'Les populations des territoires d'outre-mer' [No date]; also Blanc (1962).

‡ During the revision of the present paper, data from Barrai *et al.* (1962) were substituted for the partial results published earlier from the Italian study conducted by Cavalli-Sforza and his colleagues. The data given in Barrai *et al.* (1962) are by far the most extensive body of material now available on the distribution of consanguineous marriages by pedigree type. Unfortunately it was not possible to take account of the discussion in this paper.

absence of knowledge about the facts, the most important constant (the difference between the ages at paternity and maternity) used in computing the theoretical frequencies was selected in such a way as to produce a good fit to the data. Correction for population growth would have improved the fit further. The Brazilian data on the distribution of consanguineous marriages by pedigree type are rather different from those so far reported elsewhere. Some of the special problems of the Brazilian data were mentioned in § 3 above. The theoretical frequencies for the Brazilian data in table 29 are intended to demonstrate, by contrast with those for Italy, that very different distributions of uncle–niece and aunt–nephew marriages over the various pedigree types can result under random mating. The distribution is highly sensitive to the mean difference between ages at paternity and maternity (this is essentially the same quantity as the mean excess of groom's over bride's age at marriage). If this mean difference is large, aunt–nephew marriages will be rare.

TABLE 29. OBSERVED AND THEORETICAL NUMBERS OF UNCLE–NIECE AND AUNT–NEPHEW MARRIAGES IN TWO STUDIES

	Italian study		Brazilian study	
relationship	observed numbers	theoretical numbers	observed numbers	theoretical numbers
man marries his:				
brother's daughter	36	39·4	28	25·0
sister's daughter	67	65·6	27	27·5
woman marries her:				
brother's son	3	4·1	5	7·0
sister's son	9	5·9	6	6·5
total	115	115·0	66	66·0

Theoretical numbers computed as described in Appendix II(a) for closed stationary population with constants as follows:
Italy $\mu_M = 30$, $\mu_P = 35$, hence $\mu(a) = 5$ [see Barrai *et al.* (1962) p. 367].
Brazil: $\mu_M = 30$, $\mu_P = 33$, hence $\mu(a) = 3$.

Sources. Barrai *et al.* (1962), table 1; Freire-Maia (1961), table 2.

Table 30 compares observed numbers from Italian data with theoretical frequencies for marriages between $1\frac{1}{2}$ cousins (i.e. first cousins once removed). There are 16 pedigree types of $1\frac{1}{2}$-cousin marriages; they fall into two groups according to whether there are two persons in the ancestry of the bride and only one in the ancestry of the groom, or vice versa. (The term 'ancestry' denotes the ancestors through whom the spouses trace their common descent to one couple; it does not include that couple.) Apart from the total numbers in the ancestry of bride and groom, the sex of the ancestors enters into the computation of the frequencies expected under random mating. The computation requires classification of pedigree types by: (1) the total number of males in the combined ancestry of bride and groom, and (2) the difference between the numbers of males in the ancestry of the bride and in the ancestry of the groom. When two pedigrees are identical in these two respects their theoretical frequencies are necessarily the same.

The theoretical numbers in table 30 were obtained by distributing the total of $1\frac{1}{2}$-cousin marriages in accordance with theoretical proportions which were computed without use of any information derived from the observations. The agreement between observation and theory is very satisfactory. The theory correctly predicts the distribution between (i) those types where the bride, and (ii) those where the groom belongs to the younger generation (i.e. has an ancestry of two). The total numbers for the former are in all countries much more frequent than the latter for the same reason that uncle–niece marriages are much more frequent than aunt–nephew marriages. However, the theory also correctly predicts that some types

TABLE 30. OBSERVED AND THEORETICAL NUMBERS OF $1\frac{1}{2}$-COUSIN
MARRIAGES* IN AN ITALIAN STUDY

no. of males in combined ancestry, α_ξ	excess of males in longer line of ancestry,† ϵ_ξ or $-\epsilon_\xi$	two in ancestry of bride		two in ancestry of groom	
		observed	theoretical	observed	theoretical
0	0	144	163·6	32	37·9
1	+1	98	103·4	13	23·2
		100	103·4	34	23·2
1	−1	262	299·6	97	103·4
2	0	222	203·3	60	65·5
		200	203·3	61	65·5
2	+2	64	65·5	19	14·4
3	+1	195	136·3	52	41·5
total		1285	1278·4	368	374·6
grand total				1653	1653·0

Theoretical numbers computed as in Appendix II(a) for closed stationary population with constants: $\mu_M = 30$, $\mu_P = 35$, hence $\mu(a) = 5$. Data from Barrai et al. (1962).

* $1\frac{1}{2}$ cousins = first-cousins once removed.

† column gives ϵ_ξ where there are two in ancestry of bride, $-\epsilon_\xi$ where there are two in ancestry of groom.

where the groom is in the younger generation will be as frequent as some types where the bride is in the younger generation though the pedigrees suggest no resemblance. Each side of figure 12 shows three pedigrees for which the theoretical frequencies are identical. It is to be noted also that some of the features in which the observed and theoretical frequencies agree derive from taking into account variances, as well as means, of the intervals between the dates of birth of potential spouses. The excesses of the observed over the theoretical frequencies in the bottom line of the table, and the deficiencies in the top line can scarcely be regarded as chance fluctuations in view of the situation with respect to second cousins.

The data from Austria (Orel 1932) and Japan (Morton 1955) are similar, as far as can be seen from small numbers, to those from Italy in table 30. The Brazilian data show some differences, as well as considerable similarities; they can be 'fitted' by means of suitable constants for ages at parenthood, such as those used for table 29.

There are sixteen types of marriages between second cousins. Their frequencies in the Italian study are compared in table 31 with theoretical frequencies expected under random mating. It is clear at a glance that there are systematic discrepancies between the observed and theoretical values. The variation in the theoretical values does not account for much of the variation in the observations. The observed numbers fall short of the theoretical where the ancestry is mostly female; the discrepancies are in the other direction when the ancestry is mainly male. The same

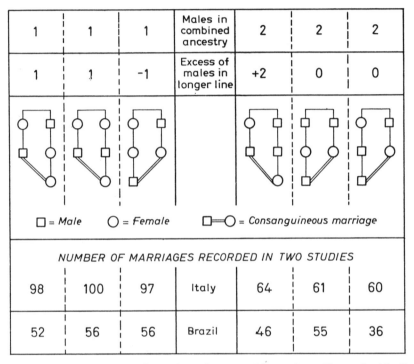

1	1	1	Males in combined ancestry	2	2	2
1	1	-1	Excess of males in longer line	+2	0	0

□ = Male   ○ = Female   □━○ = Consanguineous marriage

NUMBER OF MARRIAGES RECORDED IN TWO STUDIES						
98	100	97	Italy	64	61	60
52	56	56	Brazil	46	55	36

FIGURE 12. Selected pedigrees of marriages between $1\frac{1}{2}$ cousins. (On each side there are 3 pedigrees for which the theoretical frequencies are identical.) *Sources.* Barrai *et al.* (1962, table 1), Freire Maia (1961, table 3).

phenomenon is observed in data from Austria, Japan and Brazil. In all cases marriages of second cousins when all intermediate ancestors are male are over twice as numerous as expected under random mating. A smaller discrepancy of the same sort was noted above for $1\frac{1}{2}$ cousins.

There may be systematic departures from random mating, though one would expect second-cousin marriages to be less governed by 'non-random' circumstances than those between closer relatives. Various forces which might explain the distribution of second-cousin marriages by type may be suggested (for example, special migration patterns). However, it is tempting to attribute at least a part of this phenomenon to under-reporting. There is other evidence, which is reviewed in §5 below, that the recording of second-cousin marriages is very incomplete. It seems likely that some types are less completely recorded than others. As already noted by Orel (1932), and recently confirmed by the extensive analyses of Barrai *et al.*

(1962), the observed frequencies of the types vary not only with the number of male ancestors, but also according to whether female ancestors occur among the parents or grandparents of the spouses. Marriages between the grandchildren of two sisters are conspicuously rare. The explanation put forward by Orel seems plausible, at least for European countries. His theory is that in such cases there is an increased chance that the spouses may not be aware of their relationship. There must be many people who do not know both their grandmothers' maiden names. A few decades

TABLE 31. OBSERVED AND THEORETICAL NUMBERS OF MARRIAGES OF SECOND COUSINS IN AN ITALIAN STUDY

no. of males in combined ancestry, α_ξ	excess of males in ancestry of bride, ϵ_ξ	observed	theoretical
0	0	252	412
1	−1	229	341
		239	341
1	+1	260	408
		253	408
2	+2	310	373
2	0	398	373
		312	373
		323	373
		328	373
2	−2	278	268
3	−1	443	319
		426	319
3	+1	507	370
		432	370
4	0	774	343
total		5764	5764

Theoretical numbers computed as in Appendix II(a), with $\mu(a) = 5$. Data from Barrai et al. (1962), table 1.

ago the proportion of grandparents who survived until their grandchildren could know them well was much lower than it is now. Even if the spouses suspect the relationship between them, there may be no circumstances likely to arouse the suspicion of others. They may then prefer, Orel suggests, to avoid complications by concealing their relationship.

The argument just sketched is not inconsistent with the supposition, which was stated above, that in the case of first-cousin marriages it is the paternal parallel type which is most inadequately reported. For one's links with one's parents are generally based on personal acquaintance and in this respect the link with the mother is stronger. Links with remoter ancestors are (except in some pre-literate societies) more likely to be traced through the paternal line and, in Europe, to be remembered in particular through the identity of surnames.

5. Some implications of random mating

One consequence of the assumption of random mating is that the rate (per unit of time) at which consanguineous marriages of a given type occur is independent of the size of the isolate, as illustrated in § 2. For example, we might calculate that in a population with given demographic characteristics there will be, on average, two first-cousin marrriages per decade. This result will hold whether there are ten marriages per year in our isolate or 1000. The reason is that while in a larger isolate there will be more potential marriages between cousins, the probability that a potential marriage actually takes place will be less because the number of potential partners available for marriage to any one person is larger. While the frequency of consanguineous marriages of any type under random mating does not depend on the size of the isolate, it does depend on its demographic characteristics. For example, if every marriage produces exactly three births, there will be fewer consanguineous marriages than if nine out of ten marriages produce only one birth, but one in ten marriages produces twenty-one births.

Table 32. Rates of first-cousin marriage in isolates

	period	total no. of marriages	no. of first-cousin marriages	rate of first-cousin marriage per decade
Swiss villages:				
I	1874–1922	479	5	1·0
II	1844–1932	249	12	1·4
Hungarian village	1866–1916	157	5	1·0
North Swedish villages:				
I	1890–1947	843	8	2·0*
II	1890–1947	281	8	2·0*
III	1890–1947	191	13	3·25*

Sources. Brenk (1931), Ruepp (1935), Nemeskéri & Thoma (1961), Böök (1958).

* Swedish data relate only to marriages where both partners survived to 1947. The cousin marriages among them have been taken as roughly equal in number to the cousin marriages of four decades.

Calculations described in Appendix I(*b*) suggest that in the demographic conditions of early eighteenth-century Europe there would under random mating have been something like one first-cousin marriage per decade in an isolate. Presumably this is the right order of magnitude for most human populations before the modern era. As a result of declining mortality there would probably have been in late nineteenth- and early twentieth-century European conditions something between one-and-a-half and four first-cousin marriages per decade in an isolate. These figures should only be taken as broad indications of orders of magnitude. Where mortality was high and population growth slow, something near the lower limit of the range should apply.

There are not many populations in Europe which can convincingly be regarded as isolates and for which we have data on the frequency of cousin marriage; but some figures are shown in table 32. They confirm theoretical expectation surprisingly well.

The figures for the Swiss and Hungarian villages are straightforward. The Swedish ones are not strictly comparable with them because they are based, not on a recording of marriages as they occurred, but on a retrospective survey of all marriages where both spouses were alive in 1947. Rough adjustment has been made for this difference in computing the rate of first-cousin marriage per decade. The figures relate to three neighbouring parishes 60 miles north of the Arctic Circle, on the Finnish border. The third of these parishes has a small population spread over a considerable area and the report (Böök 1958) suggests that it perhaps should not be regarded as a single isolate. In the demographic conditions of these Swedish villages in the relevant period, the rate of first-cousin marriage per decade would probably be considerably higher (under random mating) than in the conditions of the Swiss and Hungarian villages covered by the data in table 32, and the figures are in agreement with this presumption. The Swedish figures also strikingly demonstrate how the number of first-cousin marriages is of the same order of magnitude in isolates of very different size. Thus, the largest of the Swedish isolates is five times as large as the smallest. The rates of consanguineous marriages in a European isolate would always be expected to be somewhat smaller than those calculated for a completely closed population, for there has been some emigration, from isolated rural communities, in all parts of Europe in the recent past. Some additional data are available for Swiss villages based on retrospective surveys, like the Swedish figures of table 32 (Grob 1934; Egenter 1934). They are in agreement with the figures shown.

Given the number of cousin marriages in a large area, it is possible to compute how many isolates it consists of, if one is willing to assume that it is made up of isolates and if one knows the number of cousin marriages per isolate. In this way the mean size of isolates may be estimated. The details are given in Appendix II (c). Estimation of isolate size was the main purpose for which the Dahlberg formulae were devised.

The result that under random mating the number of consanguineous marriages is independent of the size of an isolate may be shown (Appendix I (b), equation (31)) to hold not only for the comparison of stationary populations of different size, but also for the same population increasing or decreasing over time, provided its vital rates remain constant. In an increasing population, the number of consanguineous marriages of any type per year will remain the same under random mating, though the frequency of consanguineous marriages among all marriages will of course decline as the population increases. Data to illustrate this theoretical result are difficult to find; but the figures in table 33 provide an illustration, though not an ideal one.

Table 33 relates to marriages among the descendants of a single Swiss couple who migrated with some of their relatives to the Western United States in the early nineteenth century. They have remained a close community because of their religion ever since. They have multiplied very rapidly, as is shown by the tenfold increase of the total number of marriages in 50 years. The number of marriages between first cousins per decade has remained roughly constant. This number is higher than in the European isolates, as expected on theoretical grounds.

In the nineteenth century few second-cousin marriages could occur because there were hardly any second cousins yet among the descendants of the original Swiss couple. However, the last line of the table shows that more recently second-cousin marriages have been several times more numerous than those among first cousins. This is unusual. In almost all the available data (table 25 shows examples) second-cousin marriages are less numerous than those between first cousins. This is not in agreement with theoretical expectation under random mating.

TABLE 33. CONSANGUINITY IN A MIDWESTERN UNITED STATES ISOLATE

period	total marriages	no. of marriages between first cousins	no. of marriages between second cousins	rate of first-cousin marriage per decade
1875–99	42	8	3	3·2
1900–24	128	4	34	1·6
1925–49	446	13	72	5·2

Source. Hammond & Jackson (1958).

TABLE 34. RELATIVE FREQUENCIES OF CONSANGUINEOUS MARRIAGES OF VARYING DEGREE UNDER RANDOM MATING

(ratios to frequency of first-cousin marriage)

population	uncle–niece and aunt–nephew	first cousins	$1\frac{1}{2}$* cousins	second cousins
closed stationary	0·06	1	0·6	3
with growth, no migration	0·05	1	0·75	4·5
with migration, no growth	0·07	1	0·5	2
with migration and growth	0·06	1	0·6	3

Growth: $1\frac{1}{2}$ % per year. Migration: 16 % emigrate before 25. For method of computation, see Appendix II(b).

* $1\frac{1}{2}$ cousins = first cousins once removed.

The theoretical ratios which the numbers of consanguineous marriages of different degrees bear to each other under random mating vary with several factors. In the case of uncle–niece and aunt–nephew marriages and those of $1\frac{1}{2}$, $2\frac{1}{2}$ etc. cousins important influences are the average ages at maternity and paternity. More exactly, the relative frequency of marriages between cousins who are not of the same generation varies with the ratio which mean ages at parenthood bear to the variances of the ages at parenthood. In all calculations in the present paper the same variances (appropriate to populations of high fertility) have been assumed. For illustration let us take ages of 27 and 31 years as extreme mean ages at maternity. If fathers are on average 5 years older than mothers, calculation shows that under random mating the number of uncle–niece and aunt–nephew marriages together would be some 9 % or 5 % respectively of the number of first-cousin marriages. The number of $1\frac{1}{2}$-cousin marriages would be 80 % or 50 % respectively of the number of first-cousin marriages. These figures are for a closed stationary population.

The influence of growth and migration on the relative frequencies of different degrees of consanguineous marriages is illustrated in table 34. An intermediate value has been adopted for marriages between relatives who are not of the same generation.

Population growth tends to increase, but migration to decrease, the ratio of the more remotely consanguineous marriages to the number of first-cousin marriages. In most of the populations for which we have data, both of these influences have been at work—in opposite directions. One can detect to some extent the variations to be expected on theoretical grounds, e.g. a higher ratio of the more remotely related marriages in the more isolated areas (see, for example, Cavalli-Sforza 1956) and those with faster growth. In the two Swiss isolates covered by table 32 there were seventeen marriages between $1\frac{1}{2}$ cousins and fifty-two between second cousins against seventeen first-cousin marriages. These numbers are in good agreement with the theoretical ratios. However, most of the data seem clearly at variance with theoretical calculations based on random mating; there are far too few second-cousin marriages in relation to the numbers of first-cousin marriages. Two explanations suggest themselves. One is that the assumption of random mating is violated. There may be special factors which increase the number of first-cousin marriages above the level expected under random mating, while leaving the second-cousin marriages unaffected. Such factors are believed to have great influence in many non-European populations. Secondly, consanguineous marriages between remoter relatives may be under-reported. This seems very likely to be at least part of the explanation. There is direct evidence both in Germany (Wulz 1925) and in Japan that this is so. In Japan it is possible to check statements of relationship against the official family registers. Out of a sample of 201 cases reported to be unrelated, the spouses were actually second cousins in eight cases (Yanase 1962, p. 129). If the proportion of unrecorded second-cousin marriages is typically anything like 4 % of all marriages, the statistics on second-cousin marriages in Japan are highly incomplete. As shown above, the distribution of second-cousin marriages by type in the studies so far available also suggests that many spouses who are in fact second cousins have been unaware of their relationship. $1\frac{1}{2}$-cousin marriages also appear to be somewhat under-reported, though much less than marriages between second cousins. Statistics of marriages of $2\frac{1}{2}$ cousins and third cousins are available from Catholic marriage dispensations up to 1917, since which year dispensation has no longer been required for such marriages; the numbers reported seem in most cases far too low.

Migration and under-reporting probably go together. When people move about they are less likely to marry their distant relatives and also, if they do so, less likely to be aware of it. The frequencies of marriages between more distant relatives than first cousins may well be substantially under-reported; mean coefficients of in-breeding calculated from such data may be too low especially if, as has been frequent practice, remoter relationships than those of second cousins have been ignored. Moreover, if the extent of under-reporting increases with the breakdown of isolates, the decrease in the coefficient of inbreeding over time has been over-estimated.

All the calculations of theoretical frequencies presented so far relate to populations whose vital rates are constant over time. However, vital rates over much of the world have not been constant. The Western world has undergone a 'demographic revolution' in which both death rates and birth rates have been lowered to a fraction of their former values.

It has been widely believed that the frequencies of consanguineous marriages must have been declining over the Western world for say the past century as a result of the breakdown of isolates. The same tendency has been reinforced by the fall in family size. However, the effect of the fall in mortality in tending to increase the frequency of consanguineous marriages has been neglected.

It would be possible to work out roughly the sort of changes in the frequency of consanguineous marriages which would have occurred if the recorded changes in mortality and fertility had been the only influences at work. Such calculations could be attempted for any country which has good demographic statistics going back for a sufficient period.

We may consider what would have happened as a result of the fall in death rates if the population were in fact living in rigidly separated isolates. As already pointed out the average annual number of, say, first-cousin marriages would have risen rapidly as a result of the fall in death rates; at the same time the total population and, of course, the annual number of marriages would also have been increasing within each isolate. However, the number of consanguineous marriages would undoubtedly have increased more rapidly than the total number of marriages, and the frequency of consanguineous marriages would have risen. The same conclusion holds if we assume, not the existence of rigidly separated isolates, but a fixed rate of interchange of population between isolates.

Increasing frequencies of consanguineous marriages have indeed been recorded quite widely for the nineteenth century (Wulz 1925; Alstrom 1958; Cavalli-Sforza *et al.* 1960; Sutter & Tabah 1955), contrary to the decrease which many have expected to flow from the 'breakdown of isolates'. Over much of Europe it seems likely that the decline in death rates would have created an extremely strong tendency for such an increase until well into the twentieth century (when the effects of the fall in family size would be expected to show themselves). The breakdown of isolates was probably not a force strong enough, at least until recently, to counterbalance by itself the consequences of declining mortality and cause a net decline in the frequency of consanguineous marriages.

The fall in family size has occurred too recently in most countries to have much effect on numbers of consanguineous marriages until the last few decades. This fall, unlike the decline in mortality, reinforces the effect of the breakdown of isolates in reducing the frequency of consanguineous marriages. However, this is not the whole story. The transition from large to small families is likely to have temporary and somewhat paradoxical effects. Falling family size like increasing migration (i.e. the breakdown of isolates) reduces the expected ratio of second- to first-cousin marriages. But during the transition there will be a period when marriages are contracted by persons whose parents were mostly born when families were already small, but their grandparents were still born in an era of large families. Now the number of one's first cousins is determined by the number of one's parents siblings, whereas the number of one's second cousins is determined by the number of one's grandparents' siblings. As the fall in family size begins to affect the frequency of consanguineous marriages one would expect, therefore, an *increase* in the ratio of second-cousin to first-cousin marriages even though the long run effect of declining fertility, as of the breakdown of isolates, is in the opposite direction. Something of this kind may well be the explanation of the rise in the relative frequency of second-cousin marriages shown in the Belgian figures in table 25. The numbers of marriages of varying degrees of consanguinity expressed as a ratio to first-cousin marriages present the following picture:

	first cousins	1½ cousins	second cousins
1920–24	1	0·3	1·3
1955–59	1	0·5	2·8

It may thus well be a coincidence that the figures for 1955–59 agree with the theoretical ratios of table 34.

6. Discussion

In §§ 3 to 5 data on consanguineous marriages have been confronted with the theoretical frequencies of such marriages which would be observed under random mating. The theoretical frequencies were not obtained by 'fitting' constants to data on consanguineous marriages (except in one instance in table 29). The theoretical frequencies were computed using only the general demographic characteristics of the human species together with information on certain demographic variables (growth, migration, mean ages at marriage and parenthood) in particular populations. No information derived from studies of consanguineous marriages is involved.

In many respects there is a surprising agreement between the observed and theoretical frequencies under random mating. Does this mean that consanguineous marriages occur at random? It is difficult to prove, at least from European data, that they do not. The features which have been interpreted (e.g. Morton 1955) as evidence of non-randomness, such as differences between the frequencies of the four types of first-cousin marriage, the comparative scarcity of marriages between 1½ cousins, etc., become consequences of random mating under the definition of the present paper. A geneticist from another planet with no experience of the workings of human society might be tempted to conclude that consanguineous marriages do indeed occur at random.

Such a conclusion is scarcely possible for human beings. We know, for example, that even in Western society there are special influences favouring first-cousin marriage (such as the greater likelihood of intimate contact between their families) and also special influences (such as disapproval) inhibiting such marriages. It seems impossible that the balance between the two kinds of influence should be everywhere and at all times the same and precisely such as to cause first-cousin marriages to occur at the frequency required by random mating.

One reason why discrepancies between the observed numbers of consanguineous marriages and the theoretical values have not emerged more strikingly is that there has been no emphasis on data which bring out such discrepancies; for example the Hindu groups in the study by Sanghvi *et al.* (1956) where only matrilateral cross-cousin marriages occur. If they were available, statistics for many non-Western societies might well reveal strikingly 'non-random' features. For European populations the basic difficulty is that the methods have been too crude. To bring out accurately departures from random mating would require calculations taking careful account of the demographic characteristics of particular populations and changes in these characteristics over time. The rough methods of the present paper are adequate to show that, in some European communities at some periods, the order of magnitude of first-cousin marriages was in accordance with that expected under random mating. Far more elaborate investigations would be needed to establish whether cousin marriages occurred at half or twice the rate expected under random mating. Such investigations will probably depend on improvements in the available data both on consanguineous marriages and on demographic characteristics (especially migration and correlations between relatives in fertility, mortality, etc.).

Our formulae are based on the concept of the isolate. Through much of human history and in many societies even today it seems reasonable to apply this concept, with allowance for migration, because the majority of people marry within small, fairly well-defined groups. The boundaries of the isolate may be set by barriers of geography, caste, class, religion, etc. However, under the conditions of modern, industrialized, highly mobile and largely urban populations it scarcely makes sense to speak of isolates. While preserving the basic idea that random mating implies that the chance of mating between two individuals depends upon the interval between their dates of births, the analysis could be reformulated in various ways to do without the isolate. For example, one can imagine a population continuously spread over a large area, but not subdivided into isolates, individuals born near each other being more likely to marry. Some of the results would probably remain unaltered on any formulation; for example the argument concerning the four types of first-cousin marriage. Such formulations may also be useful in dealing with species other than man. Whether any moderately simple model can give useful results for modern industrialized populations seems doubtful.

The observed frequencies of consanguineous marriages reflect rather strikingly several of the features expected under random mating presumably because the availability of potential spouses of suitable age is an important determinant of such marriages. If differences in the opportunities for various types of marriage are great enough, they will be reflected in the numbers of marriages even if the extent to which the opportunities are taken up is not independent of the relationship between the potential spouses. For example, there are fewer opportunities for marriage between $1\frac{1}{2}$ cousins than between first cousins because the age difference in the case of $1\frac{1}{2}$ cousins is likely to be unsuitable. This is an important effect, since the total number of $1\frac{1}{2}$ cousins of the opposite sex which an individual has is on average several times the number of first cousins. A calculation of the theoretical ratio of $1\frac{1}{2}$-cousin marriages to first-cousin marriages is an assessment of the effect of demographic conditions on these two kinds of marriage. In cases where the demographic factors alone would cause smaller differences between the frequencies of different types, these differences may nevertheless be detectable in the observations if the non-demographic influences are not powerful enough to override them. The comparison of the four types of first-cousin marriage seems to provide an illustration, at least in societies where there are not strong rules or preferences for one or other type.

Calculation of the frequencies to be expected under random mating is therefore a tool for studying the effects of demographic factors on the extent to which inbreeding occurs. For example, changes in mortality or differences between the mean ages at paternity and maternity may have substantial effects on consanguinity. This is not only a subject of interest in itself, but important for the understanding of the actual data on consanguineous marriages, as shown by several examples above. The demographic revolution would be transforming the mating patterns of human societies, even apart from the 'breakdown of isolates'.

The extent to which the data on consanguineous marriages conform to the theoretical frequencies under random mating suggests that the latter might profitably be used as a substitute where data are defective or totally lacking. Of course, only broad orders of magnitude could be obtained in this way. For example, mean coefficients of inbreeding might be computed by taking account of matings between second cousins and remoter relatives in the proportions expected under random mating. Calculations for species other than man would be feasible in so far as reasonable estimates of demographic characteristics can be made. The study of man whether in palaeolithic times or even in the Middle Ages provides a similar problem in that direct information on inbreeding is no longer obtainable. Almost all the direct data we have, or can hope to collect, are influenced by the entirely novel demographic conditions of modern times. Modern data form slippery foundations for arguments concerning long-term genetic equilibria.

APPENDIX I. THE DERIVATION OF THE FORMULAE

(a) *The stationary case*

Our assumptions may be stated under four headings.

We consider (1) a closed community (an isolate) with (2) a stationary population. 'Stationary population' for this purpose means not merely that the total number of people does not change, but also that all relevant demographic characteristics (age-specific death rates, first marriage rates, re-marriage rates, birth rates, intervals between births, etc.) are constant over time.

Stationary population models have generally been applied in demography to whole countries, i.e. populations numbered in millions. In such cases random fluctuations in numbers of births, deaths, etc., would be negligible in proportion to the totals involved. However, we wish to apply our formulae to communities of a few hundred persons. It is not obvious that fluctuations from year to year in numbers of births or marriages can be neglected. Fluctuations over time are taken into account in part (d) below. In parts (a), (b) and (c) only the expected values of the births, marriages etc., of the total population are involved. The word 'expected' has sometimes been omitted.

Our formulae are, moreover, based on (3) a special type of assumption about random mating, as explained in § 2 above. Consider a newborn boy and a newborn girl. Let the time of the boy's birth be subtracted from the time of the girl's birth and let the resulting difference be denoted z (measured in years). Random mating will be taken to mean that the probability that the baby boy and baby girl will eventually marry each other is a function of z only. Finally (4) we assume that there are no correlations between the vital rates experienced by relatives, i.e. no correlations between parents and their children or between sibs in chances of survival, of marriage or having various numbers of children, in ages at marriage, intervals between births etc. In the present state of ignorance about these matters, it seems impossible to dispense with this assumption, though it is certainly unrealistic.

Given the assumptions, it is possible to calculate the average rate at which consanguineous marriages of any type ξ would occur in our isolate. The average annual number of such marriages will be denoted Y_ξ.† A type ξ of consanguineous marriages is most easily specified by a pedigree diagram. Each pedigree diagram represents a particular type; i.e. a type is characterized not only by a number of ancestors by which the bride and groom trace their descent from a single couple, but also by the sequence in which males and females occur among these ancestors. Values of ξ include: man marrying sister's daughter, marriage between the children of two brothers, etc.

Marriage between relatives descended from one common ancestor (e.g. the children of half brothers), and not from a single couple, will not be considered. Formulae for such marriages could easily be written out. It is a defect of many of the statistical studies of consanguineous marriages that no indication is given of the way in which marriages of persons descended from half-sibs have been classified. For separate data on such marriages in Austria see Orel (1932). Such data are likely to be incomplete. There must be many pairs of half-sibs born where one or both are illegitimate. They and their descendants may be unaware of the relationship. Even if such cases could somehow be included in the statistics, the number of consanguineous marriages between descendants of half-sibs would in most societies be very greatly outnumbered by consanguineous marriages where the spouses trace their common descent from a couple.

Illegitimate births have not been explicitly introduced into the formulae. It may be said that, so far as genetic effects are concerned, it makes no difference what sanction a mating may have, and that the word marriage in the following pages should be taken to refer to any union. This is satisfactory for theoretical deductions; but the statistics, both on consanguineous marriages and on the demographic characteristics of the population, are likely to be affected in many ways by the distinctions between legal or recognized unions and others. Also permanent unions require rather different treatment in an analysis of consanguinity from casual ones. It is quite consistent to take 'marriages' in the formulae to refer to legal marriages while basing the computation of reproduction rates or rates of growth on all births, including illegitimate ones.

† For list of symbols used in this paper, see p. 171.

We may write
$$Y_\xi = M \cdot N_\xi \Pi_\xi, \tag{1}$$

where Y_ξ is the expected number of marriages of type ξ per year, M is the expected total number of marriages per year, N_ξ is the average number of potential marriages of type ξ among the descendants of a couple, i.e. the average number of pairs of live-born children (taking one of each sex) related in the manner ξ among the descendants of a couple, Π_ξ is the chance that a potential marriage of relationship ξ actually take place, i.e. the chance that a newborn boy and newborn girl between whom there exists relationship ξ will marry.

To be specific, consider marriages between paternal parallel cousins, i.e. children of two brothers. We proceed at first without the restrictive assumptions set out above. Suppose that a couple has j live-born children. Among them it is possible to form $j(j-1)/2$ pairs of whom on average $p^2 j(j-1)/2$ will be pairs of two brothers (where p is the proportion of boys among total births).

Now consider the children of a pair of brothers. Suppose that the first brother has b_1 boys and g_1 girls and the second brother has b_2 boys and g_2 girls. The number of potential marriages among them is $b_1 g_2 + b_2 g_1$. Let R_P be the mean number of live-born sons which a boy fathers in the course of his life and let σ^2 denote the variance of the number of sons born to one person. The variance of the number of his daughters is taken to have the same value. Let ρ be the correlation between the number of girls born to one brother and number of boys born to another.

We then have
$$E(b_1) = E(b_2) = R_P, \tag{2}$$

and
$$E(g_1) = E(g_2) = R_P q/p, \tag{3}$$

where p and q are the proportions of male and female births.

Now
$$(E(b_1 g_2) = E(b_2 g_1) = R_P^2 q/p + \rho \sigma^2.$$

We now use the restrictive assumptions explained at the outset. We suppose first that $\rho = 0$. Secondly, in a stationary population each generation replaces exactly the births from which it sprang, i.e. $R_P = 1$.

Thus
$$E(b_1 g_2 + b_2 g_1) = 2q/p. \tag{4}$$

Thus, among the descendants of a marriage with j children, the expected number of potential paternal parallel cousin marriages is
$$pq j(j-1). \tag{5}$$

If n_j is the chance that a marriage produce j live births and, if there is no correlation between the vital rates of parents and those of their children, we have
$$N_\xi = pq \sum_{j=2}^{\infty} j(j-1) n_j. \tag{6}$$

If we write
$$\sum_{j=2}^{\infty} j(j-1) n_j = \bar{n}, \tag{7}$$

then
$$N_\xi = \bar{n} pq. \tag{8}$$

In the derivation of formula (8) ξ was taken to denote paternal parallel cousins. The same argument may be repeated for any other type of relationship and it is soon realized that, if stationarity is assumed, the result is always the same. N_ξ is not a function of ξ.

Thus if we consider the categories of relationships normally regarded as distinct in Western society, namely, 'uncle–niece', 'aunt–nephew', 'first cousin', etc., we can easily obtain the average number of potential marriages of each category among the descendants of one couple. There are two types of uncle–niece marriages and two types of aunt–nephew marriages, four types of first-cousin marriages, sixteen types of marriages between first cousins once removed, sixteen types of second-cousin marriages, etc. For simplicity we may at this point take $p = q = \frac{1}{2}$ which is not far from the truth.

category of relationship	av. potential number of marriages among descendants of one couple
uncle–niece and aunt–nephew	$\bar{n}$
first cousins	$\bar{n}$
first cousins once removed	$4\bar{n}$
second cousins	$4\bar{n}$

We now need to evaluate Π_ξ, i.e. the chance that a particular boy should marry a particular girl who is related to him in relationship ξ. Again for concreteness we shall deal with the case when ξ represents paternal parallel cousins.

To evaluate Π_ξ we begin by defining two functions. We first consider the conditional probability that, given that a boy has a girl relative of relationship ξ, she should be born in an interval of time z to $(z+dz)$ after his birth. Let $h_\xi(z)$ be a function such that this probability may be written $h_\xi(z)\,dz$.

Secondly, consider the probability that a marriage takes place between a particular boy and a particular girl born an interval z apart. Suppose that this probability be written $m(z)$.

With these definitions we have

$$\Pi_\xi = \int_{-\infty}^{+\infty} m(z)\,h_\xi(z)\,dz. \tag{9}$$

Now it may be shown (Appendix I (d) below) that, if $a(z)\,dz$ is the proportion of marriages when the girl is between z and $(z+dz)$ years younger than the boy, we may write

$$m(z) = \frac{1}{pq}\frac{U^2}{M}a(z). \tag{10}$$

Here U is the ratio of the number of marriages to the number of births. Equation (10) holds only under random mating in the sense defined above.

Substituting in (9) we have

$$\Pi_\xi = \frac{1}{pq}\frac{U^2}{M}\int_{-\infty}^{+\infty} a(z)\,h_\xi(z)\,dz, \tag{11}$$

and, from (1) and (8), $$Y_\xi = \bar{n}U^2\int_{-\infty}^{+\infty} a(z)\,h_\xi(z)\,dz. \tag{12}$$

The number of marriages M cancels out in the derivation of formula (12). Under the assumptions made the average annual number of consanguineous marriages of any particular type ξ (or of various types put together) is independent of the size of the isolate.

We now return to equation (11). The function $a(z)$ can be determined directly when adequate data on the distribution of marriages by the age of the bride and groom exist. Such data are available for many areas.

Very little is known about $h_\xi(z)$. $h_\xi(z)$ specifies the distribution of a random variable, say z_ξ, the difference between the dates of birth of a boy and girl between whom relationship ξ holds. z_ξ is the sum of a number of ages at parenthood (those in the boy's ancestry being given a negative sign), and the difference between the dates of birth of the pair of sibs through whom they are related. We shall assume that each of the variables whose sum is z_ξ is subject to a probability distribution and that all these distributions are independent. We can then find the distribution $h_\xi(z)$ for any ξ provided we know the following distributions: (i) the distribution of ages at fatherhood; (ii) the distribution of ages at motherhood; (iii) the distribution of the intervals between the dates of birth of pairs of live-born children of a marriage. (We need to consider not only the intervals between successive births, but between all possible pairs which may be formed by selecting two children from all the children of a marriage.) It will be assumed that z_ξ being a sum of independent random variables has an approximately normal distribution By means of the independence assumption one can derive from the three distributions mentioned formulae for the mean and variance of $h_\xi(z)$.

For this purpose additional notation is required. To simplify the following definitions we use the term 'ancestry' of the potential spouses to mean their ancestors up to, but not including, the original couple from whom both are descended.

Let $\alpha_{\xi,1}$ denote the number of males in the ancestry of the boy;
　　$\alpha_{\xi,2}$ denote the number of males in the ancestry of the girl;
　　$\beta_{\xi,1}$ denote the number of females in the ancestry of the boy;
　　$\beta_{\xi,2}$ denote the number of females in the ancestry of the girl.

Also, let　　　　　　$\alpha_{\xi,1}+\alpha_{\xi,2} = \alpha_\xi$　and　$\beta_{\xi,1}+\beta_{\xi,2} = \beta_\xi$.

Let $\mu_\xi(h)$ and $V_\xi(h)$ be the mean and variance of $h_\xi(z)$. Let μ_M and μ_P, V_M and V_P denote the means and variances of the distributions of maternal and paternal ages at the births of children. Finally, let V_i denote the second moment about the origin of the distribution of intervals between the births of pairs of sibs. It is the second moment about the origin, and not the variance, which is needed. The complication arises because z is positive if the girl is born later than the boy and negative if her birth occurs before his. Hence the interval between the birth of her ancestor and the birth of his ancestor in the original sib-pair must be given the appropriate sign. This is achieved by using the second moment about the origin in formula (14).

With these definitions we can write:

$$\mu_\xi(h) = (\alpha_{\xi,2}-\alpha_{\xi,1})\,\mu_P+(\beta_{\xi,2}-\beta_{\xi,1})\,\mu_M \tag{13}$$

and
$$V_\xi(h) = \alpha_\xi V_P+\beta_\xi V_M+V_i. \tag{14}$$

Given that z_ξ has been assumed normally distributed, it is convenient to take the distribution $a(z)$, i.e. the distribution of age differences between the spouses, to be normal also. Some experimental calculations suggest that once $h_\xi(z)$ is taken to be normal, practically the same results are obtained if $a(z)$ is assumed normal as if actual distributions of age differences derived from marriage statistics are used.

Let $\mu(a)$ and $V(a)$ be the mean and variance of $a(z)$, the distribution of age differences. We may now write, using normal distributions for $a(z)$ and $h_\xi(z)$ in (11),

$$\Pi_\xi = \frac{1}{pq}\frac{U^2}{M}\frac{1}{2\pi[V_\xi(h)\,V(a)]^{\frac{1}{2}}}\int_{-\infty}^{+\infty}\exp\left[-\frac{(z-\mu_\xi(h))^2}{2V_\xi(h)}-\frac{(z-\mu(a))^2}{2V(a)}\right]dz. \quad (15)$$

Upon integration, this becomes

$$\Pi_\xi = \frac{1}{pq}\frac{U^2}{M}[2\pi\{V_\xi(h)+V(a)\}]^{-\frac{1}{2}}\exp\left[-\frac{\{\mu_\xi(h)-\mu(a)\}^2}{2\{V_\xi(h)+V(a)\}}\right]. \quad (16)$$

If we write

$$\mu_\xi(h)-\mu(a) = \Delta_\xi \quad (17)$$

and

$$V_\xi(h)+V(a) = S_\xi, \quad (18)$$

we obtain

$$\Pi_\xi = \frac{1}{pq}\frac{U^2}{M}\frac{1}{\sqrt{(2\pi S_\xi)}}e^{-\Delta_\xi^2/2S}, \quad (19)$$

and finally

$$Y_\xi = \frac{\bar{n}U^2}{\sqrt{(2\pi S_\xi)}}e^{-\Delta_\xi^2/2S_\xi}. \quad (20)$$

(b) The non-stationary case

The formulae which have been developed above can be generalized to cover the case of any stable population, i.e. the population need not be stationary. The terms 'stable' population and 'stationary population' are here used in their technical demographic senses. A stable population is one whose demographic characteristics (age specific death rates, marriage rates, etc.) are constant over time. It has a constant age structure. The total population increases or decreases at a constant rate (denoted by r). A stationary population is a special case of a stable population where $r = 0$.

The abandonment of the restriction of stationarity brings the model a good deal nearer to reality. Most of the populations for which data have been and can be collected on the frequency of consanguineous marriages are not stationary, but more or less rapidly increasing. It has been found in other types of demographic analysis that not infrequently populations in which fertility is high (i.e. family limitation is not widely practised) can for many purposes be treated as stable populations. Moreover, the extension to the stable case is the essential step in showing how our model can be applied in a yet more general situation, i.e. when the demographic characteristics of a population change with time.

The extension of the argument to cover the general stable case, though somewhat intricate, adds nothing essentially new. Use will be made of various known results concerning marriages and reproductivity in a stable population.† The main

† In demographic literature stable populations have generally been defined in terms of constant age specific mortality and fertility rates. The extensions to marriage and reproductivity were mainly developed by Karmel (1948). The treatment in the present section largely follows chapter VI of the paper by Hajnal (1950). The notation is, however, slightly different.

conclusions from the present section can easily be summarized for those who do not wish to follow the argument in detail. They are, first, that *the number of marriages of any given relationship is constant over time in a stable population*; secondly, that this number can be calculated from the formulae already developed for the stationary case except for small adjustments shown in formula (44) below.

In a stable population the expected numbers of births, marriages, etc., are functions of time. They increase or decrease at the same rate as the expected total population. If we write $B(t)$ and $M(t)$ for births and marriages at time t respectively,† we have

$$B(t+s) = B(t)\,\mathrm{e}^{rs} \tag{21}$$

and

$$M(t+s) = M(t)\,\mathrm{e}^{rs}. \tag{22}$$

In a stable population when $r \neq 0$ each generation does not exactly replace the number of births from which it sprang. Let R_P and R_M be the paternal and maternal reproduction rates, i.e. R_P is the average number of sons born in the course of their lives to a randomly selected group of newborn boys, and R_M is similarly defined as the average number of daughters per newborn girl.

We now consider the number of potential marriages of type ξ among the offspring of marriages occurring at time t. We write

$$N_\xi = pq\,\Sigma[j(j-1)\,n_j]\,R_P^{\alpha_\xi}R_M^{\beta_\xi}, \tag{23}$$

where α_ξ is the number of male ancestors and β_ξ the number of female ancestors of the potential spouses up to, but not including, the original couple. Formula (23) is a generalization of formula (6) and subject to the same restrictive assumptions, except that it is no longer assumed that $R_P = R_M = 1$. Let $\alpha_\xi + \beta_\xi = \delta_\xi$ say. δ_ξ is then a measure of the closeness of the relationship ξ, as follows:

category of relationship	δ_ξ
uncle–niece or aunt–nephew	1
first cousins	2
first cousins once removed	3
second cousins	4
etc.	

The subscript ξ will be omitted when α_ξ, β_ξ or δ_ξ occur as exponents.

Let $\Pi_\xi(t)$ be the probability that a potential marriage of persons in relationship ξ, descended from a couple married at time t, should actually take place. In the stationary case, Π_ξ did not vary with t; now it does because, as the population of the isolate increases, the probability that a potential marriage should occur decreases since the number of potential spouses for any person increases with the population.

In order to proceed we must, therefore, chose a system of dating a potential marriage. We may locate it at the point of time when the boy is born, or when the girl is born; when the earlier of the two is born, etc. The equations come out simplest if one locates each potential marriage at a point mid-way between the birth of the two potential spouses. We therefore consider potential marriages where the boy is

† It would be more accurate to say: 'Let $B(t)$ and $M(t)$ be functions such that the expected numbers of births and marriages between time t and $(t + \mathrm{d}t)$ are $B(t)\,\mathrm{d}t$ and $M(t)\,\mathrm{d}t$ respectively.' It seems more convenient to keep to simpler language.

born at $s - \frac{1}{2}z$, and the girl at $s + \frac{1}{2}z$ after t, the time of the marriage of the original couple from whom they are both descended.

We may now write

$$\Pi_\xi(t) = \int_0^{+\infty} \int_{-\infty}^{+\infty} h_\xi(s, z) \, m(t + s, z) \, dz \, ds. \tag{24}$$

Here, $h_\xi(s, z)$ is the probability density function of the distribution of potential marriages of relationship ξ with respect to s (the midpoint between the dates of birth of the spouses) and z (the age difference between them). $m(t + s, z)$ defines the probability that a marriage should take place between a boy born at $(t + s - \frac{1}{2}z)$ and a girl born at $(t + s + \frac{1}{2}z)$.

$m(t + s, z)$ can be expressed in a form analogous to (10) above if the function $a(z)$ is defined in a form suitable to the stable population. In a stable population various distributions of marriages by the age of the spouses must be distinguished. The distribution we get by considering all marriages occurring simultaneously will differ from that obtained by following the marriages of a group of men born at the same time, and this distribution in turn will differ from that yielded by the marriages of a group of women born together.

Let $u(x, y)$ be the ratio of the expected number of marriages at any time t of brides aged x and bridegrooms aged y to the expected number of births at $t - y$. Similarly, let $\bar{u}(x, y)$ be the ratio of the number of such marriages to births at $t - x$. Then $u(x, y)$ may be said to represent the distribution of marriages in a 'generation of men', i.e. the distribution obtained by following a cohort of men through life, while $\bar{u}(x, y)$ is the distribution of marriages in a generation of women.

Thus

$$\bar{u}(x, y) = u(x, y) \, e^{r(x - y)}. \tag{25}$$

We define

$$u^*(x, y) = [u(x, y) \, \bar{u}(x, y)]^{\frac{1}{2}}. \tag{26}$$

$u^*(x, y)$ may be said to represent the distribution of marriages in a joint generation. We now define $a(z)$ for the stable case by

$$a(z) = \frac{1}{U} \int_0^{+\infty} u^*(x, x + z) \, dx, \tag{27}$$

where

$$U = \int_0^{+\infty} \int_0^{+\infty} u^*(x, y) \, dx \, dy. \tag{28}$$

For the stationary case U was defined as the ratio of the number of marriages to the number of births. When $r = 0$ equation (28) reduces to this definition.

We can now write

$$m(s, z) = \frac{1}{pq} \frac{U}{B(s)} a(z). \tag{29}$$

This formula is based on the assumption of random mating. The derivation, which will not be given here, is exactly analogous to that of formula (10) above.

Hence

$$\Pi_\xi(t) = \frac{1}{B(t)} \frac{U}{pq} \int_0^{+\infty} \int_{-\infty}^{+\infty} h_\xi(s, z) \, e^{-rs} a(z) \, dz \, ds \tag{30}$$

and

$$Y_\xi = \frac{M(t)}{B(t)} U \bar{n} R_P^\alpha R_M^\beta \int_0^{+\infty} \int_{-\infty}^{+\infty} h_\xi(s, z) \, e^{-rs} a(z) \, dz \, ds. \tag{31}$$

Y_ξ is independent of t since the ratio $M(t)/B(t)$ is constant. Just as the number of potential consanguineous marriages increases with population growth, so the chance that a potential marriage take place decreases. This is analogous to a result obtained for the stationary case, namely that the number of consanguineous marriages is independent of the size of the isolate.

If $r = 0$, $R_P = R_M = 1$ and formula (31) reduces to formula (12) for the stationary case, since for a stationary population $M(t)/B(t) = U$. The effect of departures from stationarity may be more easily envisaged by replacing part of formula (31) with suitable approximations. For this purpose we introduce two averages θ and $\bar{s}_\xi$.

We first define θ by the relation

$$\theta = \frac{1}{U} \int_0^\infty \int_0^\infty \tfrac{1}{2}(x+y)\, u^*(x,y)\, dx\, dy. \tag{32}$$

Thus θ is the mean age at marriage in the $u^*(x,y)$ distribution.

Then, to a good approximation, for any r likely to occur in practice

$$U = \frac{M(t+\theta)}{B(t)} = \frac{M(t)}{B(t)} e^{r\theta}. \tag{33}$$

Secondly, we define $\bar{s}_\xi$ by the relation

$$\bar{s}_\xi = \int_0^\infty \int_{-\infty}^{+\infty} s \cdot h_\xi(s,z)\, dz\, ds. \tag{34}$$

Thus $\bar{s}_\xi$ is the mean value of s_ξ, the interval from a marriage to the point midway between the dates of birth of two descendants of the marriage between whom there is relationship ξ. Putting θ and $\bar{s}_\xi$ into (31) we obtain

$$Y_\xi = U^2 \bar{n} R_P^{\alpha_\xi} R_M^{\beta_\xi} \exp\left[-r(\theta+\bar{s}_\xi)\right] \int_0^\infty \int_{-\infty}^{+\infty} h_\xi(s,z) \exp\left[-r(s-\bar{s}_\xi)\right] a(z)\, dz\, ds. \tag{35}$$

$\bar{s}_\xi$ may be regarded as the sum of (1) the mean interval between the marriage of a couple and the dates of birth of their children, and (2) the mean interval elapsing between the birth of these children and the birth of those of their descendants between whom there is relationship ξ. Thus

$$\bar{s}_\xi = \phi + \tfrac{1}{2}[\alpha_\xi \mu_P + \beta_\xi \mu_M], \tag{36}$$

where ϕ is the mean duration of marriage at which births occur, and μ_P and μ_M are, as before, the means of paternal and maternal ages at the birth of children. Then, to a good approximation

$$\theta + \phi = \tfrac{1}{2}(\mu_P + \mu_M), \tag{37}$$

$$\theta + \bar{s}_\xi = \tfrac{1}{2}[(\alpha_\xi + 1)\mu_P + (\beta_\xi + 1)\mu_M] \tag{38}$$

$$\exp[r\mu_P] = R_P \quad \text{and} \quad \exp[r\mu_M] = R_M. \tag{39}$$

By means of these relationships we obtain

$$Y_\xi = \exp\left[\tfrac{1}{2}r\{(\alpha_\xi - 1)\mu_P + (\beta_\xi - 1)\mu_M\}\right] U^2 \bar{n} \int_0^\infty \int_{-\infty}^{+\infty} h_\xi(s,z) \exp\left[-r(s-\bar{s}_\xi)\right] a(z)\, dz\, ds, \tag{40}$$

or $\qquad Y_\xi = R_P^{\frac{1}{2}(\alpha_\xi - 1)} R_M^{\frac{1}{2}(\beta_\xi - 1)} U^2 \bar{n} \int_0^\infty \int_{-\infty}^{+\infty} h_\xi(s,z) \exp\left[-r(s-\bar{s}_\xi)\right] a(z)\, dz\, ds. \qquad (41)$

Let the expression (41) which is a function of r be denoted by $Y_\xi(r)$. Then

$$Y_\xi(0) = U^2 \bar{n} \int_0^\infty \int_{-\infty}^{+\infty} h_\xi(s, z)\, a(z)\, \mathrm{d}z\, \mathrm{d}s. \qquad (42)$$

This expression is the same as (12) if we write, as in the earlier notation for the stationary case

$$\int_0^\infty h_\xi(s, z)\, \mathrm{d}s = h_\xi(z). \qquad (43)$$

For moderate values of r such as occur in practice (i.e. $|r| < 0.03$) we can neglect the overall effect of $\exp[-r(s - \bar{s}_\xi)]$ in (41). Thus approximately

$$Y_\xi(r) = R_P^{\frac{1}{2}(\alpha-1)} R_M^{\frac{1}{2}(\beta-1)} Y_\xi(0). \qquad (44)$$

If R_M and R_P are close together (as will be the case unless both $(\mu_P - \mu_M)$ and r take abnormal values) the correction factor

$$R_P^{\frac{1}{2}(\alpha_\xi-1)} R_M^{\frac{1}{2}(\beta_\xi-1)}$$

depends essentially on the value of $\alpha_\xi + \beta_\xi = \delta_\xi$.

If we write $R = [R_P R_M]^{\frac{1}{2}}$ we have approximately

$$Y_\xi(r) = R^{(\frac{1}{2}\delta-1)} Y_\xi(0). \qquad (45)$$

R is a reproduction rate per generation (intermediate between the paternal and maternal reproduction rates). The value of R is governed by the proportion of births who survive to marriageable age, the average number of marriages contracted by such survivors and the average number of births per marriage. If, as above, n_j denotes the probability that a marriage produce j live births (and if there are no illegitimate births)

$$R = U\Sigma j n_j. \qquad (46)$$

By means of the approximations described, the effect of rate of growth $r \neq 0$ can be envisaged without considering the detailed shape of the function $h_\xi(s, z)$. We calculate $Y_\xi(O)$ by the formula for the stationary case, but using constants $U, \bar{n}$, etc., appropriate to the population in question. We then apply the correction factor indicated by (44). The effect of this correction factor will be negligible for marriages of first cousins, for marriages of first cousins once removed the effect is essentially multiplication by $\sqrt{R}$, for marriages between second cousins multiplication by R, etc.

(c) *Migration*

It is possible to adjust our formulae in a very simple fashion in order to illustrate the major effects of migration. We retain the framework of an isolate with a stable population except that we relax the assumption that there is no movement of population in or out of the isolate (the term 'isolate' will be retained, though this is perhaps not quite logical). It is convenient to begin by assuming that migration has no effect on population growth, i.e. that every emigrant is replaced by an immigrant of the same sex and age who is subject to the same chances of death, marriage, etc. We further assume for simplicity that migration takes place before the marriageable age range is reached. No related migrants enter the isolate, i.e. consanguineous marriages can occur only between persons born in the isolate. Suppose that the proportion of men reaching marriageable age who have not been replaced by immigrants is λ_P and the proportion of girls not replaced is λ_M.

The chance that a boy and a girl both born in the isolate should marry there is $\lambda_P \lambda_M$ times what it would be if there were no migration. The expected number of sons born within the isolate to a newborn male child is now $\lambda_P R_P$ in place of R_P; the expected number of daughters born within the isolate to a newborn male child is $\lambda_P R_P(q/p)$ in place of $R_P(q/p)$, and so on.

Let $\hat{Y}_\xi$ be the average yearly number of marriages between relatives of type ξ in the isolate when migration occurs. If Y_ξ is defined by equation (31) we have

$$\hat{Y}_\xi = \lambda_P^{(\alpha+1)} \lambda_M^{(\beta+1)} Y_\xi. \tag{47}$$

Putting in the approximation for Y_ξ given by formula (44) one obtains

$$\hat{Y}_\xi = \lambda_P^{(\alpha+1)} R_P^{(\frac{1}{2}\alpha-\frac{1}{2})} \lambda_M^{(\beta+1)} R_M^{(\frac{1}{2}\beta-\frac{1}{2})} Y_\xi(0). \tag{48}$$

Formulae (47) and (48) remain valid under more general conditions about migration than those so far discussed. The restriction that every immigrant replaces an emigrant of the same sex and age, is not necessary, i.e. it may be supposed that migration *can* affect the growth of population in the isolate. We assume that the expected numbers of emigrants and immigrants of each sex and age per unit of time at any point of time bear constant ratios to the expected total population of the isolate at that point. The properties of a stable population which are required for our formulae will then hold (i.e. the numbers of births, marriages and deaths increase, or decrease, at a fixed rate, etc.) and formulae (47) and (48) will remain valid provided the symbols occurring in them are appropriately interpreted.

For this purpose R_P may be redefined as follows: Consider a group (or 'cohort') of boys born in the isolate within a short period of time. As they pass through life all the sons born to them may be added up; the average number of sons will be R_P under the definition for the closed stable population. For the case where migration occurs, the required modification is to exclude from the sons counted all those born outside the isolate, but to include the sons of immigrant fathers if the father was born at the same time as the cohort under consideration. R_M must be redefined on the same lines. λ_P is the proportion of bridegrooms who were born within the isolate and λ_M the proportion of brides born within the isolate. $Y_\xi(0)$ retains essentially the same meaning as before.

(d) The chance of marriage between a newborn boy and a newborn girl

The line of argument illustrated in table 24 may be applied under more general conditions. Consider a stationary population with time divided into discrete time units (say 'seasons') and age correspondingly measured in the number of seasons elapsed since the season of birth. For both males and females probabilities are given of marrying at each age persons of each particular age of the opposite sex. Random mating obtains in the sense that in any season any of the pairs which can be formed of the marrying males and females with a given combination of ages are equally likely to occur. The population is assumed large, there are B births per season of whom a proportion p are male and $1-p = q$ female. Denote the number of marriages per season by M and let $M/B = U$; of the marriages a proportion $a(z)$ are such that the groom was born z seasons before the bride.

The argument set out with a numerical example in table 24 above may be used[†] to find a formula for $m(z)$, the probability that a newborn boy marry a newborn girl born z years later. The result is

$$m(z) = \frac{1}{pq}\frac{U^2}{M}a(z). \tag{10}$$

This formula is identical with formula (10) of part (a), except that in continuous time $a(z)$ must be taken to be the density function of the distribution of marriages with respect to z, the excess of groom's age over bride's age. This formula is entirely satisfactory. Unfortunately, however, if time is taken to be continuous it seems difficult to justify its use for finite populations, even large ones, by an argument along the lines just indicated.

A justification can, however, be provided by an entirely different route. We now simply *assume* that there is a fixed probability, denoted by $m(z)$, that a newborn boy shall eventually marry a newborn girl born z years later. It can be shown, without entering further into the events taking place between birth and marriage, that $m(z)$ must under wide conditions satisfy formula (10).

We begin by setting out the argument for the case of discrete time units. We assume that the population is stationary. Let the numbers of boys and girls born in season t be denoted by $b(t)$ and $g(t)$ respectively. Then if B is the expected total of births and p, q the proportions of male and female births as before, we obtain for the expected values

$$E[b(t)] = pB, \tag{49}$$

$$E[g(t)] = qB. \tag{50}$$

The number of pairs, one male and one female, which may be formed among the male births of season t and the female births of season $t+z$ is $b(t)g(t+z)$. Now

$$E[b(t)g(t+z)] = pqB^2 + \text{cov}\,(z), \tag{51}$$

where cov (z) is the convariance between $b(t)$ and $g(t+z)$. This convariance is assumed by virtue of stationarity to depend only on z, not on t.

On account of stationarity also, the expected number of marriages contracted in one season will be constant over time and equal to the expected total number of marriages contracted by the males born in one season. Let this number be denoted, as before, by M. Then

$$M = \sum_z E[b(t)g(t+z)m(z)] = pqB^2\sum_z m(z) + \sum_z \text{cov}\,(z)m(z). \tag{52}$$

We now assume that the last term on the right which involves cov (z) can be neglected. The justification for this assumption is discussed below. If it is granted, it follows that

$$\sum_z m(z) = \frac{1}{pq}\frac{M}{B^2} = \frac{1}{pq}\frac{U^2}{M}, \tag{53}$$

where $U = M/B$.

† The argument is given, in a different notation, in Hajnal (1960 Note II, p. 190).

Finally if we write

$$a(z) = \frac{m(z)}{\sum\limits_{z} m(z)}$$

the result is formula (10) as given above.

If cov (z) is small, $a(z)$ may be treated as the proportion of marriages in which the bride was born z seasons after the groom. This argument, unlike the earlier one, translates without difficulty into continuous time, if the functions involved are interpreted as functions of continuous parameters for time and age.

The corresponding formula (29) for the stable population may similarly be derived by both lines of argument.

It remains to discuss the assumption made in passing from equation (52) to equation (53), namely that terms in cov (z) can be neglected. This is certainly plausible, provided the population is of moderate size, if we make two assumptions generally made in mathematical treatments of population growth, namely that vital rates are constant over time and that all members of the population are independent of each other so far as probabilities of vital events are concerned. Turning to reality, a cursory examination of the fluctuations observed in some eighteenth-century European villages in actual numbers of births and deaths suggests that the assumptions in the derivation of formula (10) are a sufficient approximation to the facts, at least outside periods of catastrophe (famine, epidemics).

To complete the discussion, we return to the first line of argument for deriving formula (10), that implied in table 24. Both in that table and at the beginning of this section the argument was developed on the assumption that the population is large. This is not necessary. For a simple organism a fully probabilistic treatment is not too difficult. For example, consider the organism of table 21. For simplicity of exposition, we suppose in addition that an individual dies at the end of the season in which it reproduces. Each male has one male offspring, so that there will be a constant number of males at all times. We start the consideration of the system at an arbitrary point of time. Every male alive at that time will be represented at all future times by one and only one male descended from him in an unbroken male line.

A line of descent from male to male may be regarded as a two-state Markov chain. At the beginning of each season there are two possibilities: either the male in the line under consideration was born in the previous season and hence this will be the first season in which he can mate, or the male in this line did not mate in the previous season and hence is certain to mate in the current season. If there are n males and hence n male lines of descent in the population, we may regard them as forming a Markov chain with $n+1$ states, according to whether $0, 1, ..., n$ matings occurred in the previous season. The transition probabilities for such a chain are easily set up. Additional assumptions are needed to assign probabilities for matings between particular males and particular females. One way is to suppose that the number of matings is determined solely by the probabilities of matings for males governing the Markov chain described. When the males who are to mate in a season have been selected an equal number of females is chosen by selecting, first, those who are already in their second season and have not yet mated and then selecting at random the necessary additional number from those one season old. Finally all ways of pairing the males and females are assumed equally likely. With this scheme it is possible to work out probabilities (which now depend on the state of the system) that two individuals with a given age difference will mate, and also the probabilities (also dependant upon the state of the system), that a pair of relatives of a particular type be born with a given age difference. These probabilities may be combined to determine the probabilities of consanguineous matings.

APPENDIX II. THE COMPUTATIONS

Once the constants to be used have been decided upon, computations using the theory presented are very simple.

The basic equation is formula (20) of Appendix I, namely

$$Y_\xi = \frac{\bar{n}U^2}{\sqrt{(2\pi S_\xi)}} \exp[-\Delta_\xi^2/2S_\xi], \tag{20}$$

where Y_ξ is the number of consanguineous marriages of pedigree type ξ per year (if Δ_ξ and $S_\xi^{-\frac{1}{2}}$ are measured in years) in a closed stationary population.

(a) Relative frequencies of marriages of different type

In the above formula only Δ_ξ and S_ξ vary with the type of consanguineous marriage. Thus if, as in tables 26, 29–31 and 34 we are concerned only to compare the frequencies of different types of marriage we need compute only the expression $S_\xi^{-\frac{1}{2}} \exp[-\Delta_\xi^2/2S_\xi]$ which will be denoted κ_ξ.

General expressions for Δ_ξ and S_ξ may be found from equations (13), (14), (17) and (18) of Appendix I. Values of S_ξ depend on the second moments of certain time intervals. For all the calculations in the present paper the same values of the second moments have been used, namely,

$$V_P = 70, \quad V(a) = 30;$$

$$V_M = 40, \qquad V_i = 55.$$

The justification for these values will be found in Appendix III. V_P and V_M are the variances of the ages of parents at paternity and maternity respectively, $V(a)$ is the variance of the excess of groom's age over bride's age and V_i is a quantity depending upon the intervals between the births of children in a family. To compute S_ξ for a pedigree type ξ one needs to know α_ξ, the total number of males in the ancestry of the bride and groom up to, but not including, the original couple from which they are both descended.

With the values of the second moments given above, S_ξ works out as follows:

degree of relationship ξ	S_ξ
uncle–niece and aunt–nephew	$125 + 30\alpha_\xi$
first cousins	$165 + 30\alpha_\xi$
$1\frac{1}{2}$ cousins	$205 + 30\alpha_\xi$
second cousins	$245 + 30\alpha_\xi$

Δ_ξ is a function of the mean ages of paternity and maternity (denoted by μ_P and μ_M respectively) and of the mean excess of groom's age over bride's age ($\mu(a)$). For practical purposes† one may write

$$\mu(a) = \mu_P - \mu_M. \tag{54}$$

If both spouses belong to the same generation, i.e. for first cousins, second cousins, third cousins, etc., Δ_ξ is a function of $\mu(a)$ only. For this case:

$$\Delta_\xi = (\epsilon_\xi - 1)\,\mu(a), \tag{55}$$

where

ϵ_ξ = (no. of males in ancestry of bride) − (no. of males in ancestry of groom).

$$= \alpha_{\xi\,2} - \alpha_{\xi,1}. \tag{56}$$

For uncle–niece marriages and $1\frac{1}{2}$, $2\frac{1}{2}$, etc. cousin marriages where the groom is of the older generation,

$$\Delta_\xi = (\epsilon_\xi - 1)\,\mu(a) + \mu_M. \tag{57}$$

For aunt–nephew marriages and $1\frac{1}{2}$, $2\frac{1}{2}$, etc. cousin marriages where the bride is of the older generation,

$$\Delta_\xi = (\epsilon_\xi - 1)\,\mu(a) - \mu_M. \tag{58}$$

By way of illustration, for first-cousin marriages S_ξ and Δ_ξ are as follows:

type ξ	α_ξ	S_ξ	ϵ_ξ	Δ_ξ
paternal parallel	2	225	0	$-\mu(a)$
maternal parallel	0	165	0	$-\mu(a)$
patrilateral cross-cousin	1	195	-1	$-2\mu(a)$
matrilateral cross-cousin	1	195	$+1$	0

For $\mu(a) = 5$, the computation of relative expected frequencies of the four types of first-cousin marriages may be set out thus:

type ξ	$\Delta_\xi^2/2S_\xi$	$\exp[-\Delta_\xi^2/2S_\xi]$	$S_\xi^{-\frac{1}{2}}$	κ_ξ	$\kappa_\xi/\Sigma\kappa_\xi$
paternal parallel	0·056	0·946	0·0667	0·0631	0·241
maternal parallel	0·076	0·927	0·0779	0·0722	0·275
patrilateral cross-cousin	0·256	0·774	0·0716	0·0554	0·211
matrilateral cross-cousin	0	1·000	0·0716	0·0716	0·273
total	—	—	—	0·2623	1·000

where

$$\kappa_\xi = S_\xi^{-\frac{1}{2}} \exp[-\Delta_\xi^2/2S_\xi]. \tag{59}$$

If κ_ξ is computed for the 16 types of $1\frac{1}{2}$ cousins using $\mu_M = 30$ and $\mu(a) = 5$, and the results summed for all types, the total is 0·1536. The ratio 0·1536/0·2623 gives the ratio of $1\frac{1}{2}$-cousin to first-cousin marriages in a closed stationary population (table 34).

Adjustments for migration are made by means of formula (47) of Appendix I. For first-cousin marriages, this involves multiplying each κ_ξ by $\lambda_P^{(\alpha+1)}\lambda_M^{(3-\alpha)}$, where λ_P is the proportion of boys and λ_M the proportion of girls who do not emigrate before reproducing. For illustration we take $\lambda_P = 0·81$ and $\lambda_M = 0·87$. These values

† In fact ($\mu_P - \mu_M$) will be slightly greater than $\mu(a)$; see Hajnal (1950, p. 358).

correspond to Swedish emigration rates in 1901–10 between birth and age 25 (Hajnal (1953), p. 128, table C).

type ξ	$\lambda_P^{(\alpha+1)} \lambda_M^{(3-\alpha)}$	$\kappa_\xi \lambda_P^{(\alpha+1)} \lambda_M^{(3-\alpha)}$	proportionate distribution
paternal parallel	0·462	0·0292	0·223
maternal parallel	0·533	0·0385	0·294
patrilateral cross	0·497	0·0275	0·210
matrilateral cross	0·497	0·0356	0·272
total		0·1308	0·999

The distribution between different types is affected only by the ratio λ_P/λ_M. The rate of migration assumed would halve the number of first-cousin marriages in an isolate. The overall effect of migration on the sum of the four types depends mainly on $\frac{1}{2}(\lambda_P + \lambda_M) = 0.84$. Adjustments to the ratios given in table 34 were obtained by multiplying by 0·84 for $1\frac{1}{2}$ cousins, and by $(0.84)^2$ for second cousins.

The growth adjustment is made by formulae (44) and (45) of Appendix I. If growth is fixed in terms of annual rates, formula (39) must first be used.

(b) Absolute numbers

To use formula (20) given above for evaluating Y_ξ itself $\bar{n}$ and U need to be evaluated. They will vary, in accordance with the demographic characteristics of the population. We may first consider values appropriate to European populations in the eighteenth century and earlier. Similar values would probably have applied for most human populations before modern influence began to reduce mortality from its former high level.

$\bar{n}$ is defined by (7) of Appendix I. It depends on the mean and variance of the distribution of marriages by the number of births they produce. If the mean number of births per marriage is 4, $\bar{n}$ must be of the order of 20. For the direct determination of $\bar{n}$ from formula (7) one preferably requires figures based on listing the total number of children born to marriages contracted in a given period. Data of this sort compiled from the parish registers of the French village of Crulai in the eighteenth century (Gautier & Henry 1958, p. 124) give $\bar{n} = 20.8$.

U is the ratio of marriages in a generation to the births from which they sprang, i.e. $2U$ is roughly the proportion of persons surviving to marriageable age (multiplied by the average number of marriages per survivor). For eighteenth-century Europe with something like one half the children born surviving to age 20, U must be of the order of 0·25. Also if a marriage produces 4 children on average, U will be about 0·25 in a population where growth is very slight. (See Appendix I, equation (46)).

Finally from § (a) above the factor κ_ξ summed over the four types of first-cousin marriage gives 0·26 for $\mu(a) = 5$. Smaller values of $\mu(a)$ give a slightly bigger, and larger ones a slightly smaller total. The sum of the four Y_ξ for first cousins is thus of the order of 0·1 or 0·15, i.e. there would be something like 1 or $1\frac{1}{2}$ first-cousin marriages per decade in a closed isolate with eighteenth-century demographic characteristics.

The fall in mortality which took place over much of the Western world in the nineteenth century would certainly have increased this figure several-fold over much of Europe. In the second half of the nineteenth century $U = 0.35$ and $\bar{n} = 30$ would probably not be untypical. For such values also the number of first-cousin marriages per year may be computed, with negligible error, from formula (20) without any correction for growth (see formula (45) above).

(c) *Mean isolate size*

It may be desired to apply the assumptions (i) that isolates are rigidly separated, and (ii) that random mating obtains, and with these assumptions to estimate the mean size of isolates in a region. This is one of the main traditional uses of the Dahlberg formulae. Our formulae may be used for the same purpose.

Let M be the average annual total number of marriages in the region in the period studied. Let M_ξ of these be marriages involving persons of relationship ξ (either of a single type or several types together, e.g. all marriages of first cousins). Assume that Y_ξ has been calculated from the demographic characteristics of the region. Thus M_ξ/Y_ξ is an estimate of the number of isolates into which the region is divided, and (MY_ξ/M_ξ) is an estimate of the average annual number of marriages per isolate.

Let us write
$$f_\xi = M_\xi/M \times 100. \tag{60}$$

Thus f_ξ is the frequency, expressed as a percentage, of marriages of relationship ξ. Moreover, let m be the crude marriage rate, i.e. the annual number of marriages per 1000 total population. Then we can write $10^5 Y_\xi/mf_\xi$ for the estimated average total population of an isolate.

Estimates of isolate size by the Dahlberg method have generally been based on first-cousin marriages. This is reasonable since statistics of first-cousin marriages are likely to be most complete and their frequency less affected by migration than in the case of marriages of more distant cousins. Different and smaller estimates of isolate size would usually be obtained if $1\frac{1}{2}$ or second cousins were used since such marriages are usually recorded in smaller numbers, relative to first-cousin marriages, than would be expected under random mating. (Uncle–niece and aunt–nephew marriages are too few to be of practical use.) If first cousins constitute 2 % of all marriages, and $m = 8$, the mean isolate size would be something like 600 or 1000 in eighteenth-century European conditions.

Such estimates of isolate size relate to total population in the ordinary sense. They can, of course, be adjusted to give mean isolate size in terms of some other desired quantity, such as the number of married women aged 20 to 40, which is believed to measure the 'breeding population'. However, mean isolate size in any of these senses is not directly comparable to estimates obtained by the Dahlberg formulae. It is doubtful if any sensible meaning can be attributed to the Dahlberg estimates (see Sutter & Goux 1961).

Appendix III. The estimation of certain quantities used in the calculations

It would have been difficult to attempt calculations using data for the particular populations for which figures about consanguineous marriages have been collected. In many cases the studies of consanguineous marriages have not covered the total population of a precisely defined geographical area in a defined period of time. When they do relate to such populations, there are still the difficulties which arise when one attempts to study foreign demographic data for units smaller than national states (lack of detailed tabulations, difficulty in obtaining publications, unfamiliarity with pitfalls in sources, etc.).

TABLE 35. CALCULATIONS FOR ESTIMATING V_i

(1) no. of children in family *less one*, i	(2) no. of marriages producing $(i+1)$ births	(3) no. of births occurring after i previous births	(4) no. of intervals of 'length' id	(5) figure in column (4) multiplied by i^2
1	35	226	935	935
2	42	191	709	2836
3	27	149	518	4662
4	26	122	369	5904
5	32	96	247	6175
6	26	64	151	5436
7	15	38	87	4263
8	9	23	49	3136
9	7	14	26	2106
10	3	7	12	1200
11	3	4	5	605
12	1	1	1	144
total	226	—	3109	37402

It seems more important in any case to explore some of the general deductions from our formulae which would apply in a wide variety of situations. Fortunately it is likely that many of the quantities needed have broadly the same orders of magnitude in most populations of substantially uncontrolled fertility, i.e. almost all human populations up to the latter part of the nineteenth century. For populations which are not stable the appropriate approximations from our formulae are obtained by putting in constants relating to the time of birth of the spouses whose marriages are studied (or even, for marriages between more remotely related persons, constants relating to one or two generations earlier). We may therefore use values of V_i, V_P, V_M appropriate for populations of high fertility for comparison with most of the data on consanguineous marriages so far published. However, for studies on Western populations where widespread family limitation has been practised for several generations, it will soon be desirable to use constants which take account of that fact; probably sequences of constants will be needed reflecting the development of the demographic transition over time.

The most difficult quantity to estimate is V_i, the variance of the distribution of intervals between the births of pairs of children of the same marriage. As is explained in connexion with equation (14) in Appendix I above, all possible pairs are considered and each interval is taken both positively and negatively, i.e. the mean of the distribution is zero by definition. There appear to be no direct data on this kind of distribution. Various methods for estimating the variance required can be suggested. The method actually adopted has at least the merit of requiring little labour. It can best be explained by setting out the calculation (table 35).

The data required are the distribution, by the total number of children produced, of an unbiased sample of all marriages (this requirement excludes, for example, data restricted to marriages where the wife survived to the end of the child-bearing period). Table 35 relates to marriages which occurred in the village of Crulai in France in the period 1674–1742 (see Gautier & Henry 1958, p. 124). The data are reproduced in column (2) (the numbers of families with fewer than two children are not given, since they are irrelevant to this calculation). Column (3) is obtained by summation of column (2) from the bottom. Column (4) is similarly computed by summing column (3) from the bottom. The idea is that a birth to a couple who have had i earlier births creates i intervals. Let us suppose that the interval between two successive births is always the same, say d years. A birth occurring after i previous births would then create one interval of length d (with the immediately preceding birth), one of length $2d$ (with the last birth but one) and so on up to an interval of length id which separates the birth under consideration from the first birth of the marriage. (The existence of multiple births is neglected.) Column (4) thus gives the distribution of intervals by length in units of d.

We require to assign a negative sign to half the intervals. Intervals of any length are as likely to be positive as negative, i.e. if we consider intervals separating the birth of children of differing sex the boy will have been born earlier in half the cases. The variance of the distribution of intervals when due account is taken of sign is, therefore, the same as the second moment about the origin of the distribution of intervals if all are treated as positive. The calculations for computing this second moment are in column (5). From the sums of columns (4) and (5) we have

$$V_i = d^2 \times \frac{37\,402}{3\,109}.$$

The major defect of this calculation is probably that it neglects the fact that marriages which are to produce many children must produce children in relatively rapid succession. Large families are responsible for an overwhelming proportion of the intervals. The longest intervals, namely those between the first and last children of large families, are overestimated by a calculation which assumes that all intervals between successive births are identical in length.

It, therefore, seemed best to take for d the mean interval between first and second births, namely 2·1 years (op. cit. p. 339). The mean of all intervals (op. cit. p. 155) was slightly larger, namely 2·3 years. We thus obtain

$$V_i = (2\cdot1)^2 \times 12 = 53.$$

The variation in the length of intervals between successive births within the same family adds an additional component to the variance. It is probably of the order of 1 year. For calculations $V_i = 55$ was adopted as a convenient round number.

This value is probably still too high for eighteenth-century conditions, but may well be too low for the period after the decline in mortality had set in. This decline increased the percentage of large families and hence of long intervals.

Estimates of V_P or V_M may be obtained by computing the variances of the distributions of the births in a year (or years) by age of father or by age of mother. Distributions of births by age of mother are very frequently available; tabulation by age of father is much rarer. Data for recent years may be found collected in United Nations, *Demographic Yearbook, 1954*.

The data for Western countries or Japan are, however, not relevant to our purpose since in these countries family limitation is nowadays widely practised and this may considerably reduce V_P and V_M. Distributions of births by age of father are available for only a few underdeveloped countries (mainly in Latin America) and the reliability of the data is often suspect. It is desirable to supplement these materials by Western data for a time before the spread of family limitation. Calculations were made using data for Scotland in 1855 (these are given in Lewis & Lewis 1905) and Norway in 1881–85 (see Norway, Statistical Office, 1890, table 22). A calculation was also made using data from a survey in India (Dandekar & Dandekar 1953). These various computations suggest that variation in V_P and V_M between populations of high fertility is not very great and that $V_P = 70$ and $V_M = 40$ are reasonable round values.

In strictness V_M should not be computed from the distribution by age of mother of the births occurring in a given calendar interval of time, but the distribution by age at maternity of the births of a cohort of women followed throughout life. The distribution of births by age of mother in a calendar period is normally distorted by the inequality in the original size of the different cohorts making up the female population of child-bearing age. It is not difficult to correct for such distortion. One way of doing so is to divide the number of births to mothers of an age group by the total female births which occurred when the mothers in that age group were born; if ratios are formed for all age groups at maternity and divided by their sum, the resulting distribution may be used for calculating V_M. Another way is to use the distribution obtained by multiplying the age-specific fertility rates by the number of women in the corresponding age-group of the life-table population (as is done for the computation of a net reproduction rate). However, it turns out that the effect of such corrections on V_M is small (though the effect on estimates of μ_M may be greater). Similar remarks, of course, apply to V_P.

For $V(a)$ the data given by Karmel (1948) suggest that 30 is a reasonable figure. Only the sum $V_i + V(a)$ enters into the calculations. Proportionately quite large variations in $V(a)$ thus make little difference.

LIST OF FREQUENTLY USED SYMBOLS

symbol	formula where first used	short definition, where possible
$a(z)$	10	Proportion of marriages where groom is z time units older than bride [see also formula (27)]
$\alpha_{\xi,1}$	13	Number of males in ancestry of groom
$\alpha_{\xi,2}$	13	Number of males in ancestry of bride
α_ξ	14	Number of males in combined ancestry of groom and bride
B	49	Expected number of births per unit time in stationary population
$B(t)$	21	Expected number of births at time t in stable population

$\beta_{\xi,1}$	13	Number of females in ancestry of groom
$\beta_{\xi,2}$	13	Number of females in ancestry of bride
Δ_ξ	17	—
δ_ξ	45	Number of persons in combined ancestry of groom and bride [see also (23)]
e_ξ	56	Excess of males in ancestry of bride over males in ancestry of groom
$h_\xi(z)$	9	Density function of distribution of intervals between births of spouses in stationary population
$h_\xi(s,z)$	24	Same as $h_\xi(z)$, but for stable population
κ_ξ	59	—
λ_P	47	Proportion of bridegrooms born within isolate
λ_M	47	Proportion of brides born within isolate
M	1	Expected number of marriages per unit time in stationary population
$M(t)$	22	Expected number of marriages at time t in stable population
$m(z)$	9	Probability of marriage between newborn boy and girl born z time units later
$\mu(a)$	15	Mean of $a(z)$ distribution
$\mu_\xi(h)$	13	Mean of z in $h_\xi(z)$ or $h_\xi(s,z)$ distribution
μ_P	13	Mean age at paternity
μ_M	13	Mean age at maternity
N_ξ	1	Mean number of potential marriages of type ξ among descendants of a couple
n_j	6	Probability that a marriage produce j live births
$\bar{n}$	7	—
p	3	Proportion of males among the newborn
Π_ξ	1	Probability that potential marriage of relationship ξ actually take place
$\Pi_\xi(t)$	24	Same as Π_ξ, but for stable population
q	3	Proportion of females among the newborn
r	21	Rate of growth (or decline) of stable population
R_P	2	Expected number of live born sons for newborn boy
R_M	23	Expected number of live born daughters for newborn girl
S_ξ	18	—
U	10	Ratio of marriages to births [see also formula (28)]
$V(a)$	15	Variance of $a(z)$ distribution
$V_\xi(h)$	14	Variance of $h_\xi(z)$ distribution
V_i	14	Second moment about zero origin of distribution of intervals between births of siblings
V_P	14	Variance of ages at paternity
V_M	14	Variance of ages at maternity
Y_ξ	1	Expected number of marriages per unit time between persons of relationship ξ

REFERENCES (Hajnal)

Agarwala, S. N. 1957 The age at marriage in India. *Population Index*, **23**, 96–107.

Alström, C. H. 1958 First-cousin marriages in Sweden 1750–1844 and a study of the population movement in some Swedish sub-populations from the genetic-statistical viewpoint. *Acta, Genet.* **8**, 295–369.

Barrai, I., Cavalli-Sforza, L. L. & Moroni, A. 1962 Frequencies of pedigrees of consanguineous marriages and mating structure of the population. *Ann. Hum. Genet., Lond.*, **25**, 347–376.

Bell, J. 1940 A determination of the consanguinity rate in the general hospital population of England and Wales. *Ann. Eugen., Lond.*, **10**, 370–391.

Blanc, R. 1962 Le mariage en Afrique; concepts et aspects démographiques. To be published in *Proceedings of International Population Conference, New York*, 1962.

Böök, J. A. 1958 Some aspects of practical applications of the theory of population genetics to man. *Bull Inst. Int. Stat.* **36** (3), 277–283.

Brenk, H. 1931 Über den Grad der Inzucht im einem innerschweizerischen Gebirgsdorf. *Arch. Julius Klaus Stift.* **6**, 1–39.

Cantrelle, P. 1960 L'endogamie des populations du Fouta Senegalais. *Population, Paris*, **15**, 665–676.

Cavalli-Sforza, L. L. 1956 Some notes on the breeding pattern of the human population. *Acta Genet.* **6**, 395–399.

Cavalli-Sforza, L. L., Moroni, A., Zalaffi, C. & Zei, G. 1960 Analisi della consanguineita osserrata in alcune diocesi dell' Emilia. *Atti Ass. Genet. Ital.* **5**, 305–316.

Chandra Sekar, C. 1948 Some aspects of Parsi demography. *Hum. Biol.* **20**, 47–89.

Dahlberg, G. 1929 Inbreeding in man. *Genetics*, **14**, 421–454.

Dahlberg, G. 1948 *Mathematical methods for population genetics.* Basle: S. Karger; New York: Interscience Publishers.

Dandekar, V. M. & Dandekar, K. 1953 *Survey of fertility and mortality in Poona District.* Poona: Gokhale Institute of Politics and Economics, publication no. 27.

Egenter, A. 1934 Über den Grad der Inzucht in einer Schwyzer Berggemeinde und die damit zusammenhängende Häufung rezessiver Erbschäden. *Arch. Julius Klaus. Stift.* **9**, 365–406.

Freire-Maia, N. 1957 Inbreeding in Brazil. *Amer. J. Hum. Gen.* **4**, 127–138.

Freire-Maia, N. 1961 The structure of consanguineous marriages and its genetic implications. *Ann. Hum. Genet., Lond.*, **25**, 29–39.

Frota-Pessoa, O. 1957 The estimation of the size of isolates based on census data. *Amer. J. Hum. Gen.* **2**, 9–16.

Gautier, E. & Henry, L. 1958 La population de Crulai, paroisse normande. *Institut National d'Études Démographiques; Travaux et Documents*, Cahier no. 33.

Gibson, G. D. 1958 Herero marriage. *The Rhodes-Livingstone Journal*, **24**, 1–37.

Grob, W. 1934 Aszendenzforschungen und Mortälitatsstatistik aus einer st. gallischen Berggemeinde. *Arch. Julius Klaus Stift.* **9**, 237–264.

Hajnal, J. 1950 Births, marriages and reproductivity. In: *Papers of the Royal Commission on Population*, Vol. II. *Reports and Selected Papers of the Statistics Committee.* London: H.M. Stationery Office.

Hajnal, J. 1953 Age at marriage and proportions marrying. *Popul. Stud.* **7**, 111–136.

Hajnal, J. 1960 Artificial insemination and the frequency of incestuous marriages. *J. R. Statist. Soc. A*, **23**, 182–194.

Hammond, D. T. & Jackson, C. E. 1958 Consanguinity in a mid-western United States Isolate. *Amer. J. Hum. Gen.* **10**, 61–63.

Kang, Y. S. & Cho, W. K. 1959 Data on the biology of Korean populations. *Hum. Biol.* **31**, 244–251.

Karmel, P. H. 1948 The relations between male and female nuptiality in stable populations. *Popul. Stud.* **1**, 353–387.

Lewis, C. J. & Lewis, J. N. 1905 *Natality and fecundity.* Edinburgh: Oliver and Boyd.

Ludwig, W. 1949 Die Häufigkeit der 4 Typen von Vetternehen. *Genus*, **6–8**, 259–266.

Mortara, G. 1948 Determinaçao de nupcialidade feminina, segundo a idade, no Brasil com base na apuraçao censitaria do estado conjugal, e aplicaçoes ao calculo da tasea de nupcialidade. *Revista Brasileira de Estatistica*, **9**, 56–82. (English version in United Nations, Department of Social Affairs, *Population Studies*, no. 7, New York 1949.)

Morton, N. E. 1955 Non-randomness in consanguineous marriages. *Ann. Hum. Genet.* **20**, 116–124.

Nemeskéri, J. & Thoma, A. 1961 Ivád: an isolate in Hungary. *Acta Genet.* **11**, 230–250.

Norway, Statistical Office. 1890 Folkemaengdens Bevaegelse 1866–1885. *Norges Officielle Statistik, Tredie Raekke* No. 106. Kristiania: H. A. Schehoug and Co.

Orel, H. 1932 Die Verwandtenehen in der Erzdiözese Wien. *Arch. Rass- u. GesBiol.* **26** (3), 249–278.

Ruepp, G. 1935 Erbbiologische Bestandesaufnachme in einem Walserdorf der Voralpen. *Arch. Julius Klaus Stift.* **10**, 193–218.

Sanghvi, I. D., Varde, D. S. & Master, H. R. 1956 Frequency of consanguineous marriages in twelve endogamous groups in Bombay. *Acta Genet.* **6**, 41–49.

Schapera, I. 1957 Marriage of near kin among the Tswana. *Africa*, **27**, 139–159.

Schull, W. J. 1958 Empirical risks in consanguineous marriages: sex ratio, malformation and viability. *Amer. J. Hum. Genet.* **10**, 294–343.

Serra, A. & Soini, A. 1959 La consanguinité d'une population. *Population, Paris,* **14**, 47–72.

Sutter, J. & Goux, J. M. 1961 L'aspect démographique des problèmes de l'isolat. *Population, Paris,* **16**, 447–462.

Sutter, J. & Tabah, L. 1955 L'evolution des isolats de deux départements français: Loir-et-Cher et Finistere. *Population, Paris,* **10**, 645–674.

Twiesselmann, F., Moreau, P. & François, J. 1962 Évolution du taux de consanguinité en Belgique de 1918 à 1959. *Population, Paris,* **17**, 241–268.

Wulz, G. 1925 Ein Beitrag zur Statistik der Verwandtenehen. *Arch. Rass- u. GesBiol.* **17**, 82–95.

Yanase, T. 1962 The use of the Japanese family register for genetic studies. In: *The use of vital and health statistics for genetic and radiation studies* (*Proceedings of the seminar sponsored by the United Nations and World Health Organisation, held in Geneva, 1960*). New York: United Nations.

[*Editors' Note:* The discussion has been omitted.]

Reprinted from *Stochastic Models in Medicine and Biology,* J. Gurland, ed., The University of Wisconsin Press, Madison, 1964, pp. 179–196

Monte Carlo Simulation: Some Uses in the Genetic Study of Primitive Man

W. J. SCHULL and B. R. LEVIN

Through most of his existence man has been a hunter and gatherer of food -- shaped by rather than shaping his environment. This way of existence imposed a substantial number of restrictions, social and biologic, upon his evolution. Clearly these restrictions have varied with time, but it seems reasonable to suppose that the first major change occurred with man's recognition that he could exercise some control over his source of food either through farming or herding, or both. The earliest evidence of this important step is to be found in the Shanidar valley culture of Northern Iraq of some 80 centuries before Christ (Solecki, 1963).

As a hunter and gatherer man's numbers must have been sharply limited -- a response to the limitations of food and the obvious need for mobility. The smallness of the groups in which he lived was both an asset and a handicap. It restricted rather markedly the number and kind of his diseases and by so doing lessened the role played by disease in his evolution . (For a discussion of the role of disease in evolution, see Haldane, 1948). But this same smallness exposed the individual group to the constant threat of extinction either from beasts or marauding fellow men. With the advent of agriculture and a more sedentary way of life, man's culture changed dramatically, as did the selective forces which operated upon him. Of particular importance was the changed role of disease. Some have held that contemporary man is largely, if not wholly, a product of events since the advent of agriculture. If this is so, then the exploration of the manner in which primitive man's way of life influenced his genetic structure has only heuristic interest, but we believe that it holds more.

Stochastic models in genetics

The processes which shape the genetic composition of a population are generally regarded as either systematic or dispersive. The former consist of those directional processes such as mutation, migration, and selection; whereas the latter, the dispersive forces, are the fluctuations in gene frequencies due to random sampling of gametes, generally referred to as "genetic drift," random variation in selection intensities, and random variation in migration rates. The effects of most of these processes are dependent upon the mating behavior of the population under scrutiny. The study of the relationship of these processes to the persistence and spread of a gene or group of genes forms the core of population genetics. Much of the present theory we owe to Sewall Wright, the late R. A. Fisher, J. B. S. Haldane, and more recently, Motoo Kimura. Attention has, in general, centered upon those situations where mathematically explicit statements were possible, but even here it has been frequently necessary to make simplifying assumptions which limit the generality of the result. Clearly, difficult or not, an explicit, comprehensive, and preferably stochastic statement of evolutionary theory is desirable. Such a statement does not now exist and because of the complexity of the problem progress toward it is slow. It is tedious to separate the relevant from the irrelevant, the important from the unimportant. However, digital computers and Monte Carlo solutions now afford us the opportunity to seek numeric answers to some of the problems which have thus far proven to be mathematically intractable. Simulation methods admittedly lack the appeal of explicit mathematical statements, but if one is pragmatic, these methods hold great promise for an early insight into a variety of interesting and important problems. In fact, at this juncture, it may well be that numeric analysis is more rewarding than an analytic, mathematical approach. What, now, are some of these problems which have proven intractable to formal methods?

Most population genetic theory has been until quite recently largely of a deterministic nature -- a matter of limited concern, perhaps, if emphasis is upon expected or average behavior in large populations. Clearly, in small populations, such as those which must have characterized primitive man, the stochastic element can not be ignored. Until 1954 and Kimura's work on the role of random fluctuations in selection intensities in the quasi-fixation of genes, the only random element prominently incorporated into the theory was that associated with the sampling of gametes. Earlier work, notably that of Fisher and Wright, had concentrated on the steady state distributions. Kimura has subsequently incorporated several other factors into the stochastic model, but to do so he has generally been obliged to assume a constancy of population size, randomness of mating, frequency independent selection intensities, etc. Thus, the case which he analyzes, though undoubtedly biologically meaningful for a variety of animals, may be an exceptional situation in man. Many of the cases which he as well as others have studied fail to take into account the simultaneous action of all of the processes, systematic and dispersive, which shape gene frequencies, although there are notable and important exceptions to this statement. The program which we propose to discuss shortly admits of some 50 different parameters which may be varied singly or in groups.

Among the important unsolved problems which can be readily studied numerically are the following: (1) The effect on the extinction of a gene of frequency-dependent selection intensities. (2) The effects on the genetic processes of age-specific birth and death rates. (3) The effects of a variety of kinds of phenotypic selection on gene frequencies in a natural population. (4) The rate of approach to homozygosity in the case where a population is subdivisible into two or more interbreeding subunits or demes in which the selection coefficients differ among the units. (5) Some years ago, the chairman (Crow, 1955) suggested on intuitive grounds, as nearly as we can tell, that in nature non-constant selective values must more often lead toward fixation than toward maintenance of variability in a panmictic, that is, in a randomly breeding, population. If our surmise that this was intuition is correct and there is no formal proof of the statement, then the problem could be answered by numeric means. (6) Finally, virtually every problem which has thus far been solved could be reinvestigated treating population size as a random variable. Situations in man can be cited which approximate virtually all of the problems we have enumerated. For example, in the case of the interbreeding subunits with different selection intensities, we know that the Birhor of India (Williams, 1963) live in bands, or demes of 25 to 35 persons. They are a hunting and gathering people and each band has a strong sense of territoriality. At any given time, some groups are relatively more affluent than others. It seems reasonable to assume that selection may, as a consequence, be unevenly distributed over neighboring bands. However, each band exchanges wives in a seemingly fixed pattern with the bands which surround it; hence the frequency of a given gene in a given band may be some complex function of a variety of selective forces.

To indicate how some of these problems might be approached let us examine the simulation program we have begun to use. We begin with a few general remarks on the nature of this program and the uses to which it, as well as other programs, have been put.

Some general remarks on the program

The applicability of simulation methods to genetic problems has been recognized by many, and, as a consequence, there already exists a modest but promising literature which attests to the broad potential of these techniques. Fraser (1957, 1962), for example, has applied them to a study of the effects of linkage and epistasis on selection. Lewontin and Dunn (1960) have investigated with a stochastic

model the effect of a positive gametic selection on the t-allele polymorphism in the house mouse. Levin, Petras, and Rasmussen (1963)
have extended the Lewontin and Dunn model to include intra-demic
migration. Crosby (1963) has simulated certain evolutionary problems
in Oenothera, the primrose. Martin and Cockerham (1960) have
modeled several problems involving mass selection and linkage such
as may arise in plant and animal breeding. Finally, Brues (1963) has
used simulation methods in a study of the interaction of genetic drift
with maternal-fetal incompatibility reactions and selection in the
maintenance of the ABO blood group polymorphism in man. Ultimately
the value of the models used by these investigators as well as the
one to be proposed by ourselves is dependent upon the validity of the
assumptions made in their construction. In the case of Monte Carlo
simulations, the basic assumptions are set out in the form of a series
of simple decisions, for example, the choice of a male parent, the
segregation of two alleles, or the determination of the sex of an individual. Under these circumstances, it seems unlikely that the user
of Monte Carlo methods is apt to incorporate unwittingly many cryptic
assumptions. Be this as it may, decisions are made by comparing the
various probabilities of the different outcomes of an event to a random
number, r, where $0 \leq r \leq 1$. As an illustration, in the choice of a
parent from a population in which the various genotypic frequencies
are $f(AA)$, $f(Aa)$, and $f(aa)$, if the random number is less than or
equal to $f(AA)$, the parent is taken to be homozygous AA . If the
random number is greater than $f(AA)$ but less than or equal to the
sum $f(AA) + f(Aa)$, the parent is heterozygous. Finally, if the random number is greater than the latter sum, the parent is homozygous
aa . Quite clearly the integrity of this simulation is dependent upon
the randomness of the so-called random numbers. In our case, these
are generated by a subroutine which uses the power residue method
(IBM Reference Manual, 1959). The random number generation sequence
which is generated has a periodicity of 2^{b-2} terms when the word size is
of b-bits and the arithmetic assumes the binary point to be at the
extreme right of the word. With the IBM 7090 this means a periodicity
of 2^{33} . A difference in runs is obtainable by merely starting at a
different place in the sequence.

The various steps and decisions in the program are presented dia-
grammatically in Figure 1. In the main, the figure is self-explanatory
but a few remarks seem in order. Although all events occur in a serial fashion in the program, individuals in the same generation or age-
set are mated at the same instant in biological time. The sampling
of mates from the parental pool alters the genotypic frequencies in
these pools, that is, sampling is without replacement. Selection in
the form of differential mortality is represented by storage decisions.
Individuals selected against simply are not stored. Selection in the

form of differential fertility is represented, obviously, by the differential production of offspring. At the moment, immigration is viewed
as being a draw from an infinite source which is supposedly representative of the surrounding populations. Emigration is a random reduction
of the population under scrutiny.

As most of you are well aware, Monte Carlo methods are, by their
very nature, inefficient in terms of computer time. In this simulation,
which was programmed in MAD (Michigan Algorithm Decoder), and run
on an IBM 7090, with a binary deck, about one-third of a second is
required per generation to produce a population of 100 individuals. A
single 200 generation run requires slightly more than one minute.
Needless to say, the number of runs required depends primarily upon
the relative stability of the system being simulated. At this preliminar-
ary stage, we have usually been content with 15 runs. These are
averaged and the variation in the system at a given generation judged
by the mean of the ranges derived from successive sets of five runs.
There is as yet no consensus with respect to the optimal number of
runs. Clearly, to document with precision some specific model more
than 15 runs may be needed, but as a screening procedure this number
may be adequate.

The input parameters

One of the major justifications for simulation programs is, as we
have indicated, the opportunity they afford to study the effect on any
given genetic model of concurrent variation in a number of parameters.
In a very real sense, the limiting factor to this variation is the inef-
ficiency of the conventional digital computer where one is obliged to
achieve stochastic operations through machine elements which are
essentially deterministic in behavior and are programmed serially.
Recently, Connelly and Justice have advocated a special computer with
statistical switches, the latter being biasable binary random function
generators. This computer would not be obliged, presumably, to
spend most of its time generating or examining random numbers.

As we have previously stated, we can vary some 50 parameters
in the present program, and can readily extend this with the addition
of a few more statements (we presently employ about 370). We
would be remiss, however, if we led you to believe that our present
program is capable of solving all of the problems previously outlined.
It is not; in fact, it is still quite crude but we believe that it has
promise.

We turn, now, to a brief consideration of the parameters of interest to us, and the values which seem appropriate for our purposes.

Population size. -- Our notions on the probable size of the bands
of primitive man stem from contemporary hunting and gathering cultures.
The representativeness of these groups is open to challenge. This

notwithstanding we have selected the population sizes we have used
largely on the basis of information on three such groups of people.
We know that the Xavante, a primitive group living along the Rio das
Mortes in Brazil, usually live in bands of approximately 200 (Neel et
al., 1963). The Birhor of India, who like the pygmies of the Ituri
forest may represent a highly specialized case of hunting and gathering culture, travel in groups of approximately 25 during the wet season,
and in groups averaging 30-35 in the dry period of the year (Williams,
1963). Finally, the average tribal size for the Australian aboriginal,
as judged by linguistic standards, is given as 523 (Birdsell, 1950)
although the hunting band size is undoubtedly much smaller. We are,
of course, interested in the size of the effective breeding population,
and the figures just quoted are not necessarily breeding sizes. Accordingly, we have generally assumed the effective population size
to be between 50 and 200.

We can with our present program view population size as a fixed
constant or as a random variable subject to some distribution. While
most of our runs to date have been on populations of constant size,
we have attempted a few where the size was permitted to vary randomly between two limits. When the upper bound was reached, the group
automatically divided into two, fissioned so to speak. When the lower bound was attained, a fusion occurred to return the population to
the mean value. Our interest in population bounds stems from the
belief that empirically at least, the more sagacious of primitive men
must have recognized that there was an optimum population size and
that drastic departures from this optimum threatened the existence of
the group. In support of this contention, it is known that certain contemporary but primitive groups practice a form of population control
through infanticide, and that others do on occasion bud-off groups
comprising several families. The Xavante studied by Neel and his
colleagues have recently gone through such a process.

Age stratification. -- Most primitive peoples have a less quantitative sense of time than characterizes our own culture. Exact ages
are, as a rule, unknown, and moreover they are generally viewed as
unimportant. Age differentials tend to be recognized not in terms of
seasons or years but rather in terms of the age-sets to which individuals belong. The persons comprising a given age-set move together
through life much as a cohort; they attain manhood or womanhood
simultaneously; they marry in a prescribed manner; etc. As a generality, the regulations governing the males are the more rigid. While
the difference between oldest and youngest in an age-set undoubtedly
varies somewhat between age-sets, the average difference can be
estimated on the basis of dentition, bone age, etc. in those instances
where it is not a matter of tradition. The interval associated with an
age-set in the Xavante, for example, appears to amount to five years,

on the average. We have, for convenience, elected to recognize, with a given sex and generation, three age-sets rather than the five or six the Xavante data would suggest. Thus, in our simulation, each age-set corresponds to a span of approximately ten years.

Mating formula. -- Man appears to have exercised more ingenuity in the restrictions he has formulated to control his mating behavior than in ways of distributing the economic burden of his culture. He may be monogamous or polygamous, and if polygamous, either polyandrous or polygynous. His polygamy may be concurrent or serial since marriages may be arranged for some lesser period than life. Temporary arrangements such as wife lending further complicate his behavior. Moreover, he frequently favors a choice of spouse which gives rise to a correlation between uniting gametes either as a consequence of inbreeding or a phenotypic assortative mating. Finally, he generally has fixed but not inviolate notions about an acceptable age differential twixt husband and wife. With respect to man, in no other area, perhaps, are our present models more unrealistic than in their treatment of the manner in which he mates.

The occurrence of overlapping, or possibly we should say, interbreeding generations has long been a matter of concern in population genetics. When it has been possible to incorporate a sense of this overlapping into the model it is generally assumed that overlapping is a continuous phenomenon, a reasonable assumption in our culture (see Moran, 1962). However, in primitive man with his age-set notion of time, this would hardly seem to be the case. The overlapping must have been discontinuous. We can and have incorporated interbreeding generations into our runs through the specification of the manner in which mating occurs between age-sets. One simply requires the female to be drawn from an age-set younger than the male.

Again, to give some dimension to the variation in marital practices we cite the Xavante. In this group, chiefs, heads of clans, and other important members of the tribe are generally polygynous, with the number of wives varying from two to five. Lesser males are either monogamous or go unmarried. Among the Xavante studied by Neel and his colleagues, of 37 living adult males 23 were monogamous and 14 polygynous. Of the latter 14, 11 had two wives, two had three, and one, the chief, had five wives.

Selection. -- Selection intensity may be considered at several levels in the program, but in all cases selection implies that some individuals are differentially represented by progeny in succeeding generations. This may arise either as a consequence of differential survival, or differential fertility, or both. Selection to be genetically meaningful, and, therefore, of evolutionary significance, must correlate either with one's phenotype or one's genotype. In general, it is much easier to treat differential mortality than differential fertility, for, as we have stated, the former is merely a decision with respect to storage and this decision can be made in such fashion that the selection intensity is a fixed constant with or without random fluctuations. Differential fertility is, however, the more interesting. For example, among the Xavante to which reference has been made, the present chief has fathered a minimum of 25 per cent of the next generation -- the precise figure being a function of how one elects to define a generation. If this is a continuing custom, its consequences would seem large.

At present, we treat fertility as if it were an attribute of a mating. This attribute, however, is viewed as being a multiple of a linear function of a constant associated with the male and a constant associated with the female, and these latter constants are genotype dependent. This permits us to assign specific genotypic values, but still have differences in fertility between reciprocal crosses. In man an important case in point would be the probable difference in fertility between the mating of an Rh+ male with an Rh- female and the reciprocal, namely, the marriage of an Rh- male with an Rh+ female. Rucknagel and Neel (1961) have presented data which can be interpreted as suggesting that a similar situation may obtain in Africa in matings of normal individuals with those heterozygous for the gene which when homozygous gives rise to sickle cell anemia.

We can not accommodate at present a phenotypic selection in the usual sense. There are numerous interesting problems in this area, however. A case in point occurs in those groups of people where the chiefs are polygamous, contribute, therefore, disproportionately to subsequent generations, and where the selection of the chief is essentially phenotypic.

Migration and mutation. -- These two processes differ essentially only in the frequency of their occurrence. We do not at present allow for mutation, but this can be readily done. We can accommodate migration either in or out, but have, as previously stated, been more concerned with notions of a critical population size and sporadic but relatively large-scale migration rather than the case of a constant, small and perhaps fluctuating migration pressure.

Number of loci. -- At present we can only simulate models which involve one genetic locus, but we are currently extending this to two loci.

Number of alleles at a locus. -- Two alleles are routinely considered, but extension to three, four, or some other small number is a simple matter.

Some preliminary results

As frequently happens in any complex program a surprising amount of time can be invested in debugging. Our experience is no exception.

Accordingly, the results which we have to show are less exciting than we should like. They are essentially of two kinds. On the one hand, they represent tests of the program through contrasting stochastic with deterministic results for certain simple situations. On the other hand, they afford an insight, admittedly limited, into the effect of a discontinuous overlapping of generations and some kinds of fertility differentials. We propose to expose these cases by enumerating the restrictions under which the runs occurred and then commenting briefly on what we view as the salient features of each case.

Case 1. Effect of discrete overlapping of generations on genetic drift.--Kimura's model of genetic drift assumes non-overlapping generations but views the changes in gene frequency as continuous. As we have indicated, in primitive man and in time spans of tens of generations this may not be true. Our model permits us to examine this proposition. The population which we simulated was one of fixed size mating monogamously and at random. The sizes chosen were 48, 102, and 198 so that we might have six age-sets of equal size, three male and three female. The age-sets were ranked, and the mating system assumed that a male drew his spouse from the next age-set in rank to the rank of his own set. For example, males of age rank 1 would mate with females of age rank 2 and their offspring would constitute the age rank 1 (male and female) of the next generation. Thus, in each generation one-third of the males obtained their spouses from the next generation. ·All matings were assumed to produce two offspring, one male and one female. No selection, mutation, or migration operated on the system, and the initial population was assumed to be completely heterozygous. The results are to be found in Figure 2.

Case 2. A locus with incomplete selection against one homozygote, and reproductive compensation. -- Among the San Blas Indians of the Caribbean, albinism reaches extraordinary frequencies for human populations or any population for that matter. The maintenance of these high gene frequencies in the face of obvious selection has been an intriguing problem. We know that, in the past, albino infants were often destroyed, and that among those who reached adulthood the males were not permitted to marry and the females, though marriageable, were not viewed as particularly desirable spouses. Albinism is also relatively common among the Hopi Indians of the United States. Woolf and Grant (1962) have suggested that genetic drift, cultural selection favoring the albino, or a small heterozygous advantage could account for the Hopi frequency.

We have simulated the San Blas case in the following manner: We have assumed a population of fixed size, 102, breeding randomly with overlapping generations as outlined in Case 1. Monogamy was the rule. There was no mutation nor migration. However, the albino, the recessive homozygote, was assumed to have a survival frequency of only 0.25. But we assumed that all matings ultimately produced exactly two children, one male and one female. Thus, in a mating of two heterozygotes, if an albino is produced and eliminated by selection, this mating produces another child! This has been called reproductive compensation. Again, the initial population was assumed to be completely heterozygous. This choice of input parameters is, of course, a value judgment, but we have no reason to believe that the shape of these curves will be a function of the input parameters. There exists a deterministic solution for a closely related case, and this along with our results will be found in Figure 3. We note that the two models agree moderately well. Exact agreement is not to be expected for the stochastic model includes reproductive compensation which is not present in the deterministic model. Reproductive compensation can only slow down the rate of elimination of the gene, and the figure is in agreement with this.

Case 3. A balanced polymorphism with complete selection against one homozygote and 20 per cent selection against the other. -- This may be viewed as a somewhat extreme version of the situation which appears to obtain with respect to sickle cell anemia in man in malarious areas of the world. Our simulation invoked the following restrictions: The population was assumed to have a constant size, namely 102 individuals. Mating was at random between age-sets but such that overlapping generations occurred. The population was monogamous. No individuals of one homozygous class survive, and only 80 per cent of the other homozygous group. All matings, however, ultimately produced two offspring, one male and one female. Again, the initial population was taken to be completely heterozygous. The steady state frequency on a deterministic basis for a population where the fitness of the homozygotes are 0.8 and 0 can be readily computed. Our stochastic model differs from this, however, in that reproductive compensation occurs. The steady state value for the case just cited is approximately 0.166; the results of our simulation are to be found in Figure 4. It is patent on inspection that a stable equilibrium exists at a gene frequency somewhat greater than 0.1, but not as high as the deterministic value. This is undoubtedly attributable to the reproductive compensation.

Case 4. A locus in which differential fertility is a function of the mating type rather than the genotype. -- We know that in man maternal-fetal incompatibilities occur, and that some of these may take the form of hemolytic disease of the newborn, a serious and often fatal anemia. Thus, in the Rh system an Rh+ child conceived by an Rh- mother runs a risk of dying not run by an Rh- child of the same mother. Selection is not complete against the Rh+ children of Rh- mothers, however, for immunization of the mother is a necessary condition before hemolytic disease can occur. Immunization requires either a transfusion or the conception of at least one previous Rh+ child. The system is more complicated than this in that hemolytic disease as a consequence of Rh incompatibility depends, in part, upon the compatibility of mother and child with respect to another genetic locus, namely, that associated with the ABO system. This situation leads us to consider the simulation of a population in which selection occurred through ascribing differentials in fertility to matings rather than to specific genotypes. Our model does not describe the Rh system exactly, but it is a step in this direction. Specifically, we assume a population of fixed size, 100, discrete and non-overlapping generations. Mating is monogamous but at random. There is no mutation, nor migration. Of the nine possible mating types when the sex of the parents are considered, all are assumed to have equal fertility save for the case of an RR x rr, and Rr x rr . These are assumed to have only half, and three-fourths the fertility of the others. The sex ratio is not a random variable. The initial population is completely heterozygous. The results are to be seen in Figure 5. Of particular interest to us was the relatively slow rate of decay and the variability of the 15 runs; compare the range with those seen in Figure 2 and 3.

As we stated at the outset of the consideration of these cases, the data are primarily of a program proofing kind. We have specifically limited ourselves to date to cases where a deterministic solution exists or where intuition can be assumed to be a reliable guide. We have satisfied ourselves that the program contains no flaws in logic, and that we may now turn to more exciting pursuits. Specifically, and at an early date, we expect to simulate a human population of the kind the Xavante presently appear to be. That is to say, we propose to assign to the parameters values consistent with those estimated from the Xavante, and to examine questions of the persistence and spread of major genes under circumstances so defined.

Summary

In summary, despite some quite sophisticated and ingenious mathematical arguments, progress toward a stochastic theory of evolution which incorporates all of the processes which perturb gene frequencies has been slow. Most of the cases which have been investigated seem to have more relevance for subhuman forms than for man. Numerous interesting problems in the evolution of man exist, and are presently unsolved. Monte Carlo simulation offers a means of investigating many of these. We are currently engaged in developing and testing a program which would permit us to describe the genetic consequences associated with certain existing hunting and gathering cultures.

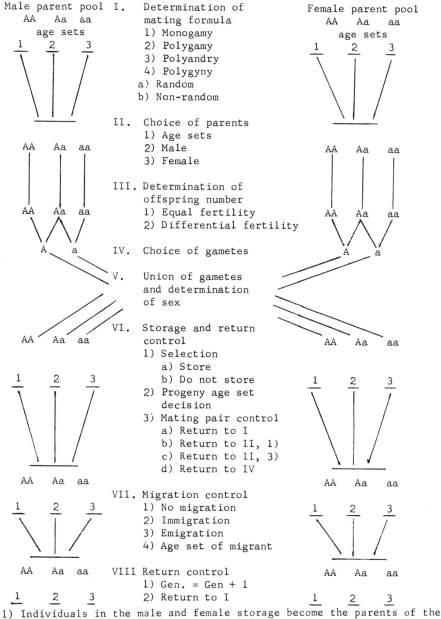

Male parent pool   I.   Determination of   Female parent pool
   AA   Aa   aa   mating formula   AA   Aa   aa
     age sets   1) Monogamy   age sets
    1   2   3   2) Polygamy   1   2   3
                         3) Polyandry
                         4) Polygyny
                         a) Random
                         b) Non-random

                   II.   Choice of parents
                         1) Age sets
   AA   Aa   aa   2) Male   AA   Aa   aa
                         3) Female

                  III.   Determination of
                         offspring number
   AA   Aa   aa   1) Equal fertility   AA   Aa   aa
                         2) Differential fertility

         A   a   IV.   Choice of gametes   A   a

                    V.   Union of gametes
                         and determination
                         of sex

                   VI.   Storage and return
                         control
   AA   Aa   aa   1) Selection   AA   Aa   aa
                            a) Store
                            b) Do not store
    1   2   3   2) Progeny age set   1   2   3
                           decision
                         3) Mating pair control
                            a) Return to I
                            b) Return to II, 1)
                            c) Return to II, 3)
                            d) Return to IV
   AA   Aa   aa   AA   Aa   aa
                  VII.   Migration control
    1   2   3   1) No migration   1   2   3
                         2) Immigration
                         3) Emigration
                         4) Age set of migrant

   AA   Aa   aa   VIII   Return control   AA   Aa   aa
                         1) Gen. = Gen + 1
    1   2   3   2) Return to I   1   2   3

1) Individuals in the male and female storage become the parents of the
   next generation.
2) Continues iterations until Gen. = desired number or fixation occurs.
   FIGURE 1. A diagrammatic representation of the various steps and
decisions represented in the simulation program under discussion.

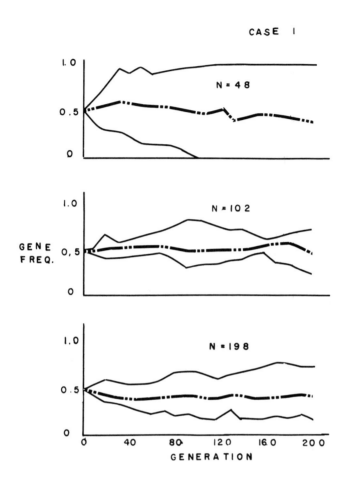

FIGURE 2.   Effects of discrete overlapping of generations
            on genetic drift. The average of six runs is
            indicated by the heavy interrupted line; the
            solid lines define the range described by the
            six runs.

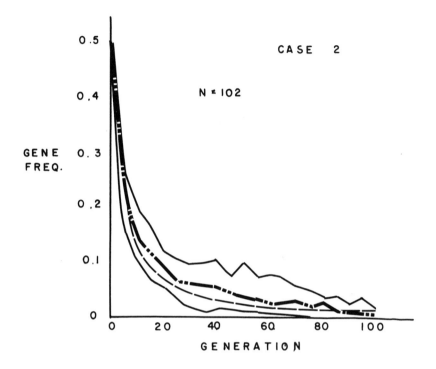

<u>FIGURE 3.</u> Change in gene frequency associated with a locus where
selection proceeds against one homozygote but reproductive compensation occurs. The average of the 15 runs is
indicated by the heavy interrupted line while the solid
lines define the mean range for each point. The deterministic solution is represented by the light interrupted
line.

389

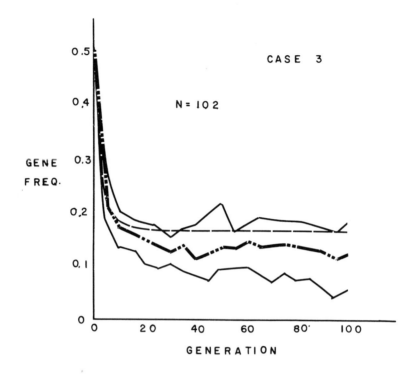

<u>FIGURE 4</u>. Change in gene frequency associated with a locus
with complete selection against one homozygote and
20 per cent selection against the other. The average
of 15 runs is indicated by the heavy interrupted line
represents the deterministic solution associated with
a similar but not identical situation. The solid lines
define the mean range for each point.

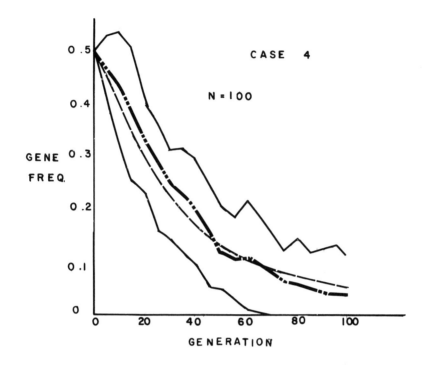

FIGURE 5. Change in gene frequency associated with a locus in
which differential fertility is viewed as a function of
the mating type rather than the genotype. The heavy
interrupted line is the average of 15 runs; the light
interrupted line is the deterministic solution. The
range at each generation is described by the uninter-
rupted lines.

REFERENCES

1. Birdsell, J. B. 1950. Some implications of the genetical concept of race in terms of spatial analysis. Cold Spring Harbor Symposium 15: 259-314.

2. Brues, A. M. 1963. Further contributions to the problem of ABO blood group polymorphism. Paper read at the Annual Meeting of the American Association of Physical Anthropologists.

3. Crosby, J. L. 1963. Evolution by computer. New Scientist No. 327: 415-417.

4. Crow, J. F. 1955. General theory of population genetics: Synthesis. Cold Spring Harbor Symposium 20: 54-59.

5. Fraser, A. S. 1957. Simulation of genetic systems by automatic digital computers. I. Introduction. II. Effects of linkage on rates of advance under selection. Aust. J. Biol. Sci. 10: 484-491, 492-499.

6. Fraser, A. S. 1962. Simulation of genetic systems. J. Theoret. Biol. 2: 329-346.

7. Haldane, J. B. S. 1949. Disease and evolution. La Ricerca Scientifica (Suppl.) 19: 3-11.

8. IBM Reference Manual 1959. Random number generation and testing.

9. Kimura, M. 1954. Process leading to quasi-fixation of genes in natural populations due to random fluctuations of selection intensities. Genetics 39: 280-295.

10. Levin, B., Petras, M., and Rasmussen, D. 1963. The effect of migration on maintaining a polymorphism in the house mouse. (In manuscript.)

11. Lewontin, R., and Dunn, L. C. 1960. The evolutionary dynamics of a polymorphism in the house mouse. Genetics 45: 705-722.

12. Martin, F., and Cockerham, C. 1960. High speed selection studies. In: Biometrical Genetics (). Kempthorne, ed.) New York: Pergamon Press. pp. 35-45.

13. Moran, P. A. P. 1962. The statistical processes of evolutionary theory. Oxford: Clarendon Press. pp. viii and 200.

14. Neel, J. V., Junqueira, P. C., Salzano, F. M., Keiter, F., and
    Maybury-Lewis, D. 1963. Studies in the Xavante Indians of the
    Brazilian Mato Grosso. (In manuscript.)

15. Rucknagel, D., and Neel, J. V. 1961. The Hemoglobinopathies.
    In: Progress in Medical Genetics (Ed.: A. Steinberg). New
    York: Grune and Stratton, Inc. pp. 158-260.

16. Solecki, R. S. 1963. Prehistory in Shanidar Valley, Northern
    Iraq. Science 139: 179-193.

17. Stout, D. B. 1946. Further notes on albinism among the San Blas
    Cuna, Panama. Amer. J. Phys. Anthrop. 4: 483-490.

18. Williams, B. J. Unpublished observations.

19. Woolf, C. M., and Grant, R. B. 1962. Albinism among the
    Hopi Indians in Arizona. Amer. J. Human Genet. 14: 391-400.

20. Wright, S. 1951. The genetical structure of populations. Ann.
    Eugen. 15: 323-354.

[*Editors' Note:* The discussion has been omitted.]

A FINAL NOTE

We have concluded an historical survey of major areas of demographic genetics as we view this developing field. We now wish to include a brief word concerning work that must be incorporated into demographic genetics in the future as it concerns large-scale questions of human populations.

From a genetic point of view much information can be gathered from studying one population at a time, one population over time, or several similar populations at one time. From present gene frequency distributions and a demographic knowledge of present populations, models of past and future genetic phenomena may be developed. Certainly, there is much of value that can, and has, been done along these lines. Yet present populations, although large in size, represent only a miniscule fraction of the number of different human populations that have existed in the course of our genetic evolution. These ancestral populations, generally called "primitive," show regularities of structure and behavior, if present representatives and archeological data may be believed, that must be understood if we are fully to characterize the demographic genetics of human populations. Recently, extensive demographic and genetic studies have been made among some of these populations; more should be undertaken.

In dealing with primitive populations, one is faced with difficulties that do not exist among literate, national populations. First, one must deal largely with anamnestic data, which varies in quality and is often difficult to check. Second, primitive societies structure their entire social life around kinship relations. Marriage is politically and economically based in such groups, and regular prescribed marriages are often specified if not fulfilled. Primitive populations are highly structured along putative genealogical lines, and their entire concept of kinship differs fundamentally from ours. Levi-Strauss (1948) has dealt at length with these questions.

Primitive societies exhibit a small number of basic types of kinship relations but with endless variations. Primitive kinship is *classificatory* in that various relatives, both genealogical and affinal, are often given the same kinship term and treated identically. Thus, although there are only four basic first-cousin types (see Paper 23), many individuals of the same age in one's own kin group ("clan") are considered to be cousins; many of these are only remotely related. Other cousins of exactly the same genealogical relationships are differentiated completely, such as cousins on mother's and father's sides. Furthermore, as Kroeber (1909) noted long ago, our Western languages are poorly equipped to translate such terms.

From this it can be appreciated that regular marriage rules among primitives, even when observed, do not necessarily imply the degree of consanguinity that our terminology would predict. In fact, the elaborate nature of human social relations may be only incidental to genetic change, and human populations may generally approximate panmixia, as do most other natural populations. Morton et al. (1971) has dealt with this question in preliminary fashion.

Work to establish algebraic approaches to kinship has been done by several authors who deserve mention since in the future such techniques may add great power to our analytic approaches. Ruheman (1945, 1967) has discussed regularity and symmetry in various kinship systems; Courrege (1974) has applied the mathematical theory of groups to this problem. Ballonoff (1973, 1974b) relates symmetries of kinship and marriage patterns algebraically to population statistics.

REFERENCES

Anderson, W. W. (1971) Genetic equilibrium and population growth under density-regulated selection. *Amer. Naturalist* 105:489–498.

——, and King, C. E. (1970) Age-specific selection. *Proc. Nat. Acad. Sci.* 66:780–786.

Ballonoff, P. A. (1973) Stability properties of marriage systems: theory of minimal structures. *Proc. IX I.U.A.E.S.*

—— (ed.) (1974a) *Genetics and Social Structure.* Stroudsburg, Pa.: Dowden, Hutchinson & Ross.

—— (1974b) Statistical theory of marriage structures. In: Ballonoff, P. (ed.), *Mathematical Models of Social and Cognitive Structures.* Urbana, Ill.: University of Illinois Press, pp. 11–27.

—— (1974c) Structural models of demographic transition. In: Ballonoff, P. A. (ed.), *Genealogical Mathematics.* Paris: Maison des Sciences de l'Homme (in press).

Bateman, A. J. (1947) Contamination in seed crops. III. Relation with isolation distance. *Heredity* 1:303–336.

—— (1950) Is gene dispersion normal? *Heredity* 4:353–363.

—— (1962) Data from plants and animals. In: Sutter, J. (ed.), *Human Displacements.* Paris: Hachette/Editions "Science Humaines," pp. 85–90.

Beiles, A (1974) A buffered interaction between sex ratio, age difference at marriage, and population growth in humans, and their significance for sex ratio evolution. *Heredity* 33:265–278.

Bernardelli, H. (1941) Population waves. *J. Burma Res. Soc.* 31:1–18.

Birch, L. C. (1948) The intrinsic rate of natural increase of an insect population. *J. Animal Ecol.* 17:15–26.

Bodmer, W. F. (1968) Demographic approaches to the measurement of differential selection in human populations. *Proc. Nat. Acad. Sci.* 59:41–50.

——, and Cavalli-Sforza, L. L. (1968) A migration matrix model for the study of random genetic drift. *Genetics* 59:565–592.

References

————, and Edwards, A. W. F. (1960) Natural selection and the sex ratio. *Ann. Hum. Genet.* 24:239–244.

Boorman, S. A., and Levitt, P. R. (1973) A frequency-dependent natural selection model for the evolution of social cooperation networks. *Proc. Nat. Acad. Sci.* 70:187–189.

Bourgeois-Pichat, J. (1968) The concept of a stable population: application to the study of populations of countries with incomplete demographic statistics. New York: United Nations Publication ST/SOA/Ser. A/39.

Brownlee, J. (1911) The mathematical theory of random migration and epidemic distribution. *Proc. Roy. Soc. Edinburgh* 31:262–289.

Cannings, C., and Cavalli-Sforza, L. L. (1973) Human population structure. In: Harris, H., and Hirschhorn, K. (eds.), *Advances in Human Genetics*, Vol. 4. New York: Plenum, pp. 105–171.

————, and Skolnick, M. H. (1975) Genetic drift in exogamous marriage systems. *Theoret. Pop. Biol.* 7:39–54.

Cavalli-Sforza, L. L. (1958) Some data on the genetic structure of human populations. *Proc. X Intern. Congr. Genet.* 1:389–407.

———— (1962) The distribution of migration distances: models and applications to genetics. In: Sutter, J. (ed.), *Human Displacements.* Hachette/Editions "Sciences Humaines," pp. 139–158.

———— (1973) Analytic review: some current problems of human population genetics. *Amer. J. Hum. Genet.* 25:82–104.

————, and Bodmer, W. F. (1971) *The Genetics of Human Populations.* San Francisco: W. H. Freeman.

————, and Zei, G. (1967) Experiments with an artificial population. In: Crow, J. F., and Neel, J. V. (eds.), *Proc. III Intern. Congr. Hum. Genet.* Baltimore, Md.: Johns Hopkins Press, pp. 473–478.

————, Kimura, M., and Barrai, I. (1966) The probability of consanguineous marriages. *Genetics* 54:37–60.

Chapman, R. N. (1931) *Animal Ecology.* New York: McGraw-Hill.

Charlesworth, B. (1970) Selection in populations with overlapping generations. I. The use of Malthusian parameters in population genetics. *Theoret. Pop. Biol.* 1:352–370.

———— (1971) Selection in density-regulated populations. *Ecology* 52:469–474.

———— (1972) Selection in populations with overlapping generations. III. Conditions for genetic equilibrium. *Pop. Biol.* 3:377–395.

———— (1973) Selection in populations with overlapping generations. V. Natural selection and life histories. *Amer. Naturalist* 107:303–311.

———— (1974) Selection in populations with overlapping generations. VI. Rates of change of gene frequency and population growth rate. *Theoret. Pop. Biol.* 6:108–133.

————, and Charlesworth, D. (1973) The measurement of fitness and mutation rate in human populations. *Ann. Hum. Genet.* 37:175–187.

————, and Giesel, J. T. (1972a) Selection in populations with overlapping generations. II. Relations between gene frequency and demographic variables. *Amer. Naturalist* 106:388–401.

————, and Giesel, J. T. (1972b) Selection in populations with overlapping generations. IV. Fluctuations in gene frequency with density-dependent selection. *Amer. Naturalist* 106:402–411.

Christian, J. J. (1970) Social subordination, population density, and mammalian evolution. *Science* 168:84–90.

Coale, A. J. (1972) *The Growth and Structure of Human Populations: A Mathematical Investigation.* Princeton, N.J.: Princeton University Press.

Cole, L. C. (1954) The population consequences of life history phenomena. *Quart. Rev. Biol.* 29:103–137.

Cotterman, C. W. (1940) A calculus for statistico-genetics. In: Ballonoff, P. A. (ed.), *Genetics and Social Structure.* Stroudsburg, Pa.: Dowden, Hutchinson & Ross, 1974, pp. 157–271.

Courrege, P. (1974) A mathematical model of the structure of kinship. In: Ballonoff, P. A. (ed.), *Genetics and Social Structure.* Stroudsburg, Pa.: Dowden, Hutchinson & Ross.

Crawford, M., and Workman, P. (eds.) (1973) *Methods and Theories of Anthropological Genetics.* Albuquerque, N.M.: University of New Mexico Press.

Crew, F. A. E. (1937) The sex ratio. *Amer. Naturalist* 71:529–559.

Crow, J. F. (1958) Some possibilities for measuring selection intensities in man. *Hum. Biol.* 30:1–13.

———, and Kimura, M. (1970) *An Introduction to Population Genetics Theory.* New York: Harper & Row.

———, and Mange, A. P. (1965) Measurement of inbreeding from the frequency of marriages between persons of the same surname. *Eugenics Quart.* 12:199–203.

D'Ancona, U. (1954) The struggle for existence. *Bibliotheca Biotheoretica,* Vol. VI. Leiden: E. J. Brill.

Dahlberg, G. (1928) Inbreeding in man. *Genetics* 14:421–454.

——— (1948) *Mathematical Methods for Population Genetics.* New York: Wiley–Interscience.

Darwin, C. R. (1896) *The Descent of Man.* New York: D. Appleton and Co.

Demetrius, L. (1969) The sensitivity of population growth rate to perturbations in the life cycle components. *Math. Biosci.* 4:129–136.

Dyke, B., and MacCluer, J. W. (1974) *Computer Simulation in Human Population Studies.* New York: Academic Press.

Eaton, J. W., and Mayer, A. J. (1953) The social biology of very high fertility among the Hutterites: the demography of a unique population. *Hum. Biol.* 25:206–264.

Edwards, A. W. F. (1962) Genetics and the human sex ratio. *Advan. Genet.* 11:239–272.

Emlen, J. M. (1970) Age specificity and ecological theory. *Ecology* 51:588–601.

——— (1973) *Ecology: An Evolutionary Approach.* Reading, Mass.: Addison-Wesley.

Etherington, I. M. H. (1939) Genetic algebras. *Proc. Roy. Soc. Edinburgh.* 59:242–258.

——— (1941) Non-associative algebra and the symbolism of genetics. *Proc. Roy. Soc. Edinburgh* 61:24–42.

Euler, L. (1760) Recherches générales sur la mortalité et la multiplication du genre humain. *Histoire de l'Académie Royale des Sciences et Belles-Lettres,* année 1760, pp. 144–164, Berlin, 1767. Reprinted in *Theoret. Pop. Biol.* 1:307–314.

References

Ewens, W. (1969) *Population Genetics.* London: Methuen.
Falconer, D. W. (1960) *Introduction to Quantitative Genetics.* New York: Ronald Press.
Felsenstein, J. (1971) Inbreeding and variance effective numbers in populations with overlapping generations. *Genetics* 68:581–597.
Fish, H. D. (1914) On the progressive increase of homozygosis in brother–sister matings. *Amer. Naturalist* 48:759–761.
Fisher, R. A. (1922) On the dominance ratio. *Proc. Roy. Soc. Edinburgh* 42:321–421.
——— (1930) The distribution of gene ratios for rare mutations. *Proc. Roy. Soc. Edinburgh* 50:205–220.
——— (1939) Stage of enumeration as a factor influencing the variance in the number of progeny, frequency of mutants and related quantities. *Ann. Eugen.* 9:406–408.
——— (1958) *The Genetical Theory of Natural Selection.* New York: Dover.
Fraser, A. S. (1962) Simulation of genetic systems. *J. Theoret. Biol.* 2:329–346.
Freire-Maia, N. (1957) Inbreeding in Brazil. *Amer. J. Hum. Genet.* 9:284–298.
Frota-Pessoa, O. (1957) The estimation of the size of isolates based on census data. *Amer. J. Hum. Genet.* 2:9–16.
Giesel, J. T. (1971) The relations between population structure and rate of inbreeding. *Evolution* 25:491–496.
Gilbert, J. P., and Hammel, E. A. (1966) Computer simulation and analysis of problems in kinship and social structure. *Amer. Anthropologist* 68:71–93.
Goodman, L. A. (1971) On the sensitivity of the intrinsic growth rate to changes in the age-specific birth and death rates. *Theoret. Pop. Biol.* 2:339–354.
Graunt, J. (1662) Natural and political observations mentioned in a following index, and made upon the Bill of Mortality, with reference to the government, religion, trade, growth, air, diseases and the several changes of the said city. In: *The Economic Writings of Sir William Petty,* Vol. II. New York: Cambridge University Press.
Hairston, N. G., Tinkle, D. W., and Wilbur, H. M. (1970) Natural selection and the parameters of population growth. *J. Wildlife Management* 34:681–690.
Hajnal, J. (1963) Random mating and the frequency of consanguineous marriages. *Proc. Roy. Soc.* B159:125–177.
Haldane, J. B. S. (1927) A mathematical theory of natural and artificial selection, IV. *Proc. Cambridge Phil. Soc.* 23:607–615.
——— (1953) Animal populations and their regulation. *New Biology* 15:9–24.
———, and Jayakar, S. D. (1962) An enumeration of some human relationships. *J. Genet.* 58:81–107.
———, and Moshinsky, P. (1939) Inbreeding in Mendelian populations with special reference to human cousin marriage. *Ann. Eugen.* 9:321–340.
Harrison, G. A., and Boyce, A. J. (1972a) Migration, exchange, and the genetic structure of populations. In: Harrison, G. A., and Boyce, A.

J. (eds.), *The Structure of Human Populations*. New York: Oxford University Press, pp. 128–145.

—— (eds.) (1972b) *The Structure of Human Populations*. New York: Oxford University Press.

Hiorns, R. W., Harrison, G. A., Boyce, A. J., and Kuchemann, C. F. (1969) A mathematical analysis of the effects of movement on the relatedness between populations. *Ann. Hum. Genet.* 32:51–60.

Hull, C. H. (ed.) (1899) *The Economic Writings of Sir William Petty*, Vol. II. New York: Cambridge University Press.

Jacob, S. M. (1911) Inbreeding in a stable simple Mendelian population with special reference to cousin marriage. *Proc. Roy. Soc.* 84:33–42.

Jacquard, A. (1970) Panmixie et structure des familles. *Population* 25:69–76.

—— (1974) *The Genetic Structure of Populations*. New York: Springer.

—— (1975) Inbreeding: one word, several meanings. *Theoret. Pop. Biol.* 7:338–363.

Jennings, H. S. (1914) Formulae for the results of inbreeding. *Amer. Naturalist* 48:693–696.

—— (1916) The numerical results of diverse systems of breeding. *Genetics* 1:53–89.

Karlin, S. (1968) *Equilibrium Behavior of Population Genetic Models with Non-random Mating*. New York: Gordon and Breach.

Kempthorne, O., and Pollak, E. (1970) Concepts of fitness in Mendelian populations. *Genetics* 64:125–145.

Keyfitz, N. (1968) *Introduction to the Mathematics of Population*. Reading, Mass.: Addison-Wesley.

Kimura, M. (1958) On the change of population fitness by natural selection. *Heredity* 12:145–167.

——, and Ohta, T. (1971) *Theoretical Aspects of Population Genetics*. Monographs in Population Biology 4. Princeton, N.J.: Princeton University Press.

——, and Weiss, G. H. (1964) The stepping stone model of population structure and the decrease of genetic correlation with distance. *Genetics* 55:483–492.

King, C. E., and Anderson, W. W. (1971) Age-specific selection. II. The interaction between r and K during population growth. *Amer. Naturalist* 105:137–156.

Kroeber, A. L. (1909) Classificatory systems of relationship. *J. Roy. Anthro. Inst.* 39:77–85.

Lalouel, J. M., and Morton, N. E. (1973) Bioassay of kinship in a South American population. *Amer. J. Hum. Genet.* 25:62–73.

Leslie, P. H. (1945) On the use of matrices in certain population mathematics. *Biometrika* 33:183–212.

—— (1948) Some further notes on the use of matrices in population mathematics. *Biometrika* 35:213–245.

—— (1959) The properties of a certain lag type of population growth and the influence of an external random factor on a number of such populations. *Physiol. Zool.* 32:151–159.

Levins, R. (1968) *Evolution in Changing Environments: Some Theoretical Explorations*. Monographs in Population Biology 2. Princeton, N.J.: Princeton University Press.

Levi-Strauss, C. (1948) *Les Structures élémentaires de la parenté*. Paris: Presses Universitaires de France.

Lewis, E. G. (1942) On the generation and growth of a population. *Sankhya* 6:93–96.

Lewontin, R. C. (1965) Selecting for colonizing ability. In: Baker, H. G., and Stebbins, G. L. (eds.), *The Genetics of Colonizing Species*. New York: Academic Press, pp. 79–94.

Li, C. C. (1955) *Population Genetics*. Chicago: University of Chicago Press.

López, A. (1961) *Problems in Stable Population Theory*. Princeton: Princeton University Press.

Lotka, A. J. (1907) Relation between birth rates and death rates. *Science* 26:21–22.

———— (1924) *Elements of Mathematical Biology*. New York: Dover, 1956.

———— (1939) *Théorie analytique des associations biologiques*, Part II. Paris: Hermann.

MacArthur, R. H. (1962) Some generalized theorems of natural selection. *Proc. Nat. Acad. Sci.* 48:1893–1897.

———— (1972) *Geographical Ecology. Patterns in the Distribution of Species*. New York: Harper & Row.

————, and Wilson, E. O. (1967) *The Theory of Island Biogeography*. Princeton, N.J.: Princeton University Press.

MacCluer, J. W. (1967) Monte Carlo methods in human population genetics: a computer model incorporating age-specific birth and death rates. *Amer. J. Hum. Genet.* 19:303–312.

———— (1975) Monte Carlo simulation: The effects of migration on some measures of genetic distance. In: Crow, J. F. (ed.), *Genetic Distance*, in press.

————, and Schull, W. J. (1970) Frequencies of consanguineous marriage and accumulation of inbreeding in an artificial population. *Amer. J. Hum. Genet.* 22:160–175.

————, Neel, J. V., and Chagnon, N. A. (1971) Demographic structure of a primitive population: a simulation. *A.J.P.A.* 35:193–208.

Malécot, G. (1948) *Les Mathématiques de l'hérédité*. Paris: Masson.

———— (1950) Quelques schémas probabilistes sur la variabilité des populations naturelles. *Ann. Univ. Lyon Sci., A* 13:37–60.

———— (1959) Les modèles stochastiques en génétique de population. *Publ. Inst. Statist., Paris* 8:173–210.

———— (1969) *The Mathematics of Heredity*. San Francisco: W. H. Freeman.

Malthus, T. (1798) *Essay on the Principle of Population as It Affects the Future Improvement of Society*. London: J. Johnson.

May, R. M. (1973) *Stability and Complexity in Model Ecosystems*. Princeton, N.J.: Princeton University Press.

Maynard-Smith, J. (1972) *On Evolution*. Edinburgh: University of Edinburgh Press.

McLaren, I. A. (ed.) (1971) *Natural Regulation of Animal Populations*. New York: Atherton Press.

Meats, A. (1971) The relative importance to population increase of fluctuations in mortality, fecundity and the time variables of the reproductive schedule. *Oecologia* 6:223–237.

Moran, P. A. P. (1962) *The Statistical Processes of Evolutionary Theory.* New York: Oxford University Press.

Morgan, K. (1969) Monte Carlo simulation of artificial populations: the survival of small, closed populations. Paper presented at the Conference on the Mathematics of Population, Berkeley, Calif.

Morton, N. E. (1955) Non-randomness in consanguineous marriages. *Amer. J. Hum. Genet.* 20:116–124.

—— (1969) Human population structure. *Ann. Rev. Genet.* 3:53–74.

—— (ed.) (1973) *Genetic Structure of Populations.* Honolulu: University of Hawaii Press.

——, Imaizumi, Y., and Harris, D. E. (1971) Clans as genetic barriers. *Amer. Anthropologist* 73:1005–1010.

——, Yee, S., Harris, D. E., and Lew, R. (1971) Bioassay of kinship. *Theoret. Pop. Biol.* 2:507–524.

Nei, M. (1965) Variation and covariation in subdivided populations. *Evolution* 19:256–258.

—— (1971) Fertility excess necessary for gene substitution in regulated populations. *Genetics* 68:169–184.

—— (1973) The theory and estimation of genetic distance. In: Morton, N. E. (ed.), *Genetic Structure of Populations.* Honolulu: University of Hawaii Press, pp. 45–54.

—— (1975) *Molecular Population Genetics and Evolution.* Amsterdam: North-Holland.

——, and Imaizumi, Y. (1966) Genetic structure of human populations. *Heredity* 21, 3 parts.

——, and Roychoudhury, A. K. (1974) Sampling variances of heterozygosity and genetic distance. *Genetics* 76:379–390.

Norton, H. T. J. (1928) Natural selection and Mendelian variation. *Proc. London Math. Soc.* 28:1–45.

Pearl. R. (1913) A contribution towards an analysis of the problem of inbreeding. *Amer. Naturalist* 47:577–614.

—— (1914) On the results of inbreeding in a Mendelian population: a correlation and extension of previous conclusions. *Amer. Naturalist* 48:491–494.

Pianka, E. R. (1972) r and K selection or b and d selection? *Amer. Naturalist* 106:581.

Pielou, E. C. (1969) *An Introduction to Mathematical Ecology.* New York: John Wiley & Sons.

Pollack, E., and Kempthorne, O. (1970) Malthusian parameters in genetic populations. I. Haploid and selfing models. *Theoret. Pop. Biol.* 1:315–345.

—— (1971) Malthusian parameters in genetic populations. II. Random mating populations in infinite habitats. *Theoret. Pop. Biol.* 2:357–390.

Price, G. R. (1972) Fisher's "fundamental theorem" made clear. *Ann. Hum. Genet.* 36:129–140.

——, and Smith, C. A. B. (1972) Fisher's Malthusian parameter and reproductive value. *Ann. Hum. Genet.* 36:1–7.

Rieger, R., Michaelis, A., and Green, M. M. (1968) *A Glossary of Genetics and Cytogenetics: Classical and Molecular,* 3rd ed. New York: Springer.

References

Robbins, R. B. (1917) Some applications of mathematics to breeding problems. *Genetics* 2:489–504.

——— (1918a) Applications of mathematics to breeding problems. *Genetics* 3:73–92.

——— (1918b) Some applications of mathematics to breeding problems. III. *Genetics* 3:375–389.

Roughgarden, J. (1971) Density-dependent natural selection. *Ecology* 52:453–468.

Ruheman, B. (1945) A method for analyzing classificatory relationship systems. *Southwestern J. Anthrop.* 1:531–576.

——— (1967) Purpose and mathematics: a problem in the analysis of classificatory kinship systems. *Bijdragen* 123:83–124.

Schull, W. J. (1972) Genetic implications of population breeding structure. In: Harrison, G. A., and Boyce, A. J. (eds.), *The Structure of Human Populations.* New York: Oxford University Press, pp. 146–164.

———, and Levin, B. R. (1964) Monte Carlo simulation: some uses in the genetic study of primitive man. In: Gurland, J. (ed.), *Stochastic Models in Medicine and Biology.* Madison, Wisc.: University of Wisconsin Press, pp. 179–198.

———, and MacCluer, J. W. (1968) Human genetics: structure of human populations. *Ann. Rev. Genet.* 2:279–304.

———, and Neel, J. V. (1965) *The Effects of Inbreeding on Japanese Children.* New York: Harper & Row.

Scudo, F. M. (1971) Vito Volterra and theoretical ecology. *Theoret. Pop. Biol.* 2:1–23.

Sharpe, F. R., and Lotka, A. J. (1911) A problem in age-distribution. *Phil. Mag.* 21:435–438.

Shaw, R. F. (1958) The theoretical genetics of the sex ratio. *Genetics* 43:149–163.

———, and Mohler, J. D. (1953) The selective significance of the sex ratio. *Amer. Naturalist* 87:337–342.

Skolnick, M. H., and Cannings, C. (1974) Simulation of small human populations. In: Dyke, B., and MacCluer, J. W. (eds.), *Computer Simulation in Human Population Studies.* New York: Academic Press.

Slobodkin, L. B. (1961) *Growth and Regulation of Animal Populations.* New York: Holt, Rinehart and Winston.

Smith, C. A. B. (1969) Local fluctuations in gene frequencies. *Ann. Hum. Genet.* 32:251–260.

Stern, C. (1973) *Principles of Human Genetics,* 3rd ed. San Francisco: W. H. Freeman.

Sutter, J. (1962) Human displacements. *Entretiens de Monaco en sciences humaines.*

———, and Goux, J. M. (1961) L'aspect démographique des problèmes de l'isolat. *Population* 16:447–462.

———, and Tabah, L. (1948) Fréquence et répartition des mariages consanguins en France. *Population* 3:608–630.

———, and Tabah, L. (1951) Effets des mariages consanguins sur la descendance. *Population* 6:59–82.

———, and Tabah, L. (1952) Effets de la consanguinité et de l'endogamie. *Population* 7:249–266.

————, and Tran-Ngoc-Toan (1957) The problem of the structure of isolates and of their evolution among human populations. *Cold Spring Harbor Symp. Quant. Biol.* 22:379–383.

Turner, J. R. G. (1970) Changes in mean fitness under natural selection. In: Kojima, K.-I. (ed.), *Mathematical Topics in Population Genetics.* New York: Springer, pp. 32–78.

van Valen, L. (1971) Group selection and the evolution of dispersal. *Evolution* 25:591–598.

Verhulst, P. F. (1838) Notice sur la loi que la population suit dans son accroissement. *Correspondance mathématique et physique publiée par A. Quetelet* (Brussels) 10:113–121.

Volterra, V. (1927) Una teoria matematica sulla lotta per l'esistenza. *Scientia* 41:85–102.

———— (1931) Variations and fluctuations of the number of individuals in animal species living together. In: Chapman, R. N. (ed.), *Animal Ecol.* New York: McGraw-Hill, pp. 409–448.

Wahlund, S. (1928) Zuzammensetzung von Populationen und Korrelations-erscheinungen vom Standpunkt der vererbungslehre aus betrachtet. *Hereditas* 11:65–106.

Ward, R. H., and Sing, C. F. (1970) A consideration of the power of the ψ^2 test to detect inbreeding effects in natural populations. *Amer. Naturalist* 104:355–366.

Watson, H. W., and Galton, F. (1874) On the probability of the extinction of families. *J. Anthropol. Inst. Gt. Brit., Ireland* 4:138–144.

Watt, K. E. F. (1968) *Ecology and Resource Management.* New York: McGraw-Hill.

Weiss, G. H., and Kimura, M. (1965) A mathematical analysis of the stepping stone model of genetic correlation. *J. Appl. Prob.* 2:129–149.

Whittaker, E. (1959) Biography of Vito Volterra, 1860–1940. In: Volterra, V., *Theory of Functionals.* New York: Dover.

Wright, S. (1921) Systems of mating. *Genetics* 6:111–178.

———— (1929) Fisher's theory of dominance. *Amer. Naturalist* 63:274–279.

———— (1931) Evolution in Mendelian populations. *Genetics* 16:97–159.

———— (1943) Isolation by distance. *Genetics* 28:114–138.

———— (1949) Population structure in evolution. *Proc. Amer. Phil. Soc.* 93:471–478.

———— (1960) Physiological genetics, ecology of populations and natural selection. In: Tax, S. (ed.), *Evolution After Darwin.* Chicago: University of Chicago Press, pp. 429–475.

———— (1969) *Evolution and the Genetics of Populations,* 2 vols. Chicago: University of Chicago Press.

————, and Kerr, W. E. (1954) Experimental studies of the distribution of gene frequencies in very small populations of *Drosophila melanogaster.* II. Bar. *Evolution* 8:225–240.

Yasuda, N., and Morton, N. E. (1967) Studies on human population structure. *Proc. III Intern. Congr. Hum. Genet., pp. 249–265.*

————, Cavalli-Sforza, L. L., Skolnick, M., and Moroni, A. (1974) The evolution of surnames: an analysis of their distribution and extinction. *Theoret. Pop. Biol.* 5:123–142.

AUTHOR CITATION INDEX

Agarwala, S. N., 373
Alström, C. H., 373
Altman, S. A., 175
Anderson, P. K., 175
Anderson, W. W., 182, 397, 401
Andrewartha, H. G., 196
Andrews, R. V., 175
Asdell, S. A., 175

Bacon, F., 50
Bailey, E. D., 175
Bajema, C. J., 213
Ballonoff, P. A., 397
Barrai, I., 373, 398
Bateman, A. J., 293, 306, 325, 397
Beiles, A., 397
Bell, J., 373
Bender, M. S., 175
Bennett, J. H., 93
Bernardelli, H., 397
Bevan, W. L., 50
Birch, L. C., 168, 397
Birch T., 50
Birdsell, J. B., 392
Blakeman, J., 294
Blanc, R., 373
Bodmer, W. F., 214, 397, 398
Bøgh, H., 294
Böök, J. A., 325, 374
Boorman, S. A., 398
Bosanquet, C. H., 293
Bossard, J. H. S., 325

Bossert, W., 168
Bourgeois-Pichat, J., 398
Boyce, A. J., 400
Brenk, H., 374
Bronson, F. H., 175
Brownlee, J., 293, 306, 398
Brues, A. M., 392
Buzzati-Traverso, A. A., 168

Calhoun, J. B., 175
Cameron, A. W., 175
Cannings, C., 398, 404
Cantrelle, P., 374
Cavalli-Sforza, L. L., 325, 373, 374, 397, 398, 405
Chagnon, N. A., 402
Chandler, J. H., 213
Chandra, Sekar, C., 374
Chapman, R. N., 398
Charlesworth, B., 182, 195, 398
Charlesworth, D. 398
Cheever, H. T., 83
Chitty, D., 175, 196
Cho, W. K., 374
Christian, J. J., 175, 398
Clough, G. C., 175
Coale, A. J., 398
Cockerham, C., 392
Cole, L. C., 399
Cotterman, C. W., 399
Courrege, P., 399
Craig, J. V., 175

Crane, M. B., 293
Crawford, M., 399
Creighton, C., 50
Crew, F. A. E., 93, 399
Crosby, J. L. 392
Crow, J. F., 196, 204, 392, 399
Currence, T. M., 293

Dahlberg, G., 263, 306, 325, 374, 399
D'Ancona, U., 70, 71, 399
Dandekar, K., 374
Dandekar, V. M., 374
Darlington, P. J., Jr., 175
Darwin, C. R., 85, 93, 399
Davidson, R. T., 213
Davis, D. E., 175
Davison, J., 196
Demetrius, L., 399
De Morgan, A., 50
Dobzhansky, T., 175, 182, 196, 204, 272,
    273, 306
Dubinin, N. P., 196
Dunn, L. C., 392
Dyke, B., 399

Eaton, J. W., 399
Edwards, A. W. F., 93, 398, 399
Egenter, A., 374
Eiseley, L., 4
Elton, C., 175, 196
Emlen, J. M., 399
Erlenmeyer-Kimling, L., 213, 214
Erlich, P. R., 175
Errington, P. L., 175, 196
Etherington, I. M. H., 399
Euler, L., 399
Ewens, W., 399

Falconer, D. S., 399
Felsentstein, J., 399
Fish, H. D., 399
Fisher, R. A., 93, 196, 204, 399, 400
Fraccaro, M., 325
Frampton, V. L., 293
François, J., 375
Fraser, A. S., 392, 400
Freire-Maia, N., 325, 374, 400
Frota-Pessoa, O., 325, 374, 400
Fuller, J. L., 175

Gadgil, M., 168

Gaines, M., 196
Galton, F., 404
Gautier, E., 374
Geist, V., 175
Gershenson, S., 196
Getz, L. L., 175
Gibson, G. D., 374
Giesel, J. T., 182, 398, 400
Gilbert, J. P., 400
Glass, B., 325
Goldfarb, C., 213
Goodman, L. A., 400
Gottesman, I. I., 214
Goux, J. M., 375, 404
Grant, R. B., 393
Graunt, J., 400
Green, M. M., 403
Greene, J. C., 4
Gregory, P. H., 293, 306
Grob, W., 374
Guhl, A. M., 175
Guilday, J. E., 175

Hairston, N. G., 400
Hajnal, J., 374, 400
Haldane, J. B. S., 102, 104, 108, 168, 175,
    182, 204, 306, 325, 392, 400
Hamilton, H. W., 175
Hammel, E. A., 400
Hammond, D. T., 374
Hansing, E. D., 293
Harris, D. E., 402
Harrison, G. A., 400
Hart, B. S., 175
Healey, M. C., 175
Henry, L., 374
Hess, C., 325
Heston, L. L., 214
Hibbard, C. W., 175
Hiorns, R. W., 400
Hooper, E. T., 175
Hovanitz, W., 273
Howell, J., 50
Hubby, J. L., 168
Hughes, E. M., 213
Hull, C. H., 400
Hultkrantz, J. V., 263

Imaizumi, Y., 402, 403
Ingram, J. K., 50
Istock, C. A., 182

Jackson, C. E., 374
Jackson, P. D., 214
Jacob, S. M., 400
Jacquard, A., 400, 401
Jahn, E. F., 325
Jayakar, S. D., 168, 400
Jenkins, J. M., 293
Jennings, H. S., 401
Jensen, I., 294
Jungueira, P. C., 393

Kalela, O., 175
Kang, Y. S., 374
Karlin, S., 401
Karmel, P. H., 374
Keiter, F., 393
Kempthorne, O., 401, 403
Kerr, W. E., 405
Keyfitz, N., 401
Kimura, M., 196, 392, 398, 399, 401, 405
King, C. E., 397, 401
Koford, C. B., 175
Krebs, C. J., 196
Kroeber, A. L., 401
Kuchemann, C. F., 400

Lahiri, D. B., 75
LaLouel, J. M., 401
Lederberg, J., 214
Legouvé, E., 51
Leslie, P. H., 182, 196, 401
Levin, B. R., 392, 403
Levins, R., 168, 401
Lévi-Strauss, C., 306, 401
Levitt, P. R., 398
Lew, R., 402
Lewis, C. J., 374
Lewis, E. G., 401
Lewis, J. N., 374
Lewontin, R. C., 168, 182, 196, 392, 401
Li, C. C., 306, 401
Lindeman, R. L., 168
Linn, M. B., 293
Lloyd, J. A., 175
López, A., 401
Lotka, A. J., 54, 70, 100, 103, 105, 401, 404
Lovejoy, A. O., 4
Lowe, C. R., 93
Ludwig, W., 374

MacArthur, R. H., 168, 401, 402
MacCluer, J. W., 399, 402, 403

McCrady, A. D., 175
McKeown, T., 93
McLaren, I. A., 402
Malecot, G., 402
Mallet, E., 51
Malthus, T., 402
Mange, A. P., 399
Martin, F., 392
Martin, P. S., 175
Master, H. R., 374
Mather, K., 293
May, R. M., 402
Maybury-Lewis, D., 393
Mayer, A. J., 399
Mayer, W. V., 175
Maynard-Smith, J., 402
Mayr, E., 175, 196
Meats, A., 402
Metzger, L. H., 175
Michaelis, A., 403
Mohler, J. D., 93, 404
Moran, P. A. P., 392, 402
Moreau, P., 375
Morgan, J., 51
Morgan, K., 402
Moroni, A., 373, 374, 405
Mortara, G., 374
Morton, N. E., 204, 306, 325, 374, 401, 402, 405
Moshinsky, P., 325, 400
Mueller, C. D., 175
Muller, H. J., 204
Murie, J. O., 175
Murray, K., 175
Mykytowycz, R., 175

Neel, J. V., 393, 402, 403
Nei, M., 402, 403
Nemeskéri, J., 374
Newcourt, R., 51
Norton, H. T. J., 182, 196, 403
Norway Statistical Office, 374

Ohba, S., 182
Ohta, T., 401
Orel, H., 374

Paradowski, W., 214
Pavlovsky, O., 182
Pearl, R., 403
Pearson, J. L., 293

Pearson, K., 294
Petras, M., 392
Petty, W., 50, 51
Philiptschenko, J., 263
Pianka, E. R., 403
Pielou, E. C., 403
Pitelka, F., 175
Pollak, E., 401, 403
Price, G. R., 403
Provine, W. B., 4

Rasmussen, D., 392
Raven, P. H., 175
Reed, T. E., 213, 214
Rieger, R., 403
Robbins, R. B., 403
Robertson, F. W., 196
Rosenthal, D., 214
Roughgarden, J., 403
Roychoudhury, A. K., 403
Rucknagel, D., 393
Ruepp, G., 374
Ruheman, B., 403

Sacks, M. S., 325
Salzano, F. M., 393
Sanghvi, I. D., 374
Scaliger, J. J., 51
Schapera, I., 374
Schmidt, W., 294
Schull, W. J., 375, 402, 403
Scobell, H., 51
Scudo, F. M., 403
Semeonoff, R., 196
Serra, A., 375
Sharpe, F. R., 404
Shaw, R. F., 93, 404
Shields, J., 214
Simpson, G. G., 175
Sing, C. F., 404
Skellam, J. G., 325
Skolnick, M. H., 398, 404, 405
Slobodkin, L. B., 404
Smith, C. A. B., 403, 404
Smith, J. M., 168, 175
Soini, A., 375
Solecki, R. S., 393
Stern, C., 404
Stewart, J. Q., 325
Stout, D. B., 393
Sutter, J., 306, 325, 375, 404
Sutton, O. G., 294

Tabah, L., 306, 375, 404
Tamarin, R. H., 196
Tast, J., 175
Terman, C. R., 175
Thoma, A., 374
Thompson, D. H., 272
Timofeef-Ressovsky, N.W., 196
Tiniakov, G. G., 196
Tinkle, D. W., 400
Traill, H. D., 51
Tran-Ngoc-Toan, 325, 404
Turner, J. R. G., 404
Twiesselmann, F., 375

Ullyett, G. C., 196
U.S. Public Health Service, 214

Van Gelder, R. G., 175
van Valen, L., 404
Van Vleck, D. B., 175
Varde, D. S., 374
Verhulst, P. F., 404
Volterra, V., 70, 404

Wadley, F. M., 294
Wahlund, S., 306, 404
Ward, R. H., 404
Watson, H. W., 404
Watt, K. E. F., 405
Weinberg, W., 263
Weiss, G. H., 401, 405
Wentworth, D. T., 70
Whitaker, J. O., 175
Whittaker, E., 405
Wilbur, H. M., 400
Williams, B. J., 393
Wilson, E. O., 168, 402
Wimer, R. E., 175
Wolfenbarger, D. O., 294
Woolf, C. M., 393
Workman, P., 399
Wright, S., 168, 175, 204, 272, 273, 294, 306, 325, 393, 405
Wulz, G., 375
Wynn-Edwards, V. C., 175

Yanase, T., 375
Yasuda, N., 405
Yee, S., 402

Zalaffi, C., 374
Zei, G., 374, 398

SUBJECT INDEX

Adriatic, fishes in, 10
Africa, 333, 341
Age at reproduction, and consanguinity, 328–332
Age distribution, 1, 52–53, 54–57, 73–76, 176–178, 381–382
  selection and, 95–145, 176–182, 183–195, 207
Age of onset, and fitness, 208–213
Age-specific fertility, 54–57, 73–76, 95, 100–106, 177, 184–193, 197, 202–203, 206–213, 378
Age-specific mortality, 13–14, 52–53, 54–57, 73–76, 95, 100–106, 177, 184–193, 197, 202, 206–213, 378
Age structure (see Age distribution)
Agression, 170
Altruism, 149
Arctic, irruptions in, 174
Associations, biological, 58–72
Australia, 138, 381
Austria, 338, 341, 343

Bees, foraging pattern of, 276–280
Behavior, and demographic genetics (see Social biology)
Belgium, 333
Birth rate
  crude, defined, 53
  English, 35
Blood groups, 320–325
Brazil, 334, 336, 341, 343, 381

Carrying capacity, 7, 148, 161–168, 179–182, 189–194
Catholic church, marriages and, 308–310, 318, 333, 336
Chance of marriage, random, at birth, 363–366
Competition
  interspecific (see Populations, competition between)
  intraspecific, 169–175
Computer simulation (see Simulation)
Consanguinity
  and effective population size, 296–297, 302–304, 318–320, 324–325, 346–351
  and family size, 350–351
  frequency of, in marriages, 297–298, 300, 302, 318–320, 324–325, 326–375
  mathematical model, 353–373
  and migration, 297, 304, 318–320, 324–325, 349, 362–363
  and population growth, 349
Continuous model of isolation (see Isolation, by distance)
Cousin, marriages (see Consanguinity)
  reasons for, 335–336
  types of, 332–345
Crop pollination, by insects, 282–288
Crow's index, 197–198, 200–204

Dahlberg model of isolate size, 295–297, 302–303, 309, 318, 323, 324–325, 326–328

Death rate, crude, defined, 53
Demographic genetics
  applications of, 197–213
  definitions of, 1
Demographic versus ecological approaches,
  9–11, 150
Demography, concepts of, 1
Density
  and isolate size, 297, 320, 324
  and population dispersal, 170–171
  social effects of high, 171
Density-dependent population processes, 7,
  147, 150, 153–168, 174–175, 179–182,
  183–184
Discrete generations, and demographic
  structure, 216–218
Dispersal, and subordinate status, 171
Dispersion
  and density, 170–171
  of genes, 216, 222–223, 320–325
  parameters of, 222–223, 274–293, 296–
  297, 302–305, 307, 310–317, 324–325
  of pollen
    by insects, 275–288
    by wind, 288–290
Dominance, 169–175
Drift, genetic, 218–219, 377, 384

Ecological genetics, 1, 146–196
Ecology, of populations (*see* Population,
  ecology of)
Effective population size, 216–217, 267, 304
  of isolate, 295, 302–304, 309, 324
  and migration, 221, 267–272, 304, 324–
  325
  and overlapping generations, 217
Endogamy, as migration measure, 297, 310,
  315–316, 324
Energetics, ecological, and *r* and *K*, 165–168
England, population of, 7, 12–51
Entropy, 143
Environment
  and endogamy, 315, 324
  and migration distance, 316, 324–325
Equilibrium
  of gene frequencies, 146–147, 150–153,
  157–167, 179–182, 185, 189–194, 378
  population size, 58–72
  of sex ratio, 90–92
Ergodicity, of population structure, 95
Ethology, and genetic theory, 149, 169–175

Europe, 302, 335, 338, 352
Evolution, mammalian, 169–175

Family size, and consanguinity (*see* Consanguinity, and family size)
Fertility
  age-specific (*see* Age-specific fertility)
  differential, 79, 198–213, 302, 383, 385
Finite population size, 218, 299–301, 378,
  380
Fitness
  measured in overlapping generations, 95–
  97, 116–131, 141, 151–152, 176, 185,
  197–202, 206–213
  variance in, and natural selection, 96–97,
  141, 197, 200–204
Foraging, distances, 278–280
France, isolate size in, 296, 303–304

Gene frequencies
  changes due to demographic factors, 152,
  169–175, 178–182, 183–195
  and environment, 159–168, 179–182,
  183–184, 189–194
Generation length, 207
Genes
  correlation between, 218, 265–272, 304
  dispersion of (*see* Dispersion)
Genotypic correlations, 224, 238–240, 260–
  263
Germany, 338, 349
Gravitational model of migration, 313–315,
  324
Group selection, 149
Growth
  concept of, 8–11
  density-dependent, 147
  of populations, 1, 8, 9, 45–50, 60–72, 349
Growth rate
  genotype-specific, 95–145, 177, 183–185,
  206–209
  intrinsic, 9, 53, 54–57, 60–72, 179–180,
  183–193, 198, 358–362

Habitat, optimal, 171–172
Hardy-Weinberg law
  and Mendelian traits, 226
  with overlapping generations, 151, 194–
  195
Heterosis, simulation of, 385

Homozygosity
  and finite populations (*see* Finite population size)
  and inbreeding (*see* Inbreeding, coefficients of)
Human populations, demographic genetics of, 295, 302–306, 307–325, 326–353, 376, 395–396
Hungary, 346

Identity, genetic, 222
Inbreeding, 216, 218, 265–267, 304, 324
  coefficient of, and genetic variance, 220–221, 304
  coefficients of, 218, 265–267, 296, 304, 318
  early studies of, 218
  and recessive traits, 219
India, 334, 340, 352, 381
Infanticide, and sex ratio, 81–84
Irruptions, periodic, 173–175
Isolate, 241, 295, 302–305, 320, 352
  and genotype frequencies, 241–263
  size distribution, 369
Isolates, breakdown of, 350–351
Isolation
  by distance, 220–223, 264–273, 296–297, 303–304, 307, 324–325
  and migration, 220–223, 251–263, 264–272, 302–305, 324–325 (*see also* Isolation, by distance)
  of populations, 1, 220–223, 241–263, 264–272, 295–297, 302–305, 307, 320, 324–325, 352 (*see also* Isolates)
Isonymy, 299
Israel, 336
Italy, 296, 308–324, 333, 341, 343, 345 (*see also* Parma Valley)

Japan, 300, 333, 335, 338, 343, 349

Kinship
  effect of different systems, 303
  types of, 332–345
K selection, 148, 151, 153–168, 176, 180–182

Leslie matrix, for population projection, 11, 73–76, 151, 177–182, 184
Lewis' matrix, 73–76
Life history (*see* Overlapping generations)

Life table, 132–134
Linkage, and genotype frequencies, 229–263
Linnaeus, biological typology of, 3
Logistic growth model, 10, 62, 150, 180–182
London
  early demography of, 12–51
  migration into, 38
  mortality in, 7, 12–51
Lotka equations, 9, 52–53, 54–57, 95, 98–100, 185–186, 206–208, 358–362

Malthusian parameter, 96, 135, 198, 201, 206–209 (*see also* Growth rate, intrinsic)
Mates, association between, 216, 310–317, 324, 326–353
Mate selection, 79
Mating structure, 216, 300, 302–305, 318–325, 382, 385
Matrix, population protection (*see* Leslie matrix)
Migration, 1, 220–223, 267–272, 296–299, 302, 304, 310–317, 324–325, 383
  to city, 38
  island model, 221, 264–265, 297, 307, 322–323, 325
  matrix, 297
  parameters of (*see* Dispersion, parameters of)
  and social rank, 169–175
  stepping stone model, 221
Monte Carlo simulation (*see* Simulation)
Mortality
  age-specific (*see* Age-specific mortality)
  causes of 7, 13–51
  and consanguinity, 350
  differential, 7, 13–51, 79, 198–213, 382
  and environment, 7, 16, 38, 171
  London bills of, 7, 12–51
  and sex, 40
  social, 149
Mutation, simulation of, 383

Net reproduction rate, 198, 207–208
Normal distribution, of migration, 305, 310–312, 324

Overlapping generations, 94–145, 176–182, 183–195, 201–213, 217, 328–332, 353–373
  and simulation, 381–382

413

Panmixia (*see* Random mating)
Parental expenditure, and sex ratio (*see* Sex ratio)
Parent–offspring, distance between, 292, 304, 318-320, 324-325
Parity, type of, and fitness, 146
Parma Valley, 296, 308-324
Plague, bubonic, 29-35, 42
Pollution, air, 7, 16, 38
Polygenic traits (polyhybrids), genotype frequencies, 225-263
Population
  concept of, 3, 6
  density (*see* Density; Density-dependent population processes)
  ecology of, 7, 10, 58-72, 148, 169-175
  genetic differentiation within, 3, 150, 176-182, 185, 197, 206-213
  regulation of (*see* Regulation, of population size)
  size (*see* Size, of populations)
Population genetics, definition of, 1
Population growth (*see* Growth, of populations)
Population size
  effective (*see* Effective population size)
  fluctuations of, 63-72, 171, 173-174, 176-179, 183-184, 378,380
  periodic changes, 63-72
Populations, competition between, 10, 58-72, 150, 169-171
Preadaptation, social, 172
Predator–prey model, 62-72
Primitive societies, 376-393, 395-396
Projection matrix (*see* Leslie matrix)
Public health, and demographic genetics, 199

Random mating
  and age-difference between mates, 328-332, 336-341, 346-372
  and gene correlations, 238-240
Random numbers, 379, 380
Rank, social, 169-170
Regulation, of population size, 7, 8, 45-50, 62-72, 147, 149, 153-168, 174-175, 179-182, 189-194
Reproductive value, 96, 137
  and sex ratio, 85-87, 89-90
r selection, 148, 151, 153-168, 176, 180-182

San Blas Indians, 384
Sandwich Islands, 82

Schizophrenia, genetics of, 198, 209-211
Selection
  and differential social dispersal, 171-175
  Fisher's fundamental theory of, 96-97, 141-145, 151-152, 197, 200-201
  and overlapping generations, 94-145, 146-147, 176-182, 183-195, 202-204, 206-209, 213, 382-383
  sex ratio, 78-93
  variation in, 378
Sex, and mortality, 40
Sex-linkage, selection and, 104
Sex ratio, 1
  as continuous variable, 79, 88-93,
  distribution of, 90-93
  evolution of, 78-93
  and parental care, 79, 85-93
Sexual selection, 78-79
Simulation
  computer, 299-301, 376-386
  program debugging, 383-384
Size, of populations, 1, 10, 45-50, 58-72, 171-175
Social biology, and genetics, 79-80, 148-149, 169-175, 298, 395-396
Species, views of, 3
Stable population theory, 9, 52-53, 54-57, 206-209, 353-373
Stationary population, and consanguinity, 353-358
Stochastic processes, 2, 218-219, 299-301, 376-386
Subdivision, of population, 2, 219-220
Subordination, social, 169-175
Surnames, in genetic studies, 299
Sweden, 339, 346
Switzerland, 346, 349

Territory, 149-169
Thermodynamics, second law of, 143

United States, 347-348

Variance, Wahlund's, 220-221, 241-251, 296, 304
Village size, distribution, 303, 304, 308-310, 324
Vital rates (*see* Age-specific mortality; Age-specific fertility)

Wahlund effect, 220 (*see also* Variance, Wahlund's)

About the Editors

KENNETH M. WEISS is Assistant Professor at the Center for Demographic and Population Genetics, Graduate School of Biomedical Sciences of the University of Texas Health Science Center in Houston. He was previously Research Associate in the Department of Human Genetics, University of Michigan Medical School. He has also been a professional meteorologist. He received his B.A. in mathematics from Oberlin College and his Ph.D. in biological anthropology from the University of Michigan. He has published numerous papers on demographic problems in human evolution, including a monograph on the subject, *Demographic Models for Anthropology*.

PAUL BALLANOFF is Assistant Professor of Population Genetics at the Center for Demographic and Population Genetics, Graduate School of Biomedical Sciences of the University of Texas Health Science Center at Houston. He also holds an adjunct appointment in the Department of Mathematical Sciences at Rice University in Houston, where he teaches a course on the mathematical foundations of social anthropology. He has held previous appointments at the University of Illinois and the University of Washington and is author of a book, *Mathematical Foundations of Social Anthropology*, as well as being editor or coeditor of three other collections, including *Genetics and Social Structure* in the Benchmark series.